Sustainable Development Goals Series

The **Sustainable Development Goals Series** is Springer Nature's inaugural cross-imprint book series that addresses and supports the United Nations' seventeen Sustainable Development Goals. The series fosters comprehensive research focused on these global targets and endeavours to address some of society's greatest grand challenges. The SDGs are inherently multidisciplinary, and they bring people working across different fields together and working towards a common goal. In this spirit, the Sustainable Development Goals series is the first at Springer Nature to publish books under both the Springer and Palgrave Macmillan imprints, bringing the strengths of our imprints together.

The Sustainable Development Goals Series is organized into eighteen subseries: one subseries based around each of the seventeen respective Sustainable Development Goals, and an eighteenth subseries, "Connecting the Goals", which serves as a home for volumes addressing multiple goals or studying the SDGs as a whole. Each subseries is guided by an expert Subseries Advisor with years or decades of experience studying and addressing core components of their respective Goal.

The SDG Series has a remit as broad as the SDGs themselves, and contributions are welcome from scientists, academics, policymakers, and researchers working in fields related to any of the seventeen goals. If you are interested in contributing a monograph or curated volume to the series, please contact the Publishers: Zachary Romano [Springer; zachary.romano@springer.com] and Rachael Ballard [Palgrave Macmillan; rachael.ballard@palgrave.com].

Yin Lun

Climate Change and Traditional Knowledge

Community Based Adaptation in Eastern Himalayas, China

Springer

Yin Lun
Southwest Forestry University
Kunming, Yunnan, China

ISSN 2523-3084 ISSN 2523-3092 (electronic)
Sustainable Development Goals Series
ISBN 978-3-032-12014-4 ISBN 978-3-032-12015-1 (eBook)
https://doi.org/10.1007/978-3-032-12015-1

This Springer imprint is published by the registered company Springer Nature Switzerland AG
The registered company address is: Gewerbestrasse 11, 6330 Cham, Switzerland

Foreword by Pierre Léna

Traditional Knowledge is the Light of Wisdom for Climate Change Adaptation

The Global Climate Crisis and the Responsibility of Scientists

Throughout my long career as a scientist, whether in exploring the vastness of the universe or reflecting on the future of our planet, I have always been deeply aware of the inseparable bond between humanity's destiny and the natural environment. In the twenty-first century, climate change has emerged as one of the greatest challenges confronting all humankind. This challenge is reflected not only in the constant flow of new scientific data but also in the profound transformations occurring in every aspect of human life. It is not merely an abstract curve of rising temperatures or a complex model; rather, it is a tangible and pressing force that is reshaping human lifestyles, social structures, and even the trajectory of civilization itself.

On a global scale, the impacts of climate change are profoundly uneven. Some regions may find temporary buffers through technological advancement, while others, constrained by ecological fragility and weak economic foundations, have become the earliest and most direct victims. The Himalayas—particularly their eastern range—are a vivid microcosm of this global crisis. This is one of the most rapidly warming areas on Earth, second only to the polar regions. The accelerated melting of glaciers and snowpacks has placed river systems and water security under severe stress; grassland degradation and changes in forest composition have sharply diminished ecosystem services; and floods, landslides, and mudslides occur with increasing frequency and intensity as climate instability escalates.

In such an ecologically sensitive region, community livelihoods remain almost entirely dependent on natural resources. Agriculture and pastoralism are the foundations of survival. Forests provide fuel and timber; grasslands sustain livestock; rivers and wetlands supply water for drinking and irrigation; and medicinal plants remain an indispensable source of traditional healthcare. This profound reliance on the natural world makes Himalayan communities especially vulnerable to the impacts of climate change. They are among the first to sense the risks and among the least able to escape their consequences.

Yet it is precisely because of this long and intimate coexistence with nature that these communities have, through generations of practice, accumulated rich knowledge and adaptive strategies. This knowledge not only enables them to sustain livelihoods and social continuity under harsh conditions but has also crystallized into a unique system of wisdom. From the management of water resources to the sustainable use of grasslands, from the preservation of medicinal plants to strategies for disaster preparedness, local communities demonstrate creativity and resilience that transcend mere survival. These practices are not only the foundation of their endurance but also offer valuable lessons for other regions of the world.

Thus, Tibetan villages in the eastern Himalayas stand at the very frontline of the global climate crisis while also serving as vital windows into the relationship between humanity and nature. Their stories concern not only the fate of one mountainous region but also the broader question of how humankind might seek pathways forward in an era of global warming. In this sense, the publication of this book holds exceptional significance. It is not only a comprehensive study of a particular region but also a profound reflection on the relationship between human wisdom, traditional knowledge, and climate adaptation. Through these accounts and analyses, we are able to see that the creativity and resilience displayed by human civilization in times of crisis may well be the key to navigating an uncertain future.

The Complementarity of Traditional Knowledge and Science

In the ongoing discussions on global climate governance, traditional knowledge has been gaining increasing recognition. It is not a relic of the past, but rather a complex system formed through centuries of interaction between communities and their natural environment. This knowledge embodies accumulated observations, experiences, and beliefs across generations, recording the details of environmental change. It represents a "long-term archive," written jointly by time and by life itself.

Compared with modern science, the strength of traditional knowledge lies in its local specificity and practical application. Scientific models excel at illustrating global trends, yet they often struggle to provide precise predictions for local variations. Traditional knowledge, by contrast, is deeply rooted in the land and in daily practices. It enables communities to perceive subtle ecological changes: the timing of glacial retreat, early signs of grassland degradation, or the shifting growth and migration of medicinal plants. Such knowledge is not in opposition to science; rather, it serves as an indispensable complement and reference point for scientific inquiry.

At the same time, traditional knowledge carries distinct gender dimensions. Men and women assume different roles within communities, and their ways of interacting with the natural environment differ accordingly, giving rise to diverse sets of knowledge and skills. Women play a central role in agricultural cultivation, the collection of medicinal plants, household health, and the management of water resources. Their observations and wisdom directly influence food security, disease prevention, and the sustainable development of families and communities. Yet these perspectives have often been overlooked in research and policymaking. In reality, women's knowledge and men's experiences complement one another, together forming a more complete system of adaptation.

Thus, traditional knowledge is not only a reservoir of ecological wisdom but also a reflection of social, cultural, and gendered dimensions. It reminds us that effective adaptation to climate change cannot rely solely on scientific models and technological tools; it must also incorporate the voices and experiences of local communities. The knowledge systems of Tibetan communities exemplify this integration. They highlight human creativity and resilience in the face of uncertainty. As we reflect on the future of global climate governance, this local wisdom holds lessons of universal significance.

The Author's Contributions and Academic Standing

The author of this book, Prof. Yin Lun, is Scholar and Activist who has closely integrated "traditional knowledge" with "climate change." Over the past two decades, he has established a strong presence in the rural and mountainous regions of western China, particularly focusing on communities that are most directly impacted by the effects of climate change. His research is not confined to academic journals and papers, but is deeply rooted in practical collaboration with local communities, especially female villagers, on extensive and enduring projects.

In terms of **climate change mitigation**, Prof. Yin Lun has developed practices based on three levels of biodiversity:

- At the genetic resources level, he has promoted afforestation with native tree species to restore ecosystem functions.
- At the species diversity level, he has assisted communities in establishing carbon-neutral forests to enhance the carbon sequestration capacity of the forest.
- At the ecosystem diversity level, he has focused on the protection and restoration of rivers, wetlands, grasslands, and alpine meadows to improve their ability to absorb carbon dioxide and regulate climate.

In terms of **climate change adaptation**, he has collaborated with local female villagers, following the technical specifications outlined by the Ministry of Ecology and Environment, to implement various innovative practices:

- Using traditional knowledge of agricultural genetic resources, he has promoted the diversification and sustainable development of farming and livestock systems.
- Applying traditional medical knowledge, he has explored methods for diagnosing and treating common local diseases.
- Utilizing traditional technologies, he has facilitated the establishment of vegetable greenhouses, solar energy systems, and water diversion irrigation systems.

- Based on cultural traditions, he has set up small-scale funds and loan mechanisms to strengthen the economic resilience of local communities.
- Leveraging knowledge of geographical indications, he has developed unique agricultural and pastoral products and expanded market opportunities.

These practices are not only about preserving and revitalizing traditional knowledge but also about integrating scientific methods with community wisdom. They have made rural villages in vulnerable regions more resilient to climate risks, providing real-world examples of the intersection between "local knowledge" and "global challenges."

In the academic realm, Prof. Yin Lun has published several influential books, including *Climate Anthropology*, has edited a series of books, and has published more than 60 academic papers. He has led major research projects funded by the National Social Science Fund, significant national commissions, and provincial research initiatives. Additionally, he has collaborated with international organizations, such as the United Nations Development Programme (UNDP) and the World Bank, on over ten projects. His research has not only had a profound impact in China but has also made significant contributions to the international academic and policy spheres.

On the international stage, his projects have been featured multiple times at the Paris Peace Forum and have garnered attention from international leaders, including UN Secretary-General António Guterres, former US Vice President Al Gore, former French Prime Minister Jean-Pierre Raffarin, and many others. He has served as Lead Expert for the Intergovernmental Science-Policy Platform on Biodiversity and Ecosystem Services (IPBES) and as Senior Expert at the United Nations Office for Disaster Risk Reduction (UNDRR), playing an active role in global climate governance.

Professor Yin Lun's academic achievements and societal practices demonstrate that he is not only a rigorous researcher but also a practitioner who transforms scientific knowledge into social action.

The Value and Innovation of This Book

The value of this book lies first in the integrity and logic of its overall structure. It is composed of 12 chapters, forming a progressive logical sequence: Chaps. 1 and 2 focus on the background and action research, laying the foundational groundwork. Chapters 3–9 delve into specific dimensions of traditional knowledge and climate change adaptation, including water resources, medicinal plants, agro-pastoral livelihoods, forests, genetic resources, gender, and disasters. Chapters 10–12 then expand to explore issues of belief, national strategies, and global governance, demonstrating the integration of traditional knowledge with modern frameworks. This structure is both systematic and comprehensive, rooted in field practices while oriented toward global governance. Its unique value lies in its ability to bridge disciplinary boundaries, connecting anthropology, ecology, policy studies, and social practices, while also transcending geographical and cultural borders to make local wisdom a key component of global discussions.

Secondly, the value of this book is reflected in its interdisciplinary nature. It is not merely an anthropological field study, but also encompasses systematic analysis from ecology and forward-thinking reflections from policy studies. By employing anthropological methods to uncover the wisdom embedded in daily community practices, ecological methods to assess the environmental benefits of traditional knowledge, and policy research to consider how this knowledge can be integrated into national strategies and international mechanisms, this book builds a bridge of communication across disciplines.

Finally, the most prominent contribution of this book lies in its innovation. Globally, studies on traditional knowledge and climate change have largely remained fragmented case studies. This book, however, takes the Tibetan community as a systematic case study, comprehensively mapping the relationship between traditional knowledge and climate change adaptation. It not only highlights the rich diversity of local knowledge but also emphasizes its universal value in global climate governance. This integration from the local to the global not only provides a unique case study for Chinese academia but also offers a new research paradigm and practical experience for the international community.

Implications for the Future

The significance of this book far exceeds the academic realm. For China, it demonstrates the vibrant practice of the "ecological civilization" concept in ethnic regions. By combining traditional knowledge with modern science, it provides valuable case studies and insights for the nation's efforts in addressing climate change, rural revitalization, and biodiversity conservation. For the international community, this book emphasizes the importance of listening to local voices and respecting traditional wisdom in global climate governance. While science and technology are undoubtedly crucial, without the involvement of community practices and cultural foundations, global governance will lose its essential social support.

For young scholars and future scientists, Prof. Yin Lun's research is an inspiration: scientific research is not just about accumulating theories and data, but about deeply interacting with real-world issues, with people, and with nature. Only by combining academic knowledge with action can science truly possess the power to change the world.

For the general reader, this book reminds us to reconsider our relationship with nature. Climate change is not a distant scientific issue, but a real-world problem that permeates daily life, cultural traditions, and social structures. Understanding and drawing upon traditional knowledge may be a crucial step toward finding our path to the future.

Conclusion

As a scholar who has long been engaged in scientific research and education, I am acutely aware of the gap between science and society, and I understand the necessity of bridging this gap. This book represents such an attempt. It is not only the result of Prof. Yin Lun and his team's over two decades of dedicated effort but also a bridge in humanity's collective response to climate change. I offer my heartfelt congratulations and gratitude to Prof. Yin Lun and his collaborators. I also hope this book will inspire more individuals—whether scientists, policymakers, or ordinary readers—to reflect on how they can contribute their wisdom and actions toward mitigating and adapting to climate change in their own lives and work.

Under the concept of a "community of shared human destiny", we need not only the power of science but also the wisdom of culture and the courage of morality. May this book serve as a beacon, guiding us toward a more sustainable and harmonious future.

Pierre Léna
Membre de l'Académie des sciences, Paris
Academia Europaea, Pontifical Academy of Sciences
Professeur émérite, Université Paris Cité and Observatoire de Paris
Président d'honneur, Fondation La main à la pâte et Office for Climate Education

Foreword by Prof. Joachim von Braun

Traditional Knowledge: A Cornerstone of Climate Resilience in Himalayan Tibetan Communities

Climate change is one of the defining challenges of our time—its impacts reverberate across every continent, ecosystem, and community. Yet the burden of this crisis is not shared equally. Mountain regions, particularly the Eastern Himalayas, are among the most vulnerable: warming at rates three times the global average, glaciers retreating at unprecedented speeds, alpine meadows shrinking, and river systems—lifelines for billions downstream—experiencing increasingly erratic flows. For Tibetan communities who have inhabited these landscapes for millennia, climate change is not an abstract scientific concept but a daily reality: it disrupts crop cycles, dries pastures, shifts the ranges of medicinal plants, and increases the frequency of disasters such as landslides and flash floods. It is within this urgent context that the present book takes shape. Authored by Prof. Yin Lun, it brings together two decades of immersive fieldwork, community collaboration, and interdisciplinary research—a testament to the power of centering Indigenous voices in climate action.

As President of the Pontifical Academy of Sciences, I have long advocated for a holistic approach to global challenges—one that honors the interdependence of humans and nature, and affirms that solutions to climate change must be rooted not only in scientific innovation but also in the accumulated wisdom of local communities. The Academy's mission, as articulated in *Laudato Si'*, emphasizes that "care for our common home" requires respecting the rights and knowledge of Indigenous peoples, who have long served as stewards of fragile ecosystems. Professor Yin's work embodies this ethos. It moves beyond abstract debates over climate policy or isolated accounts of traditional practices, offering instead a systematic, evidence-based framework that connects Tibetan traditional knowledge to tangible climate resilience. For scholars, policymakers, practitioners, and all those committed to equitable climate action, this book stands as an indispensable resource.

The strength of this work lies, first and foremost, in its unwavering commitment to community-led research. Professor Yin does not position himself as an "external expert" observing Tibetan communities from a distance; instead, he serves as Collaborator and Facilitator, working alongside local villagers, Tibetan doctors, traditional leaders, and women's groups to document, analyze, and activate traditional knowledge. This approach—rooted in community-based action research (CBAR)—ensures that the book's insights are not only academically rigorous but also directly relevant to the people most affected by climate change. For instance, in Chap. 2, Prof. Yin describes how community meetings—where villagers used pebbles to vote on climate priorities—identified drought, premature crop maturation, and pest outbreaks as the most pressing threats. This participatory process did more than collect data: it empowered villagers to take ownership of climate adaptation, transforming "research subjects" into "agents of change." Such methods challenge the top-down paradigms that have long dominated climate research, reminding us that the most effective solutions often emerge from those closest to the problem.

A second defining contribution of this book is its nuanced, dynamic understanding of traditional knowledge. Professor Yin rejects the simplistic view of traditional knowledge as a static "relic of the past"—a collection of outdated proverbs or rituals. Instead, he frames it as a living, adaptive system that evolves in response to environmental change. He categorizes this knowledge into two interconnected pillars: knowledge related to climate (encompassing climatological and phenological insights) and knowledge related to biodiversity (spanning genetic, species, and ecosystem diversity). Each pillar is explored through rich, ground-level case studies that reveal the depth and practicality of Tibetan wisdom.

Consider, for instance, the book's exploration of traditional climate knowledge. Tibetan villagers in Deqin County do not rely on thermometers or satellite data to predict weather; instead, they use a sophisticated system of observations: red clouds at dusk signal clear skies, while dark clouds foretell rain; yaks stamping their hooves indicate impending storms; and the flowering of walnut trees marks the start of spring planting. These indicators, passed down through generations, are not just "folk wisdom"—they are empirically grounded. Professor Yin's team validated many of these observations with modern meteorological data, finding that local predictions of rainfall timing and drought periods were often as accurate as regional climate models. This alignment underscores a critical point: traditional knowledge is not an alternative to science, but a complementary tool—one that captures the fine-grained, local nuances of climate change that global models often miss.

Equally compelling is the book's exploration of traditional biodiversity knowledge. Professor Yin divides this into three layers: genetic diversity (e.g., the selection of drought-tolerant barley varieties), species diversity (e.g., the classification and use of medicinal plants), and ecosystem diversity (e.g., the management of forests, meadows, and rivers). Chapter 4, which focuses on Tibetan medicinal plant knowledge, is particularly insightful. Local Tibetan doctors—who learn their craft through monasteries, family lineages, or folk traditions—have classified hundreds of medicinal plants based on altitude, slope, and growing conditions. Over the past decade, these doctors have documented a clear shift in medicinal plant ranges: species once found at 3000 m now grow at 3500 m, a direct response to warming temperatures. Professor Yin uses this traditional classification to quantify species migration—turning indigenous knowledge into a measurable indicator of climate change. This innovation is groundbreaking: it demonstrates that traditional knowledge can not only inform adaptation but also contribute to scientific understanding of ecological shifts.

What makes this work truly transformative, however, is its focus on bridging divides. It bridges natural and social sciences: ecological data on glacier retreat is paired with ethnographic accounts of how "sacred mountain" beliefs drive forest conservation. For example, Tibetan villagers consider Mount Kawagebo—a 6740-meter peak—as a protector, and they prohibit logging or grazing in its surrounding forests. This cultural practice, Prof. Yin shows, has not only preserved biodiversity but also enhanced carbon sequestration, making sacred forests effective carbon sinks. The book also bridges traditional and modern practices: Tibetan communities have integrated solar water heaters and greenhouses into their livelihoods, using traditional knowledge to select optimal locations for these technologies (e.g., placing greenhouses on south-facing slopes identified in local phenological wisdom). And crucially, it bridges local action and global governance. The final chapters (Chaps. 11 and 12) situate Tibetan traditional knowledge within the United Nations Framework Convention on Climate Change (UNFCCC) and China's "ecological civilization" framework, arguing that indigenous wisdom must be integrated into international climate policy to achieve equity and effectiveness. Too often, global climate negotiations prioritize technological fixes and macro-level targets, while sidelining the knowledge of communities that have adapted to environmental change for centuries. Professor Yin's work challenges this imbalance, showing that traditional knowledge is essential to making climate action more inclusive, resilient, and just.

As someone who has spent decades researching economic and technological change in the context of sustainable development, I am particularly struck by the book's focus on livelihood resilience. The Tibetan communities studied here are semi-agricultural and semi-pastoral, relying on highland barley, yaks, forests, and medicinal plants for survival. Climate change threatens all these resources: droughts reduce crop yields, rising temperatures alter pasture quality, and melting glaciers disrupt water supplies. Yet, through traditional knowledge, communities adapt. They select drought-tolerant barley varieties like "Chaguo barley" for high altitudes and "Red-awned barley" for dry valleys. They practice rotational grazing—moving livestock between summer alpine pastures (4000 m), autumn transitional pastures (3000 m), and winter valley pastures (2000 m)—to prevent overgrazing and ensure forage availability. They build community-managed reservoirs, using customary laws to allocate water equitably between upstream and downstream households. These practices are not just "adaptations" in the technical sense; they are acts of cultural preservation. By maintaining their traditional livelihoods, Tibetan villagers preserve their identity, their social structures, and their sacred relationship with the land. This is a powerful reminder that climate adaptation is not just about surviving change, but about thriving while honoring what makes communities unique.

The book also addresses a pressing question: How can traditional knowledge be protected and amplified in an era of rapid environmental and social change? Professor Yin offers concrete, community-driven solutions. He highlights the role of traditional organizations like the "Sisterhood" (a women's group) and the "Arrow Association" (a men's group) in managing natural resources. The Sisterhood, for instance, leads forest restoration efforts, planting native species like Himalayan cypress and monitoring saplings to prevent livestock damage. The Arrow Association oversees watershed management, organizing villagers to clean rivers and reinforce embankments during the rainy season. Professor Yin also advocates for policy integration: he calls for traditional knowledge to be included in China's Nationally Determined Contributions (NDCs) and for indigenous voices to be heard in UNFCCC negotiations. These recommendations are rooted in the principle of ownership—communities must lead in defining how their knowledge is used, shared, and protected.

In closing, I wish to emphasize the timeliness and global relevance of this work. As the world grapples with the urgent need to implement the Paris Agreement and achieve the Sustainable Development Goals, Traditional Knowledge Systems offers a roadmap for centering equity and sustainability in climate action. It shows that the solutions to climate change are not just in laboratories or boardrooms—they are in the fields where Tibetan farmers select drought-tolerant seeds, in the pastures where herders practice rotational grazing, and in the forests where communities protect sacred groves. For scholars, this book provides a model for interdisciplinary, community-engaged research. For policymakers, it offers actionable insights into designing climate policies that are both effective and inclusive. For all readers, it is a testament to the resilience of human communities and the wisdom of traditional ecological knowledge.

Professor Yin Lun's dedication to this work—two decades of living, learning, and collaborating with Tibetan villagers—shines through every page. This book is not just a scholarly achievement; it is a tribute to the power of partnership between researchers and communities. It is also a call to action: to recognize that indigenous knowledge is a global public good, one that can help us build a more sustainable, equitable, and resilient future.

As we face an uncertain climatic future, we would do well to listen to the voices of the Tibetan communities in this book. They remind us that humans are not masters of nature, but part of it—and that our survival depends on honoring the wisdom of those who have lived in balance with the earth for generations. May this book inspire more scholars, policymakers, and practitioners to embrace traditional knowledge, and to work alongside indigenous communities in building a world where people and nature thrive together.

Prof. Joachim von Braun
President of the Pontifical Academy of Sciences
Professor (em.) for Economic and Technological Change
Center for Development Research (ZEF), Bonn University

Foreword by Nigel Crawhall

Traditional Knowledge: A Beacon Wisdom for Climate Change Adaptation

It is with enthusiasm that we welcome this new publication on traditional knowledge systems as a foundation for understanding, responding, and adapting to climate change.

The work of UNESCO's Local and Indigenous Knowledge Systems (LINKS) program for over the last two decades has helped us understand the critical importance of mobilizing different knowledge systems to address contemporary developmental and sustainability challenges. Indigenous and local knowledge, in complementarity to formal science, can greatly enhance our solutions and develop new pathways to connect observation, knowledge, action, and policy.

Professor Lun Yin brings a wealth of experience from fieldwork into our understanding of day-to-day responses to the rapidly changing climate. It is axiomatic that climate adaptation is something that happens in situ, based on people's best understanding of their needs, the effect on natural resources and ecosystem services, and the trends that they observe. Most rural communities have been able to develop seasonal calendars and knowledge systems that take into account variability, while in our contemporary context, climatic and environmental changes are becoming more unstable with higher frequency of extreme events. Traditional systems of natural resource management, including transhumant pastoralism and various forms of horticulture, and agricultural, along with artisanal fishing emerged to cope with these variations.

As the publication highlights, the United Nations systems have come to place greater emphasis on the traditional, indigenous, and local knowledge systems as the frontline of coping with rapid climate and environmental changes, while also growing new forms of cooperation between community-based knowledge systems and formal science. The challenge of our times is to bring the multiple evidence of challenges, opportunities, and solutions to decision-makers who are able to create enabling policy contexts and learning cycles to help us navigate these unprecedented changes.

Researchers, policymakers, educators, and community leaders will find this book insightful and helpful in our common quest for sustainability.

Nigel Crawhall
Chief of Section, Local and Indigenous Knowledge Systems
(LINKS) Programme
United Nations Educational, Scientific and Cultural Organisation
(UNESCO)
Paris, France

Preface

This book is the culmination of my two decades of work on adapting to climate change by the traditional knowledge related to biodiversity in the Eastern Himalayas of China, which represents a significant milestone in my journey. As I began writing, vivid memories of my life and work in Tibetan villages across the Eastern Himalayas over the past 20 years emerged—the faces of villagers, the snow-capped mountains, forests, rivers, and lakes. Writing a Preface is of great necessity, not only to express gratitude for all that has shaped this work but also to invite readers to step into the scene and experience just as I did picturesquely.

Traditional Culture and Beliefs as Inspiration

Dating back 20 years, during my tenure on a biodiversity conservation project in the region, I traveled by vehicle between a Tibetan village situated along the Lancang River and another Tibetan settlement adjacent to Baima Snow Mountain. Although these two villages belong to the same township and are geographically proximate, our journey constituted a spectacular traverse from the river valley at an altitude of 1800 m to the alpine meadow standing at 4000 m.

Throughout the lengthy journey, we witnessed the vibrant vitality of biodiversity and profoundly experienced the movement of the climate, ranging from scorching heat to refreshing coolness and even bone-chilling cold. The devastation caused by extreme weather events was particularly alarming, especially when the winding road carved into the steep cliffs was unfortunately disrupted by flash floods triggered by heavy rains a few days prior. With no other choice, I had to disembark and wait for the road to be repaired. An elderly Tibetan Buddhist lama, who was also a passenger in the vehicle, sat beside me by the roadside. In front of us stood the renowned Mount Taizi, composed of thirteen majestic snow-capped peaks, each revered as a sacred and inviolable deity in local traditional culture. Among them, Mount Kawagebo and its companion, Mount Miancimu, stand proudly above 6000 m in altitude.

As we gazed silently at the continuous chain of snow-capped peaks, the lama suddenly whispered, "Look, it seems these sacred mountains are gradually shedding their heavy garments. What used to be adorned in silver and white is now donning a lightweight sleeveless attire." Seeing my puzzled expression, he smiled gently and continued, "The glaciers and snow covering the mountain peaks are like the finery of the gods. However, due to climate change, the glaciers are melting and the snow is decreasing, just as the gods' robes are becoming thinner and thinner." These casual remarks by the lama stirred ripples in my heart. I realized that while climate change is a global issue, at the local level, it has sparked keen observations, profound reflections, and unique wisdom among the locals. This traditional knowledge based on personal experiences and thoughts not only reflects the locals' intuitive feelings and deep understanding of climate change but also contains valuable mechanisms and effective methods for addressing it. It is these profound insights and wisdom that have served as the source of inspiration and thought for me in writing this volume.

The Traditional Knowledge Associated with Biodiversity Adapting to Climate Change as Fundamental Element

My interest in the realm of traditional knowledge and climate change is not merely grounded in the soil of academic research; rather, it stems more profoundly from personal involvement in action-oriented projects. Deeply inspired by the Tibetan Buddhist lama, in 2005, I successfully applied to the United Nations Development Programme (UNDP) for the inaugural action research project focusing on "Traditional Knowledge and Climate Change" titled "Traditional Livelihood Knowledge of Tibetan Semi-Agricultural and Semi-Pastoral Communities in Northwest Yunnan and Strategies for Climate Change Adaptation," which pioneered the field in China. Subsequently, in 2022, I was honored to receive funding from the National Social Science Fund of China to undertake a research project Study on Traditional Knowledge Systems of Ethnic Groups in the Tibetan-Yi Corridor for Climate Change Adaptation, marking a historic first step in the mainstream social science community of China in exploring traditional knowledge and climate change research.

To date, across the vast expanse of the Eastern Himalayas in China, I have had the privilege of leading and implementing over twenty action projects and research initiatives closely tied to traditional knowledge and climate change. These projects and initiatives comprehensively cover multiple dimensions of traditional knowledge, including but not limited to water resource management, utilization of medicinal plants, maintenance of livelihoods, forest ecology conservation, preservation of genetic resources, gender role considerations, disaster risk management, and the integration of religious beliefs and practices. Each research endeavor and practical application is a profound exploration and tribute to the unique value and role of traditional wisdom in addressing the global challenge of climate change.

Based on the experience and achievements of these projects, the traditional knowledge for coping with climate change had been classified into five categories: 1. Traditional phenological information for observing climate change; 2. Traditional natural resource utilization techniques for mitigating climate change; 3. Traditional livelihoods for adapting to climate change; 4. Traditional ecosystem management mechanisms for reducing climate disaster risks; 5. Traditional cultures and customs for understanding climate change. These categories form the core chapters of the volume.

Community-Based Action Research as Methodology

CBAR is unequivocally dedicated to addressing climate change. Over the span of two decades, I have immersed myself in an innovative research endeavor within Tibetan villages nestled in the eastern Himalayas of China, an endeavor that is resolutely aimed at harnessing and utilizing traditional knowledge to counter the pervasive threats posed by climate change. This project, which unwaveringly adopts CBAR as its foundational methodology, underscores the imperative of close collaboration with villagers, jointly dissecting the multifaceted challenges and potential perils introduced by climate change through a steadfast commitment to "learning by doing," fostering continuous growth and exploration in practice.

Within this series of meticulously crafted CBAR studies, I have stood in solidarity with Tibetan villagers, leveraging their time-honored traditional knowledge to collaboratively devise an array of strategies and measures for climate change adaptation. Building upon their unique traditional social organizations, such as the "Tibetan Sisterhood" and the "Tibetan Archery Association," we have innovatively established rural practice laboratories for community-based climate change adaptation, providing villagers with an unparalleled platform to exhibit and perpetuate their traditional knowledge, while also affording them the means to transform this wisdom into tangible actions. Moreover, we have implemented tree planting projects on sacred mountains, utilizing traditional knowledge to conserve these revered forests, initiatives that have not only fortified the ecological resilience of the community but have also deepened villagers' reverence and connection to nature and the Earth Mother in an unwavering manner.

It is worth emphasizing that this research has unequivocally revealed the remarkable wisdom and strength of Tibetan villagers in addressing climate change, demonstrating the immense potential of traditional knowledge in tackling the pressing issue of global climate change. Through community-led action research, we have successfully translated traditional knowledge into practical actions for climate change adaptation, furnishing villagers with locally relevant and highly effective response solutions. In stark contrast to externally driven adaptation schemes, these community-based plans are readily embraced and adopted by villagers, as they comprehensively consider the unique geographical and environmental conditions of the area, enabling villagers to adeptly apply traditional knowledge to more effectively address the challenges posed by climate change in a resolute manner.

Furthermore, this research offers a unique perspective, granting us a deeper understanding of the impacts and challenges of global climate change from the vantage point of local communities. By delving deeply into the wisdom and experience embedded within traditional knowledge, we have provided local voices and insights for the global response and adaptation to climate change, furnishing invaluable inspiration and reference for formulating fair and effective climate change adaptation plans.

In conclusion, this volume delves deeply into the methodology of community-based adaptation to climate change, leveraging traditional knowledge, while simultaneously showcasing the unparalleled value and promising prospects of this approach in the global fight against climate change, as evidenced through a series of practical cases and research findings. Our unwavering belief is that through persistent exploration and the implementation of community-based action research, we can collectively aspire to a greener and more sustainable future, a future that is resolutely within our reach.

Gender Perspective on Climate Change Stands as a Distinctive Highlight

In my in-depth collaboration with local women, we have meticulously crafted a set of climate change adaptation strategies, grounded in their ancestrally transmitted traditional knowledge. Our journey began with a comprehensive assessment of the multifaceted threats and challenges posed by climate change, followed by the meticulous documentation and organization of the treasure trove of traditional knowledge closely associated with climate change within the female community. On this foundation, we have carefully planned and implemented a series of actions to adapt to climate change: leveraging knowledge of traditional agricultural genetic resource selection, we formulated a blueprint for promoting livelihood diversification; harnessing the wisdom of traditional medicine, we conducted precise diagnoses and effective treatments for locally prevalent diseases; integrating traditional techniques and lifestyles for sustainable utilization of biological resources, we installed vegetable greenhouses, solar energy devices, and canal irrigation systems; based on traditional cultural concepts closely linked to biodiversity, we established microfinance and credit mechanisms to address climate change; and, with a profound understanding of traditional bio-geographical indication products, we successfully developed diversified local agricultural and livestock products. These practical initiatives have significantly enhanced the resilience of local women in mitigating the disaster risks and other adverse impacts of climate change, thereby playing a pivotal role in the journey of adapting to climate change. I profoundly recognize that women's traditional knowledge serves as a treasure trove, profoundly deepening our understanding of the impacts of climate change. In the global battle against climate change, women are not only indispensable forces but also crucial participants, armed with their invaluable traditional knowledge. They are far from passive recipients of climate change; rather, they are active agents in mitigating and adapting to it. Their traditional knowledge is not only a solid foundation for their survival but also provides us with novel perspectives and approaches to tackle climate change. Although these traditional knowledge and experiences may not be easily encapsulated by the term "scientific," they undoubtedly represent a brilliant beacon of wisdom, illuminating and guiding our path forward.

Act Local and Think Global as the Core Concept

Little did I anticipate, at one time, that a research project on climate change and traditional knowledge conducted in a remote Tibetan village in the eastern Himalayas of China would attract widespread international attention. This transformation stems from a series of subsequent significant events:

In 2017, I unexpectedly received an invitation from the Climate Reality Leadership Corps, led by former US Vice President Mr. Gore, to participate in the annual 24 h of Reality global climate broadcast. To this end, a film crew was specifically dispatched to China to document the vivid practices of how I, along with local Tibetan villagers, utilized traditional knowledge to tackle climate change.

In both 2019 and 2020, I consecutively received letters from the Paris Peace Forum, announcing that two of my projects—traditional knowledge related to biodiversity in addressing climate change and traditional forest knowledge in adapting to climate change—had stood out from over 800 candidates worldwide to become showcase highlights at the second and third Paris Peace Forums, respectively. This undoubtedly presented an excellent opportunity to demonstrate to the international community the unique value of traditional knowledge in conserving biodiversity and addressing climate change. During the forums, United Nations Secretary-General António Guterres exhibited keen interest and attention toward my projects. He posed two pivotal questions to me: What constitutes traditional knowledge? And what role can traditional knowledge play in addressing climate change and conserving biodiversity? After receiving my detailed responses, Mr. Guterres highly praised my projects, noting that although they were rooted in "traditional knowledge," they embodied "innovative spirit" and served as an inspiration and model for the global fight against climate change.

In 2020 and 2022, I was invited to speak at the National Tribal and Indigenous Climate Conference, delivering keynote addresses on the themes of traditional water knowledge in adapting to climate change and traditional knowledge in mitigating climate disaster risks, respectively.

In 2023, drawing on practical experience from the eastern Himalayas, I initiated a new action research project in the Nganyi Community in western Kenya, aiming to explore the application of traditional livelihood knowledge in addressing climate change and traditional beliefs in adapting to climate change.

In 2024, I was invited to the Pontifical Academy of Sciences in Vatican City to deliver a keynote speech on traditional knowledge related to biodiversity in addressing climate change. During this time, I had the honor of being personally received by Pope Francis and listening to his profound insights on traditional knowledge and climate change.

Also in 2024, I received gratifying news from the Asian Development Bank and the Central Asia Regional Economic Cooperation (CAREC). Based on my work in the field of gender and climate change, I emerged victorious from over 800 candidates worldwide to win the 2024 Gender and Climate Change Award, becoming the first male recipient since its establishment.

Therefore, the message I aim to convey through this volume is unequivocal: traditional knowledge, as a valuable knowledge system for all humanity, plays an irreplaceable role in addressing the current severe challenges posed by global climate change. In the context of global climate change, traditional knowledge should be regarded as a treasure trove of knowledge for human society in combating climate change.

Building upon this volume, I will redouble my efforts to continue initiating actions for adapting traditional knowledge to climate change at the local level and further promoting the understanding of traditional knowledge in addressing climate change globally.

Traditional knowledge embodies the wisdom that shines brightly in tackling climate change!

Kunming, China
February 2025

Yin Lun

Acknowledgements I would first like to acknowledge the generous financial support from the Project of the National Social Science Fund of China, "Traditional Knowledge Systems and Values of Ethnic Groups in the Tibetan–Yi Corridor for Coping with Climate Change within Ecological Symbiosis Relationships" (22BMZ032), and from the Yunnan Province Philosophy and Social Sciences Innovation Team Construction Project, "Research on Traditional Ecological Knowledge for Climate Change Response Among Southwest Ethnic Groups" (2022cx04). Without the generous support and intellectual nourishment from these two projects and team members, this academic journey would have been far more challenging.

I am also grateful to Southwest Forestry University in China for its consistent support. In particular, I owe special thanks to the Climate Action Research Team for their invaluable help in editing the initial draft, as well as their meticulous efforts in proofreading and refining the later versions of the manuscript.

Furthermore, I would like to offer special thanks to many scholars and experts whose insights and perspectives have consistently inspired me throughout my research on traditional knowledge and climate change. They include Prof. Pierre Léna, Prof. Joachim von Braun, Mr. Nigel Thomas Crawhall, Prof. Alexandra Mendoza-Caminade, Prof. Melissa Nursey-Bray, Ms. Ann Marie Chischilly, Prof. Marie Kvarnstrom, Prof. Jean-Marc de Grave, Prof. Simon J. Lambert, Prof. Xue Dayuan, Prof. Long Chunlin, Prof. Guo Luo, Prof. Peng Yu, Prof. Tirso Gonzales, Ms. Zhou Hao, Ms. Huang Qing, Mr. Liu Wei, Ms. Li Xunqi, Mr. Xu Weilin, and Mr. Misiani M. Zachary.

I would like to express my profound gratitude particularly to numerous Tibetan villagers who generously shared with me their traditional knowledge and ecological wisdom pertaining to climate change. Over the past two decades, we have lived and worked together, taking collective action to address the challenges of climate change. They are Tsering Zhaba, Lurong Sinan, Tsering Samdup, Dingzhu Peichu, Aru, Amaji, Songji Kangzhu, Tibuci Li, Liqing Peichu, Qunlu, Danzhen Leci, Sinan Larong, Gerong Zhuoma. In this volume, you are the local experts who possess diverse traditional knowledge, and in life, you have been my constant companions, brothers, and closest friends. This volume also belongs to you. Your ecological wisdom and traditional knowledge illuminate the richness and diversity of human ways of adapting to an ever-changing environment.

There is a kind of support that is of utmost importance yet cannot be adequately thanked with words: the meticulous care and encouragement provided by my parents. Special thanks also go to my wife, Yanyan, who is likewise a Tibetan lady, for she have always inspired and supported me to achieve greater accomplishments.

The assistance of the Springer team, particularly that of Mr. Ragavendar Mohan and Mr. Zachary Romano, is deeply appreciated.

Competing Interests The author has no competing interests to declare that are relevant to the content of this manuscript.

Contents

About the Author

Prof. Yin Lun is Dean and Professor at the School of Marxism at Southwest Forestry University in China. He is also Research Fellow at the Southwest Research Center for Eco-civilization, under the National Forestry and Grassland Administration, and Research Fellow at the Center for Ecological Civilization Think Tank in Yunnan, Chinese Academy of Social Sciences. Additionally, he serves as Director of the Center for Biodiversity and Indigenous Knowledge (CBIK).

Professor Yin Lun is Founder and Director of the Climate Action SDG Laboratory. He is Ethno-Ecologist of Bai ethnic minority background, based in Kunming, Yunnan Province, China.

Professor Yin Lun also holds positions as Indigenous and Traditional Knowledge Consultant for the Intergovernmental Science-Policy Platform on Biodiversity and Ecosystem Services (IPBES) and as Senior Expert in Indigenous Knowledge with the United Nations Office for Disaster Risk Reduction (UNDRR).

Professor Yin Lun's major work focuses on the traditional knowledge of indigenous peoples and local communities worldwide, particularly its connection to biodiversity and climate change. His efforts are dedicated to supporting the adaptation and resilience of these communities in the face of climate change, aiming to ensure sustainable livelihoods. He leads numerous action projects and scientific research in collaboration with indigenous peoples and local communities, with a focus on biodiversity, traditional knowledge, and climate change. These projects are conducted in partnership with governmental, non-governmental, and research organizations. Professor Yin Lun is committed to serving as a bridge between traditional knowledge and scientific knowledge, and between local communities and policymakers, to improve decision-making that influences climate change adaptation and biodiversity conservation. His work emphasizes the importance of grounding conservation and development interventions in the traditional knowledge and cultural assets of indigenous peoples and local communities.

Professor Yin Lun was Winner of the 2024 Central Asia Regional Economic Cooperation (CAREC) Gender Climate Award.

Abbreviations

AR4	The Fourth Assessment Report by IPCC
AR6	The Sixth Assessment Report by IPCC
CBD	Convention on Biological Diversity
CBWM	Community-Based Water Resource Management
CNCCP	China's National Climate Change Programme
COP	Conference of the Parties
DRR	Disaster Risk Reduction
FAO	Food and Agriculture Organization
FPIC	Free, Prior and Informed Consent
FWG	Facilitative Working Group
GBIP	Gender-Based Irrigation Plan
GCA	Global Commission on Adaptation
GCC	Global Climate Convention
GCF	Green Climate Fund
GHG	Greenhouse Gas
RGG	Reducing Greenhouse Gas Emissions
ICIMOD	International Centre for Integrated Mountain Development
IGBP	International Geosphere-Biosphere Programme
IGC	Intergovernmental Committee
IIRR	International Institute for Rural Reconstruction
ILCs	Indigenous Peoples and Local Communities
ILO	International Labour Organization
IPCC	Intergovernmental Panel on Climate Change
IUCN	International Union for Conservation of Nature
LCIPP	Local Communities and Indigenous Peoples Platform
LK	Local Knowledge
MEE	Ministry of Ecology and Environment
NAPA	National Adaptation Programme of Action
NAPs	National Adaptation Plans
NDCs	Nationally Determined Contributions
NGOs	Non-Governmental Organizations
NWP	Nairobi Work Programme
PGRs	Plant Genetic Resources
PMCid	PubMed Central Identifier
PMid	PubMed Identifier
PNAS	Proceedings of the National Academy of Sciences of the United States of America

PRA	Participatory Rural Appraisal
SDGs	Sustainable Development Goals
TEK	Traditional Ecological Knowledge
TK	Traditional Knowledge
TRIPS	Trade-Related Aspects of Intellectual Property Rights
UNCCD	United Nations Convention to Combat Desertification
UNCED	United Nations Conference on Environment and Development
UNEP	United Nations Environment Programme
UNFCCC	United Nations Framework Convention on Climate Change
UNISDR	United Nations International Strategy for Disaster Reduction
WCIP	World Council of Indigenous Peoples
WIPO	World Intellectual Property Organization

Climate Change and Traditional Knowledge Among Tibetan Peoples of the Eastern Himalayas, China

1

1 Introduction

The Himalayan Mountains stretch westward from the northeast of Yunnan Province, China, to the northern regions of Pakistan and Afghanistan. This region, recognized as one of the world's biodiversity hotspots, is also home to an extraordinary cultural diversity among its numerous mountain ethnic groups. In recent years, the impacts of climate change have become increasingly evident both within the Himalayan region and beyond (Byg & Salick, 2009a, 2009b). The detrimental effects of climate change are not confined to the local area; they extend into the lower reaches of the region via the major Asian rivers originating in the Tibetan highlands, such as the Mekong, Salween, and Yangtze Rivers. These rivers are critical lifelines for hundreds of millions of downstream inhabitants, both human and non-human (Yin, 2011). As a result, the climate change occurring in the Himalayan region has gradually become a global issue of common concern (Yin et al., 2013).

Climate change impacts are already evident in the Himalayas. Limited studies on temperature and precipitation in a few localized areas indicate that warming in the Himalayas is occurring at a rate three times greater than the global average. As a result, the Himalayas are among the regions most vulnerable to climate change and are experiencing rapid environmental changes (Xu et al., 2009; Yin, 2015). The upstream snow and ice reserves in the Himalayas, crucial for sustaining seasonal water availability, are likely to be significantly affected by climate change. A shift in winter precipitation from snow to rain, driven by rising temperatures, alters the timing of peak stream flows. This change results in an earlier or even the complete elimination of the spring snow melt peak, while winter flows increase. As glaciers retreat due to warming, river flows initially increase but decline once the glaciers disappear. Ongoing climate change over the coming decades is expected to exacerbate these impacts, leading to further negative effects on river flows, groundwater recharge, natural hazards, and biodiversity. It will also affect ecosystem composition, structure, and function, as well as human livelihoods. Additionally, as glaciers retreat from prominent Little Ice Age moraines in the Himalayas, new lakes are forming. These glacial lakes pose a high risk of glacial lake outburst floods (Vedwan et al., 2001; Tse-ring et al., 2010).

In recent years, climate change and natural disasters have increasingly jeopardized the lives of local Tibetan communities, affecting their agro-pastoral livelihoods, water resources, and biodiversity, including mountain pastures, valleys, and forests. Due to the fragile natural environment and the deep reliance of these communities on these resources, such challenges have severely threatened their sustainable development and livelihoods (Yin, 2011).

Y. Lun, *Climate Change and Traditional Knowledge*, Sustainable Development Goals Series,
https://doi.org/10.1007/978-3-032-12015-1_1

This chapter focuses on climate change and traditional knowledge among Tibetan Peoples of the Eastern Himalayas, China, to effectively and sustainably respond to the impacts of climate change. This chapter begins with a comprehensive literature review on traditional knowledge and climate change, drawing on available data sources. The review encompasses two main components: international treaties and conventions, and relevant academic disciplines. It includes studies from a range of related fields, such as agrarian sciences, cultural anthropology, applied geography, ethno-ecology, and ethno-botany. Secondly, building on the literature review, this chapter analyzes the relationship between traditional knowledge and climate change. It primarily draws on qualitative research methods, including case studies and observational studies. Based on insights from previous chapters, the chapter categorizes traditional knowledge into two levels:

At the first level, traditional knowledge is conceptualized as a knowledge system related to weather and climate. This encompasses the understanding of seasonal divisions, solar terms, weather and climate predictions, and early warnings for extreme weather events and climate-related disasters. This system is further divided into:

- Traditional Climatological Knowledge: Pertaining to climate and weather patterns.
- Traditional Phenological Knowledge: Concerned with the timing of biological events and environmental changes.

At the second level, traditional knowledge is defined as a knowledge system related to biodiversity. This includes knowledge about forests, water resources, plants, medicine, and genetic resources. Given that biodiversity encompasses genetic diversity, species diversity, and ecosystem diversity, this knowledge system is similarly divided into:

- Traditional knowledge related to genetic diversity.
- Traditional knowledge related to species diversity.
- Traditional knowledge related to ecosystem diversity.

The purpose of this study was to advance beyond previous research on climate change, which has largely been confined to the field of natural science. Adopting an Ethnic Ecology perspective, this study employs interdisciplinary research methods and focuses on the Tibetan people in the eastern Himalayas as a specific case. It examines the impact of climate change on traditional knowledge related to biodiversity, explores theories connecting climate change and traditional knowledge, investigates quantitative analysis and research methodologies for traditional knowledge, and identifies practical approaches for adapting to local climate change and disasters based on traditional knowledge.

The significance of this study is as follows: First, it addresses gaps in research on climate change and traditional knowledge by facilitating interdisciplinary and innovative exploration. Second, it enhances understanding and awareness among various interest groups about traditional knowledge, and investigates the integration of traditional knowledge with policy and scientific knowledge to innovate and practice climate change adaptation. Third, the research on traditional knowledge provides a foundation for future climate change responses and the prevention and management of climate-related disasters, aiming to reduce disaster risk and vulnerability. Fourth, it encourages the inclusion of traditional knowledge in the policymaking process related to climate change.

The exploration of the relationship between climate change and traditional knowledge represents an innovation of this study. Building on interdisciplinary theories and methods, this study investigates this relationship within the field of ethno-ecology. It seeks to transcend the limitations of natural science research on climate change, which often overlooks and under appreciates the traditional knowledge of indigenous peoples and local communities.

Additionally, the study aims to address the shortcomings of traditional knowledge research in the social sciences, which typically views traditional knowledge merely as indigenous cognitive frameworks and proverbs for predicting climate and phenological changes.

The exploration of the relationship between climate change and traditional knowledge represents an innovation of this study. Building on interdisciplinary theories and methods, this study investigates this relationship within the field of ethno-ecology. It seeks to transcend the limitations of natural science research on climate change, which often overlooks and underappreciates the traditional knowledge of indigenous peoples and local communities.

Additionally, the study aims to address the shortcomings of traditional knowledge research in the social sciences, which typically views traditional knowledge merely as indigenous cognitive frameworks and proverbs for predicting climate and phenological changes.

Based on the classification of traditional knowledge related to climate change and an analysis of the impacts of climate change and disasters on this knowledge, this study expands and clarifies the scope and content of traditional knowledge. It aims to establish a research framework for understanding the intersection of climate change and traditional knowledge. This framework encompasses three aspects: traditional knowledge related to climate change, the impact of climate change on traditional knowledge, and the role of traditional knowledge in adapting to climate change.

2 Traditional Knowledge

Traditional knowledge, also known as indigenous knowledge or local knowledge, is accumulated and developed by communities based on their experiences. This knowledge is validated through practical application and continually evolves to adapt to changes in natural, cultural, and social environments.

Traditional knowledge is a broad and dynamic system encompassing various dimensions of the life ways and knowledge systems of diverse indigenous peoples and local communities. It refers to the ways of knowing that both guide and emerge from community members' close relationships with, and responsibilities toward, the landscapes, waterscapes, plants, and animals crucial for cultural flourishing. Over thousands of years, indigenous peoples and local communities have refined their traditional ways of knowing and living through continuous experiences and interactions with living beings and their environments.

The definition of traditional knowledge can be categorized into three main areas:

1. International Treaties and Conventions: This definition, particularly within the Convention on Biological Diversity, frames traditional knowledge as a system related to biodiversity. It is further specified under the concept of access and benefit-sharing as knowledge pertaining to genetic resources.
2. Academic Research: Scholars in disciplines such as national ecology and ecological anthropology, including those represented by Berkes, contextualize traditional knowledge within climate change. They define it as a dynamic process of adaptation, reflecting its evolving nature in response to environmental changes.
3. Indigenous Peoples and Local Communities: Traditional knowledge is also defined from the perspective of indigenous peoples and local communities, emphasizing its role as a lived experience and cultural practice deeply embedded within their ways of life.

2.1 Traditional Knowledge in International Treaties and Conventions

The concept of traditional knowledge is addressed in the international treaties and conventions of three key UN agencies:

1. Convention on Biological Diversity (CBD): Administered by the United Nations Environment Programme (UNEP), the CBD was the first to propose frameworks for the protection and benefit-sharing of traditional knowledge (Siebenhuner et al., 2005).
2. World Intellectual Property Organization (WIPO): This organization, along with its

related conventions, engages with traditional knowledge through intellectual property rights and cultural heritage considerations.

3. Agreement on Trade-Related Aspects of Intellectual Property Rights (TRIPS): Under the World Trade Organization (WTO), TRIPS addresses traditional knowledge within the context of trade-related intellectual property rights, influencing how traditional knowledge is treated in global trade.

The Convention on Biological Diversity (CBD) addresses traditional knowledge in its foundational documents:

1. Preamble: The CBD acknowledges the deep and traditional reliance of many indigenous and local communities on biological resources. It underscores the importance of equitably sharing the benefits derived from traditional knowledge, innovations, and practices that contribute to the conservation of biological diversity and its sustainable use (CBD Preable).
2. Article 8(j): This article mandates that parties to the CBD, as far as possible and appropriate under their national legislation, should respect, preserve, and maintain the traditional knowledge, innovations, and practices of indigenous and local communities relevant to biodiversity conservation and sustainable use. It also encourages the promotion of these practices with the approval and involvement of the knowledge holders and advocates for the equitable sharing of benefits arising from their utilization (CBD Article 8(j)).

The Convention on Biological Diversity (CBD) defines traditional knowledge with a focus on biological resources, emphasizing its role in promoting biodiversity conservation and sustainable utilization. According to the CBD, traditional knowledge encompasses the knowledge, innovations, and practices that contribute to these goals, whether directly or indirectly. This definition is closely linked to indigenous peoples and local communities, recognizing that traditional knowledge is created, maintained, and handed down through generations by these groups. It represents valuable knowledge, experiences, innovations, or practices that are integral to the cultural and ecological heritage of these communities (Xue and Cai, 2006; Schüklenk & Kleinsmidt, 2006).

According to the Convention on Biological Diversity (CBD), traditional knowledge encompasses the knowledge, innovations, and practices developed over long periods and adapted to local cultures and environments. This knowledge is typically collective and transmitted across generations, primarily through oral traditions, although it can also be documented in written form. Traditional knowledge is expressed through various mediums, including stories, songs, legends, proverbs, cultural values, beliefs, rituals, customary laws, and indigenous languages. It also includes practical life aspects, such as agricultural practices for cultivating crops and livestock varieties. Therefore, traditional knowledge can be regarded as a form of practical science.

The Convention on Biological Diversity (CBD) recognizes traditional knowledge not only as a form of knowledge, innovation, and practical science, but also as a resource, specifically a unique resource related to biological and genetic resources. The Nagoya Protocol, adopted at CBD/COP-10, states in Preamble Paragraph 22 that it acknowledges the diverse circumstances in which traditional knowledge associated with genetic resources is held or owned by indigenous and local communities. Additionally, Preamble Paragraph 24 further recognizes the unique contexts in which such traditional knowledge is maintained in different countries, whether oral, documented, or in other forms, reflecting a rich cultural heritage that is relevant for conservation and sustainable use of biological diversity (http://www.cbd.int/decision/cop/?id=12267). From this, it is evident that the CBD considers that indigenous peoples and local communities have developed unique knowledge, methods, technologies, and practical sciences in the process of utilizing local biological and genetic resources. This knowledge, methods, technologies, and practical sciences are often integrated with biological and genetic resources. Consequently, the CBD aims to protect these resources and their associated traditional knowledge, while ensuring the fair sharing of the benefits arising from their use.

Since traditional knowledge involves intellectual property, the World Intellectual Property Organization (WIPO)—the foremost United Nations agency addressing global intellectual property issues—plays a crucial role. In October 2000, WIPO established the Intergovernmental Committee on Intellectual Property and Genetic Resources, Traditional Knowledge and Folklore (IGC) to collaborate with the CBD on issues concerning traditional knowledge. The WIPO-IGC's mission is to address intellectual property issues related to the access and benefit-sharing of genetic resources and associated traditional knowledge. Annex I of WIPO/GRTKF/IC/7/5, issued at the Seventh Meeting of WIPO-IGC in 2004, outlined the characteristics of traditional knowledge as follows: (i) it is created, preserved, and inherited traditionally, passed down through generations; (ii) it has a special connection to the local community and the people who have preserved and transmitted it over generations; and (iii) it aligns with the cultural characteristics of the local or indigenous communities who are recognized as its holders. Based on these characteristics, WIPO further defines traditional knowledge as encompassing literature, art, scientific works, performances, inventions, scientific discoveries, designs, trademarks, enterprise names, and symbols, as well as unpublished works that are traditional or based on tradition. This definition includes innovations and creations across industrial, scientific, literary, and artistic domains. Traditional knowledge spans agricultural, scientific, technical, ecological, and medical knowledge (including medicines and treatments), biodiversity-related knowledge, folk art (such as music, dance, songs, handicrafts, design, legends, and artworks), language elements (such as names, geographical indications, and symbols), and other forms of intangible cultural heritage (WIPO/GRTKF/IC/3/9). WIPO's broad definition highlights that traditional knowledge is not only linked to biological resources but also encompasses a wide range of cultural expressions and intellectual activities.

In response to the Convention on Biological Diversity (CBD), the World Trade Organization (WTO) has engaged in extensive discussions on genetic resources and traditional knowledge under the Agreement on Trade-Related Aspects of Intellectual Property Rights (TRIPS). Specifically, Article 27 of TRIPS addresses whether applicants for patents must disclose the genetic resources and traditional knowledge utilized in their inventions, along with any relevant benefit-sharing arrangements. However, TRIPS does not primarily focus on folk literature and art. Consequently, the concept of traditional knowledge under TRIPS aligns closely with that under the CBD, predominantly referring to traditional knowledge related to biological and genetic resources.

2.2 Traditional Knowledge in Academic Research

Traditional knowledge—also known as indigenous knowledge (Nakashima & Roué, 2002), traditional ecological knowledge (Berkes, 1999; Huntington, 2000), local knowledge (LK), and local ecological knowledge (Gilchrist et al., 2005; Olsson & Folke, 2001)—refers to knowledge that is location-specific, acquired through long-term observation and interaction with the environment, and passed down orally from generation to generation. Despite various definitions and interpretations from scholars and experts across different fields, there remains a lack of unified and widely recognized consensus on its definition. According to Mugabe, the disparity in definitions arises from the differing intellectual perspectives of various individuals. Consequently, terms such as traditional knowledge and indigenous knowledge are often used interchangeably, reflecting the diverse cultural characteristics of the local communities that hold and utilize this knowledge (Mugabe et al., 2001).

The earliest systematic studies of traditional ecological knowledge (TEK) were conducted within the fields of ecological anthropology and ethno-ecology, both of which explore the conceptions of ecological relationships held by different cultures (Hardesty, 1977; Toledo, 1992, 2001; Nazarea, 1999; Hunn, 2008; Johnson & Hunn, 2010). For instance, ethno-ecology is defined as the study of knowledge systems developed by cultures to classify objects, activities, and events within their universe (Hardesty, 1977). Research on traditional ecological knowledge initially focuses on species identification and classification and then extends to understanding how cultures perceive ecological processes and their interactions with the environment. This research encompasses three main components: (i) local and empirical knowledge of species and environmental phenomena; (ii) practices related to agriculture, hunting, fishing, and other subsistence activities; and (iii) belief systems regarding the role of people within ecosystems and their interaction with natural processes (Berkes, 2012a, 2012b).

As early as the 1990s, scholars began to define traditional knowledge in various ways. Warren characterized traditional knowledge as a form of local knowledge, specific to a particular society or culture. This type of knowledge contrasts with the international knowledge systems generated by universities, research institutions, and large corporations. Traditional knowledge serves as a foundation for local decision-making in agriculture, health care, food production, education, and natural resource management within rural communities (Warren, 1991). Flavier described traditional knowledge as the informational basis for communication and decision-making within a society, noting its dynamic nature, shaped by both internal innovations and external knowledge systems (Flavier et al., 1995). Grenier defined it as unique, traditional, and indigenous knowledge developed by individuals living in specific geographical areas and conditions, and maintained within their communities (Grenier, 1998).

Additionally, various research institutions have defined and analyzed traditional knowledge in their reports. The International Institute for Rural Reconstruction (IIRR) considers traditional knowledge to encompass aspects such as information, practices, technologies, beliefs, tools, materials, experiments, biological resources, human resources, education, and communication (IIRR, 1996). The Indigenous Knowledge and Development Monitor (IK&DM) defines it as the knowledge and technology possessed by people in specific geographical areas that allows them to derive maximum benefit from their natural environment. This knowledge is typically passed down through generations, with each generation adapting or augmenting the original knowledge to better align with environmental changes, thereby providing effective survival strategies (IK&DM, 1998).

Entering the twenty-first century, discussions on traditional knowledge have increasingly focused on its connections to biodiversity, genetic resources, and climate change within the context of global environmental changes. Swiderska defines traditional knowledge as a collective resource possessed by indigenous or local communities, shaped by their socio-ecological environment. This knowledge encompasses innovations and practices closely linked to traditional resources and territories, local economies, genetic, species, and ecosystem diversity, as well as cultural, spiritual values, and traditional laws (Swiderska, 2006). Soejarto views traditional knowledge as a resource that can be abstractly recorded and commercially utilized in the context of biological resources (Soejarto et al., 2005). Memory considers traditional knowledge to be a technical component of ongoing resource management, noting that certain traditional practices can offer technical support for conventional resource management systems. For instance, local communities might utilize plants well adapted to their environment for ecological restoration (Memory, 2007). Xue regards traditional knowledge as an integral part of the social and cultural systems of traditional communities, including traditional land-use and management practices such as rotational grazing systems and their associated knowledge transmission. This perspective highlights the close relationship between the application and protection of traditional knowledge and resource management systems (Xue, 2011).

Berkes initially linked traditional knowledge to environmental and resource contexts and later extended its relevance to climate change, with a particular focus on how indigenous peoples and local communities adapt to environmental and climatic shifts. He defines traditional knowledge as a cumulative body of knowledge, practices, and beliefs that evolves through adaptive processes and is transmitted across generations via cultural transmission. This body of knowledge encompasses the interrelationships among living beings, including humans, and their interactions with their environment (Berkes, 2009). Traditional knowledge is understood as a dynamic process rather than a static body of content; it represents a way of knowing that continually builds upon experience and adapts to change (Berkes, 2012a, 2012b). As such, traditional knowledge is marked by its ongoing evolution and responsiveness to socio-ecological conditions, rather than existing in isolation or as a fixed entity.

2.3 Traditional Knowledge in Indigenous Peoples and Local Communities

In recent years, some organizations and conferences representing indigenous peoples and local communities have expressed disagreement with definitions of traditional knowledge that are based on Western language and science. They argue that English terms such as "traditional ecological knowledge" are often developed within academic and policy circles that are frequently disconnected from indigenous perspectives. These terms may not fully capture how indigenous communities understand and refer to their own knowledge and ways of life. The term "traditional knowledge" is broad and encompasses various dimensions of dynamic knowledge systems and life ways of diverse indigenous peoples. It generally refers to the ways of knowing of indigenous communities, which both guide and arise from their members' intimate relationships with and responsibilities toward landscapes, waterscapes, plants, and animals essential to the vitality of their cultures. Indigenous peoples' traditional knowledge and practices have been honed over thousands of years through their experiences and interactions with living beings and environments (Kyle, 2015).

For instance, in the Canadian Arctic, some communities prefer the term Inuit Qaujimajatuqangit (Arnakak, 2002; Wenzel, 2004). This term encompasses all aspects of Inuit values and ways of life, which may render it too broad to be used interchangeably with "traditional ecological knowledge" (Wenzel, 2004). Similarly, some Aboriginal communities view Traditional Ecological Knowledge (TEK) as encompassing more than just a body of knowledge. While knowledge is a component, TEK also includes spiritual experiences and relationships with the land. It is described as a "way of life", not merely the knowledge of how to live, but the actual living of that life. Aboriginal perspectives on TEK are often characterized as "verb-based", or action-oriented, contrasting with non-Aboriginal views that may treat TEK as a static body of knowledge. From the Aboriginal viewpoint, TEK is a lived experience rather than a theoretical construct; it is something that one practices and embodies (McGregor, 2004).

Traditional knowledge provides valuable insights into how climate affects the environment and influences cultures dependent on local ecological systems. Many indigenous communities adhere to traditional lifestyles that are intricately linked to seasonal and interannual patterns of preparation, harvest, and utilization, as reflected in their practices, stories, and languages. Maintaining these ways of life requires the ability to observe, record, and make informed decisions based on relationships among landscapes, water scapes, plants, and animals, utilizing various indicators. Communities that practice these skills within specific territories often possess climate trend information that spans countless generations (Kyle, 2015).

For indigenous peoples and local communities, traditional knowledge represents not only a resource and technology but also a cultural and living process. In the context of global change, particularly in addressing climate change and conserving biodiversity, traditional knowledge emerges as a rights issue. It encompasses the right of indigenous peoples and local communities to participate in global governance and influence policymaking processes.

3 Traditional Knowledge Related to Climate

Indigenous peoples and local communities residing in diverse environments—such as extremely low-lying areas, mountainous regions, deserts, tropical rainforests, islands, and temperate zones—perceive climate change primarily through changes in temperature, rainfall, snowfall, seasonal patterns, and phenological events. They also observe variations in wind, waves, storms, annual climate patterns, glaciers, snow cover, ice, rivers, lakes, and species dynamics (Salick & Byg, 2007). These observations form the foundation of their traditional knowledge related to climate.

Traditional knowledge of weather and climate encompasses the ability to observe, record, and make informed decisions based on the relationships among landscapes, waterscapes, plants, and animals using a variety of indicators, including climatological and phenological phenomena. Indigenous peoples and local communities who employ these skills within specific territories often possess detailed knowledge of climate trends that extends back through numerous generations.

In this book, traditional knowledge related to climate is defined as the direct and indirect observation and understanding of local climatological and phenological phenomena that are passed down from generation to generation within a specific region. This knowledge is utilized to predict and address climate change and extreme weather events. It is categorized into two main components: climate and phenology.

3.1 Traditional Climatological Knowledge

Traditional climatological knowledge is developed and passed down by indigenous peoples and local communities through extensive observation, understanding, and practical experience with meteorological, weather, and climate phenomena and their changes over time.

For example, the Sakha people of northeast Siberia possess traditional climatological knowledge about winter snowfall and summer rainfall. Over the past two decades, they have observed climate change impacts, such as decreased snowfall, increased winter temperatures, and heightened summer rainfall (Susan and Crate 2009). Similarly, the tribal peoples of Tharparkar, Sindh, in Pakistan use their traditional climatological knowledge to predict local climate phenomena, including lightning, wind direction, cloud formation, and occurrences of rain and drought (Ahmed et al., 2016). The Inuvialuit people in the Arctic have extensive traditional climatological knowledge regarding snowfall and ice, which highlights changes in ice distribution, thickness, and breakup due to climate change (Nicolas et al., 2004). In the eastern Himalayas of China, Tibetan communities apply their traditional climatological knowledge to predict variations in temperature, snowfall, glacier dynamics, and rainfall (Byg & Salick, 2009a, 2009b). Additionally, the Kenyah Badeng farmers in Sarawak, Malaysia, use their traditional climatological knowledge to interpret sensory inputs such as sight, sound, touch, and smell to assess current and forthcoming weather conditions, including clouds, wind, temperature, humidity, and rainfall (Garay-Barayazarra and Puri, 2011).

Climate models based on modern science can offer broad projections of climate change and estimate future trends and consequences of human activities on a macro scale. However, they often lack precision in providing detailed information about regional climate changes. Traditional climatological knowledge, on the other hand, serves as a valuable source of localized climate information. Despite the diverse global distribution of indigenous peoples and local communities, there is a notable consistency in their traditional climatological knowledge. This knowledge encompasses observations and understanding of changes in temperature, precipitation, wind patterns, ice conditions, wave dynamics, storms, and annual climate variations.

3.2 Traditional Phenological Knowledge

Traditional phenological knowledge refers to the traditional knowledge formed by indigenous peoples and local communities through long-term observation of the natural environment and ecosystem, including changes in livelihood conditions such as crops and livestock, farmland and pasture environment, and their relationship with climate. One of the purposes of this knowledge is to understand the laws of natural seasonal phenomena, so as to serve the livelihood. For indigenous peoples and local communities living around the world, traditional phenological knowledge about climate change includes changes in glaciers, snow cover, ice, rivers and lakes, species, seasons, plant growth and animal behavior.

For example, mountain farmers in the Himalayas of India possess traditional phenological knowledge about climate and its changes, based on indicators such as flowering patterns, crop yields, fruit quality, and the emergence of new pests and diseases. This knowledge enables them to predict climate changes effectively (Chaudhary & Bawa, 2011; Chaudhary et al., 2011; Vedwan & Rhoades, 2001). Indigenous smallholder farmers on the Bui Plateau in the Bamenda Highlands of Cameroon use traditional phenological knowledge related to agriculture and the local ecosystem. This includes observations of bird and insect sounds, plant behavior, and star constellations, helping them predict weather patterns and extreme weather events (Tume et al., 2019). The Nez Perce people of the Columbia Valley in the northwestern United States observe climate change through alterations in major species and ecosystems in the Columbia River (Benedict et al., 2009). In Tuvalu, South Pacific Islands, traditional phenological knowledge includes observations of sea-level rise, sea area expansion, seawater acidification, coral bleaching, and coastal erosion, which provide insights into climate change (Heather, 2009). The Maori of New Zealand possess traditional phenological knowledge about the relationships between natural environments—such as plants and animals—and periodic weather and climate changes. This includes forecasting weather and climate phenomena and classifying these phenomena (King et al., 2008). In the Andes of Peru, local farmers use the appearance of the Pleiades cluster mid-year to predict precipitation and rainfall for the rainy season months later. This knowledge also includes understanding the timing and semi-regularity of El Niño events (Orlove et al., 2002). Australian Aborigines have developed extensive traditional phenological knowledge related to local weather and climate, including biophysical phenomena that reveal upcoming weather conditions. This diversity reflects their deep understanding of the links between biophysical phenomena and climate change (Bureau of Meteorology, 2003).

In Samoa, South Pacific, indigenous knowledge includes seasonal calendars based on weather and climate events, detailed classifications of local wind and cloud types, and various methods for monitoring and predicting future weather and climate conditions, including extreme weather like tropical cyclones (Lefale, 2003).

In Saurashtra, northern India, indigenous knowledge includes beliefs and practices related to climate. For example, the flowering of golden flowers and wind direction are used to predict the onset of the rainy season in June or July (Kanani & Pastakia, 2001).

Indigenous peoples and local communities frequently reside in biodiversity-rich areas, fostering a close interaction with local climate conditions, natural environments, and ecosystems. Traditional phenological knowledge forms a vital component of this relationship. This rich and diverse body of knowledge has evolved into a crucial tool for predicting weather and climate changes, with some aspects even becoming intertwined with spiritual beliefs. Unlike scientific instruments, traditional phenological knowledge is deeply embedded in the complexities of ecosystems, biodiversity, and environmental conditions. As such, it represents a valuable and irreplaceable source of regional climate change information. By providing accurate, region-specific insights, traditional phenological knowledge enables better understanding of climate dynamics and supports efforts to address the current global climate crisis from both social and cultural perspectives.

Phenological and climatological knowledge both constitute traditional systems for observing local climate change. The key distinction between the two lies in their focus: Climatological knowledge emerges from the perception and observation of meteorological phenomena such as temperature, precipitation, sunlight, and wind. In contrast, phenological knowledge is derived from the observation of the natural environment, ecosystems, and livelihoods. Specifically, through their livelihoods, indigenous peoples and local communities assess the impacts of climate change by observing changes in the growth of crops and livestock, which are directly influenced by shifting environmental conditions.

Traditional knowledge related to climate is not in opposition to scientific knowledge. Rather, as a form of localized understanding, it complements and can be integrated with scientific knowledge to enhance the accuracy of regional weather forecasts, climate change predictions, and assessments of extreme climate disaster risks. Collaboration between traditional and scientific knowledge can significantly strengthen the ability of indigenous peoples and local communities to adapt to and mitigate the impacts of climate change.

4 Traditional Knowledge Associated with Biodiversity for Climate Change

The traditional knowledge of indigenous peoples and local communities is deeply rooted in the local ecosystem and natural environment, playing a pivotal role in the utilization, management, and protection of biodiversity. This knowledge has given rise to traditional practices associated with biodiversity conservation. The Convention on Biological Diversity (CBD) mandates that Contracting Parties preserve and maintain the knowledge, innovations, and practices of indigenous peoples and local communities, particularly those embodying traditional lifestyles relevant to the conservation and sustainable use of biodiversity. According to Xue (2011), traditional knowledge related to biodiversity can be categorized into five types: (1) Knowledge of agricultural species and genetic resources; (2) Knowledge of medicinal resources; (3) Technical innovations for the use of biological resources and traditional farming and lifestyle practices; (4) Knowledge of traditional cultures related to the conservation and sustainable use of biodiversity; and (5) Knowledge embedded in geographic biological indicators.

Climate change has significantly impacted global biodiversity, and biodiversity hotspots, which are often rich in both cultural and biological diversity, are no exception. Indigenous peoples and local communities, who inhabit these areas, possess a wealth of traditional knowledge related to biodiversity. This knowledge not only reflects their sustainable practices but also provides alternative insights into climate variability and change, grounded in their long-standing use of biodiversity resources (Yin et al., 2018). Accordingly, this book categorizes traditional knowledge related to biodiversity into three levels based on the broader concept of biodiversity: genetic diversity, species diversity, and ecosystem diversity. These categories are: (1) traditional knowledge associated with genetic diversity; (2) traditional knowledge associated with species diversity; and (3) traditional knowledge associated with ecosystem diversity.

Within this framework, the book examines traditional knowledge in the context of climate change, which forms its core focus. Chapter 7, "Traditional Knowledge in the Use of Plant Species and Genetic Resources for Climate Change Adaptation", is classified under traditional knowledge associated with genetic diversity. Chapters 4 and 5, "Traditional Knowledge and the Movement of Medicinal Plant Species in Response to Climate Change" and "Traditional Knowledge in Livelihood and Lifestyle Practices for Climate Change Adaptation", are categorized as traditional knowledge associated with species diversity. Finally, Chaps. 3 and 6, "Traditional Knowledge in Water Resource Management for Climate Change Adaptation" and "Traditional Knowledge in Forest Resource Management for Climate Change Adaptation", pertain to traditional knowledge associated with ecosystem diversity.

Building on the three levels of traditional knowledge associated with biodiversity, this book further explores traditional knowledge through the lenses of gender, disaster, and belief. Chapter 8, "A Gender Perspective on Traditional Knowledge in Irrigation Management and Climate Change Adaptation", focuses on the role of gender in resource management. Chapter 9, "Traditional Knowledge in Disaster Management for Climate Change Adaptation among Tibetan People in the Eastern Himalayas", examines the ways in which local communities respond to climate-induced disasters. Lastly, Chap. 10, "Traditional Beliefs Related to Conservation for Climate Change Adaptation", delves into the role of spiritual and cultural beliefs in biodiversity conservation and climate resilience.

4.1 Traditional Knowledge Associated with Genetic Diversity

Through generations of accumulated knowledge and the long-term processes of domesticating, cultivating, and utilizing various cultivated plant and domestic animal species, as well as other biological resources, indigenous peoples and local communities have developed traditional knowledge associated with genetic diversity. A significant component of this knowledge pertains specifically to the diversity of plant genetic resources. Therefore, this section will focus primarily on discussing traditional knowledge related to plant genetic resource diversity.

Climate change is introducing new challenges and difficulties to agriculture, thereby threatening plant genetic resources. In the future, the global population facing hunger is expected to rise, further exacerbating food insecurity for indigenous peoples and local communities. On one hand, climate change negatively impacts natural habitats, endangering the genetic diversity of crop species and the associated traditional knowledge. This situation results in "enetic erosion" of plant genetic diversity and a concomitant erosion of related traditional knowledge.

On the other hand, plant genetic resources are fundamental to global food security and agriculture, particularly in the context of climate change. The ability of a species to adjust genetically and evolve in response to selective pressures—such as abiotic stresses (drought, heat, frost, etc.) and biotic stresses (emerging diseases and pests), as well as new climatic conditions—depends on the inherent levels of genetic diversity within that species (Hammer & Teklu, 2008). In other words, both genetic resources and the associated traditional knowledge are crucial for enhancing the diversity of crop varieties, thereby enabling agriculture to better cope with the impacts of climate change and ensuring food security worldwide, particularly for indigenous peoples and local communities.

The development of crop varieties that are resilient to global climate change and other environmental shifts often relies significantly on the traditional knowledge of indigenous peoples and local communities. These groups typically inhabit regions that harbor a substantial portion of the world's genetic resources. They have cultivated various agricultural and horticultural crops from local wild species, laying the groundwork for modern plant breeding and global food security. Additionally, indigenous peoples and local communities have identified and managed a range of genes through selection and hybridization, which are adapted to both biotic and abiotic stresses. Through these practices, they have accumulated a rich body of traditional knowledge related to genetic resources.

Traditional knowledge of indigenous peoples and local communities is regarded as a crucial factor in the conservation and development of genetic resources (Javaid and Steve, 2016). This knowledge plays a transformative role in managing biodiversity within genetic resources. The sustainable use of plant genetic resource diversity, guided by traditional knowledge, enhances the resilience of indigenous peoples and local communities to climate change and directly contributes to the preservation of biodiversity (Salgotra et al., 2018).

In this context, plant genetic resources and the associated traditional knowledge are invaluable assets crucial for global food security, particularly in the face of climate change. They provide essential genes and alleles that govern resistance to biotic and abiotic stresses, often found in wild species and landraces. Given the challenges and threats that climate change poses to agriculture, the role of traditional knowledge in managing seed and plant genetic resources has become increasingly recognized. Traditional knowledge is not only vital for protecting genetic diversity but also serves as a key strategy for addressing climate change both now and in the future. On one hand, genetic information from local or indigenous species can be preserved through gene banks, with these genetic traits potentially benefiting future breeding programs aimed at enhancing disease and pest resistance or tolerance to adverse weather conditions. On the other hand, establishing a traditional knowledge database is essential for preserving breeding techniques alongside genetic characteristics, as seeds and clones stored in gene banks require instructions and knowledge for successful cultivation.

4.2 Traditional Knowledge Associated with Species Diversity

Through the development of traditional livelihoods such as agriculture and pastoralism, and the long-term domestication, cultivation, and breeding of plant and animal varieties, indigenous peoples and local communities have cultivated traditional knowledge associated with species diversity. This species diversity and the related traditional knowledge play a crucial role in enabling indigenous peoples and local communities to sustain social and economic development, particularly in the context of climate change. A key factor in ensuring the sustainable management of plant resources is how to integrate the traditional knowledge of indigenous peoples and local communities, who place the highest value on these resources. Medicinal plant species diversity represents a significant aspect of species diversity; thus, this section will focus on discussing the traditional knowledge related to medicinal plant species diversity.

Global climate change poses a significant threat to plant species diversity. Climate change has markedly impacted species ranges and ecological communities worldwide (Thomas et al., 2004). It can directly affect plant species diversity through mechanisms such as drought and heat waves, which influence photosynthesis, respiration, transpiration, and phenology. Indirectly, climate change may impact plant species diversity through increased fire frequency, the spread of parasites and diseases, altered litter quality and decomposition, potentially elevating the risk of plant mortality and extinction in many regions (IPCC, 2013). Indigenous peoples and local communities frequently inhabit areas with fragile environments and vulnerable ecosystems. Consequently, climate change has exacerbated negative impacts on local plant species diversity, particularly affecting medicinal plant species and the associated traditional knowledge.

Medicinal plant species offer a rich diversity of valuable products crucial for the health care and remedies of indigenous peoples and local communities. Based on this, traditional knowledge related to medicinal plant species diversity has been developed and integrated into their cultural and belief systems. Consequently, the loss or alteration of medicinal plant species diversity due to climate change could negatively impact the health care and medical practices of these communities. Additionally, the erosion of related traditional knowledge could further undermine their cultural and spiritual practices.

On the other hand, protecting existing plant species diversity, including medicinal plants, is a crucial aspect of how indigenous peoples and local communities respond to climate change. Traditional knowledge plays a significant role in this process. It is important to note that traditional knowledge is not static but evolves over time. Consequently, it also adapts to shifts in medicinal plant species diversity. This book explores how quantitative analysis and research on traditional knowledge related to medicinal plant species diversity can serve as indicators of species movement driven by climate change. Additionally, traditional knowledge, which is integral to cultural and spiritual practices, can help conserve medicinal plant species diversity and offer a means for indigenous peoples and local communities to adapt to climate change.

4.3 Traditional Knowledge Associated with Ecosystem Diversity

Indigenous peoples and local communities inhabit diverse ecosystems worldwide, including forests, deserts, plateaus, ice fields, arid lands, seashores, and watersheds, each of which has given rise to distinct traditional knowledge related to ecosystem diversity. Even within the same geographical area, where multiple ecosystems coexist, the resulting diversity of ecosystems fosters varied traditional knowledge among different communities. This traditional knowledge associated with ecosystem diversity plays a crucial role in supporting and maintaining local ecosystem services to varying degrees.

The forest ecosystem is one of the most widespread ecosystems globally, exhibiting significant diversity based on different forest types. For instance, the Amazon hosts tropical rainforest ecosystems, while the Himalayas are home to coniferous forest ecosystems. Within a single region, such as the Himalayas, varying altitudes result in distinct forest ecosystems: tropical rainforests are found below 1,000 m above sea level; evergreen broad-leaved forests between 1,000 and 2,500 m; coniferous and broad-leaved forests between 2,500 and 3,200 m; and coniferous forests above 3,200 m. Indigenous peoples and local communities residing in these diverse forest ecosystems develop traditional knowledge specific to their environment. Given the complexity of forest ecosystem diversity, this section will explore the traditional knowledge associated with these varied forest environments.

Indigenous peoples and local communities who possess and utilize traditional knowledge related to forest ecosystem diversity are at the forefront of global efforts to address climate change and its impacts. Due to their profound connection with, and reliance on, forest ecosystems and landscapes, they are among the first to observe, comprehend, and experience the effects of climate change on forests, woodlands, and their own livelihoods and cultures (Parrotta & Agnoletti, 2012). Across various forest ecosystems worldwide, these communities leverage their traditional knowledge to observe, interpret, and respond to climate change (Parry et al., 2007b; Crump, 2008; Galloway-McLean, 2010; Vlassova, 2006).

Forest ecosystem diversity and the associated traditional knowledge, innovations, and practices have supported the livelihoods and cultures of indigenous peoples and local communities for generations. This traditional knowledge reduces their vulnerability and enhances their resilience to climate change. The resilience enhancement manifests in the following ways:

Firstly, traditional knowledge related to forest ecosystem diversity serves as a crucial indicator for indigenous peoples and local communities in predicting climate change. By observing and interpreting changes in forest ecosystems, they can effectively and adaptively manage daily life and livelihood activities.

Secondly, traditional livelihoods of these communities often heavily depend on forest ecosystems, including non-timber forest products. Climate change's impact on these ecosystems threatens their survival both directly and indirectly. In this context, traditional knowledge can help sustain and improve forest ecosystems, thereby enhancing livelihood security amidst climate change.

Thirdly, forest ecosystems maintain the natural environment and provide essential resources such as food, water, wood, and medicinal plants. Forest ecosystem services are fundamental to the survival and social development of these communities. The adverse effects of climate change on forest ecosystems can degrade these services, leading to environmental deterioration and resource depletion. Traditional knowledge related to forest ecosystem diversity can mitigate these impacts and help maintain ecosystem services, promoting sustainable development.

Fourthly, climate change-induced damage to forest ecosystems increases the risk of local disasters. Extreme weather events can lead to forest fires, floods, debris flows, and landslides, resulting in property and life losses. Traditional knowledge related to forest ecosystem diversity can aid in predicting and mitigating the effects of extreme weather phenomena, thereby minimizing disaster risks.

5 Traditional Knowledge and Coping with Climate Change

Indigenous peoples and local communities are often described as the world's "advance guard" against climate change (Galloway-McLean, 2010). Their sensitivity to environmental changes, adaptive capacity, and resilience illustrate their ability to adjust their behaviors in response to shifting climate conditions (Nakashima et al., 2012). Traditional knowledge provides crucial insights into the processes of climate change mitigation and adaptation, and its significance in informing these strategies is increasingly recognized.

Traditional knowledge assists indigenous peoples and local communities in understanding and interpreting ecological changes, including those induced by climate change. It is valuable for conducting environmental and social impact assessments, and for analyzing climate variability (Berkes, 1999; Huntington, 2000; Nakashima & Roué, 2002; Olsson et al., 2004). Such knowledge is essential for guiding these communities' strategies to prepare for, adapt to, and mitigate the effects of climate change. The strategies historically employed by indigenous peoples and local communities to navigate environmental changes are now gaining renewed attention as significant sources of information for addressing contemporary climate challenges.

Climate scientists, policymakers, and an increasing number of engaged citizens involved in monitoring global change are increasingly recognizing the value of indigenous knowledge systems in enhancing both adaptation and mitigation strategies. The role of traditional knowledge is evident in activities related to climate change mitigation and adaptation; for example, ecologically aligned practices such as reforestation can contribute to reducing greenhouse gas (RHG) emissions, while the use of drought-resistant seeds enhances adaptation capacities (Speranza et al., 2010). Similar to global responses, indigenous peoples and local communities employ two primary approaches to confront climate change: one involves mitigating climate change by addressing its root causes, while the other focuses on adapting to its inevitable impacts. For these communities, traditional knowledge represents a dynamic process of learning and an invaluable repository for successfully navigating local environmental changes, owing to their deep, intrinsic relationship with the surrounding ecosystems. Consequently, traditional knowledge is integral to both mitigation and adaptation efforts in these regions. This section explores how traditional knowledge can empower indigenous peoples and local communities to more effectively manage climate change and actively contribute to both mitigation and adaptation strategies. Additionally, it presents an approach that positions traditional knowledge as a key agent in climate response frameworks.

5.1 Traditional Knowledge and Climate Change Mitigation

Climate change mitigation refers to human interventions aimed at reducing the sources or enhancing the sinks of greenhouse gases (IPCC, 2014). For indigenous peoples and local communities, mitigation efforts have long been embedded in natural resource conservation and ecosystem management practices. These activities align with the IPCC's definition of mitigation by serving dual purposes: reducing anthropogenic greenhouse gas emissions and enhancing carbon sinks. Traditional knowledge related to biodiversity has historically played—and will continue to play—a significant role in climate change mitigation by contributing to the reduction of greenhouse gas emissions, the expansion of carbon sinks, and the strengthening of carbon substitution mechanisms.

5.1.1 Traditional Knowledge and Greenhouse Gases Emission Reduction

The most effective approach to mitigating climate change is through significant reductions in greenhouse gas (RHG) emissions. Strategies targeting RHG reduction typically focus on decreasing fossil fuel combustion by enhancing energy efficiency, increasing the use of clean energy sources, particularly solar power, and eliminating practices like gas flaring (Nyong et al., 2007). While the imperative to reduce fossil fuel consumption and develop renewable energy resources in industrialized and urban settings is well-recognized, comparatively less attention has been directed toward rural areas—specifically, arable lands, pastures, and forests (Parrotta & Agnoletti, 2012).

Traditional knowledge, particularly that related to agro-ecosystem diversity, plays a vital role in reducing greenhouse gas emissions. This knowledge, developed by indigenous peoples and local communities through centuries of agricultural practice, represents a form of practical technology that both preserves biodiversity and promotes the sustainable use of biological resources. In addition to its environmental benefits, it contributes to improving food quality and ensuring food security. Traditional ecological agricultural techniques and biological resource processing methods are key components of this knowledge system, providing valuable tools for enhancing both environmental and food sustainability.

Before the introduction of chemical fertilizers, indigenous peoples and local communities primarily practiced organic agriculture. In contrast to modern industrial agriculture, traditional organic farming methods and technologies are more effective in reducing greenhouse gas emissions. Unlike energy-intensive industrial agriculture, these traditional practices do not depend on fossil fuel-based inputs such as inorganic fertilizers, pesticides, or mechanized farming to maintain productivity. Moreover, they avoid the deforestation and associated carbon dioxide emissions that result from large-scale agricultural expansion.

At the same time, the use of these traditional technologies and methods promotes more efficient energy utilization in agriculture, as opposed to the heavy reliance on fossil fuels characteristic of intensive and specialized modern farming. For instance, traditional agroforestry and shifting agriculture, though often lower in overall productivity, employ techniques that make more effective use of solar energy, resulting in greater energy output relative to input. In contrast, modern agricultural systems consume more energy than they produce, necessitating high-energy inputs, particularly from fossil fuels, fertilizers, and pesticides, to sustain their high productivity.

Compared to traditional agriculture, modern intensive farming lacks energy efficiency and generates higher levels of greenhouse gas emissions due to its heavy reliance on non-renewable fossil fuels. Traditional agricultural practices, by contrast, are typically more energy-efficient. In fact, when considering agricultural output relative to energy input, traditional agriculture may prove more productive than modern industrialized systems, which rely on unsustainable, high-energy inputs and produce significant greenhouse gas emissions. For indigenous peoples and local communities worldwide, the technologies and methods used in agriculture serve as a mechanism through which traditional knowledge of agro-ecosystem diversity contributes to climate change mitigation. This body of knowledge will continue to play a critical role in both present and future mitigation efforts.

5.1.2 Traditional Knowledge and Carbon Sinks

Forests are widely recognized as crucial carbon sinks, capable of capturing and storing significant amounts of atmospheric carbon dioxide. As such, forests play a vital role in reducing greenhouse gas emissions and mitigating the risks associated with climate change (Karjalainen et al., 1994; Stainback & Alavalapati, 2002). Recent scientific research indicates that more than half of the world's carbon sink capacity is attributed to "young" forests, those less than 140 years old, particularly in mid- and high-latitude regions. Consequently, efforts to enhance carbon sinks often focus on forestry programs that protect existing forests and promote afforestation in marginal areas, including rangelands (Adesina et al., 1999). Recognition of the forest's role in climate change has also influenced the Kyoto Protocol, which allows countries to account for carbon sequestered in forests when calculating national emission targets.

Indigenous peoples and local communities have long recognized the significance of forests, making sacred forests and public forest reserves common in their societies. These well-managed forests not only provide essential resources, such as food and timber, but also function as critical carbon sinks. Additionally, indigenous peoples and local communities possess traditional knowledge related to forest ecosystem diversity, encompassing methods, beliefs, and cultural practices associated with forests. These traditions have played a key role in protecting primary forests and form the technical, spiritual, and cultural foundation for forest conservation, reforestation, and afforestation efforts. In this context, traditional knowledge of forest ecosystem diversity contributes to the mitigation of greenhouse gas emissions and offers a vital solution for enhancing carbon sinks.

Indigenous peoples and local communities have traditionally practiced the fallow system of cultivation, including methods such as shifting agriculture, which fosters forest development, particularly through the emergence of new forests during fallow periods. These young forests significantly enhance carbon absorption and storage. Agroforestry is another practice that has proven effective in carbon sequestration. Traditional techniques employed in agroforestry systems increase the organic matter content in the soil, thereby enhancing agricultural productivity and alleviating pressure on forests. The presence of trees and shrubs as permanent components of these systems further promotes carbon storage in both above- and below-ground biomass and soil organic matter. Consequently, traditional knowledge related to agro-ecosystem diversity provides essential ecological goods and services, including biodiversity conservation and increased carbon sequestration in biomass and soils.

5.2 Traditional Knowledge and Climate Change Adaptation

Global climate change is unlikely to be halted in the near future, even with drastic measures to reduce CO_2 emissions, due to the inertia of the global climate system (Boer et al., 2001). Consequently, adaptation is recognized as a viable strategy to mitigate the vulnerabilities associated with the anticipated negative impacts of climate change. Climate change adaptation refers to the adjustments made by natural or human systems in response to actual or anticipated changes in environmental conditions, aiming to reduce vulnerability to climate impacts while capitalizing on any favorable opportunities that arise from these changes (IPCC, 2014). For indigenous peoples and local communities, traditional knowledge serves as a dynamic information base that has supported their adaptation to continuously changing climatic conditions. The IPCC emphasizes that traditional knowledge is an essential foundation for developing effective response strategies and managing natural resources in the face of environmental changes, including climate change.

This book focuses on the theme of traditional knowledge and climate change adaptation, specifically examining traditional knowledge associated with biodiversity and community-based adaptation strategies. In the preceding ten chapters, we have explored the relationship between traditional knowledge and climate change adaptation across various levels and types of knowledge, highlighting its potential value and role in the adaptation process. Adaptation to global climate change is often framed in terms of vulnerability and resilience. Building on the discussions in the previous chapters, this section will further analyze the relationship between traditional knowledge and climate change adaptation, particularly the role that traditional knowledge can play in reducing climate vulnerability and enhancing resilience.

5.2.1 Traditional Knowledge and Climate Vulnerability Reduction

Climate vulnerability is defined as "the degree to which geophysical, biological, and socio-economic systems are susceptible to, and unable to cope with, adverse impacts of climate change" (IPCC, 2007a, 2007b, 2007c). Vulnerability is determined by three interrelated components: sensitivity, exposure, and adaptive capacity (Prno et al., 2011). According to the IPCC, "exposure" pertains to the degree of climate stress experienced by a particular unit; "sensitivity" refers to the extent to which a system is affected by or responsive to climate stimuli, whether positive or negative; and "adaptive capacity" denotes the potential or capability of a system to adjust to climate change (Rosenzweig & Hillel, 2008).

It is increasingly recognized that indigenous peoples and local communities are among the first victims of global climate change, highlighting their heightened vulnerability to its impacts (Ford et al., 2016; IPCC, 2007a, 2014). The effects of climate change on these communities and their territories are expected to be both early and severe, primarily due to their location in vulnerable environments. Additionally, climate change poses a direct threat to many indigenous peoples and local communities that rely on resource-based livelihoods. Furthermore, the distinct social and cultural characteristics of many indigenous groups can render the decisions, policies, and actions implemented by dominant societal groups inadequate, inflexible, or even inappropriate for their needs. This disparity contributes significantly to the heightened vulnerability of indigenous peoples and local communities to climate change. For these communities, exposure refers to the presence of potentially harmful conditions, while sensitivity encompasses the occupancy and livelihood characteristics that render individuals and communities susceptible to these exposures (Prno et al., 2011; Smit & Wandel, 2006). Adaptive capacity, on the other hand, involves both local determinants and the broader context in which the community operates (Prno et al., 2011).

Indigenous peoples and local communities are not only potential victims of global climate change; they also possess the capacity to actively respond to climate change and reduce their vulnerability. This capacity is rooted in their traditional knowledge. For these communities, traditional knowledge—particularly climatological and phenological insights—can significantly mitigate vulnerability to climate change. Drawing on this knowledge, they have developed various adaptation measures that enable them to cope with climate variability and extremes. For instance, to reduce vulnerability to climate hazards, establishing an early warning system to forecast such events is essential (Ajibade & Shokemi, 2003).

The traditional climatological and phenological knowledge possessed by indigenous peoples and local communities plays a vital role in forecasting weather and climate. Drawing on this traditional knowledge, these communities assess the likelihood of extreme weather events and climate hazards, such as droughts, floods, landslides, and snowstorms, allowing them to predict the scale and impact of these events. This proactive approach enables them to implement preventive measures to reduce vulnerability to climate-related disasters. Additionally, utilizing their traditional climatological and phenological insights, indigenous peoples and local communities can forecast and interpret changes in weather and climate, ultimately informing their livelihood decisions. This includes determining and adjusting the timing of grazing, planting, crop varieties, livestock breeds, and related technologies. Consequently, climate prediction systems grounded in traditional knowledge can significantly aid indigenous peoples and local communities in mitigating the vulnerabilities of their livelihoods caused by climate change.

5.2.2 Traditional Knowledge and Climate Resilience Enhancement

As the impacts of climate change on ecosystems and human societies intensify, resilience has emerged as a crucial concept for adapting to these challenges. Following the Paris Agreement's emphasis on the global adaptation goal of "reducing vulnerability, enhancing adaptability, and fostering resilience", the notion of resilience has gained significant prominence in climate change adaptation discourse. In September 2019, during the United Nations Climate Summit in New York, the Global Commission on Adaptation released the report "Adapt Now: A Global Call for Leadership on Climate Resilience." This report underscores the profound and pervasive negative effects of climate change, asserting that adaptation has become essential for humanity, the environment, and the economy (GCA, 2019). Consequently, climate resilience has become a vital framework for scientific institutions, policymakers, international organizations, and governments in designing effective responses to climate change.

In studies and reports associated with the Convention on Biological Diversity, resilience is defined as the capacity of an ecosystem to return to its original state following a perturbation, while maintaining its essential taxonomic composition, structural integrity, ecosystem functions, and process rates. Specifically, resilience denotes a system's ability to absorb disturbances and still retain its fundamental functions and structure, thereby preserving its identity (Holling, 1973; Thompson et al., 2009; Walker & Salt, 2006). The Intergovernmental Panel on Climate Change (IPCC) defines climate resilience as the ability of a system and its components to anticipate, absorb, accommodate, or recover from the effects of hazardous events in a timely and efficient manner. This includes ensuring the preservation, restoration, or enhancement of its essential structures and functions (IPCC, 2012). Thus, resilience encompasses not only the ability of a system to maintain or return to its original state but also its capacity to adapt and learn in the face of uncertainty.

Consequently, climate resilience necessitates that social-ecological systems possess the following capabilities: (1) Absorptive Capacity—the ability to absorb external pressures brought about by climate change while maintaining the normal functioning of the system; (2) Adaptability—the capacity to adjust and respond to external shocks or stresses to enhance the sustainability of the system; and (3) Transformative Ability—the capability to prepare for future impacts of climate change (Béné et al., 2013; Folke et al., 2010; Nelson et al., 2007; Vayda & McCay, 1975).

Traditional knowledge provides valuable insights for assessing the resilience of indigenous peoples and local communities to the impacts of climate change. For these groups, their cultures and traditional knowledge are intrinsically resilient, grounded in profound knowledge of the land, familiarity with their territories, and essential skills for safety and survival. This foundation contributes significantly to their resilience. Furthermore, traditional knowledge possesses vitality and inherent adaptability in the face of climate change; it is not merely a static repository of information but a dynamic tool for transmitting social values and attitudes. Thus, traditional knowledge plays a critical role in enhancing the resilience of indigenous peoples and local communities as they adapt to climate change.

In the context of biodiversity, resilience is an emergent property of ecosystems, conferred at multiple scales by genes, species, functional groups, and processes within the system (Drever et al., 2006; Gunderson, 2000). Therefore, for the ecosystems and social systems of indigenous peoples and local communities, traditional knowledge associated with biodiversity plays a crucial role in enhancing climate resilience.

Traditional knowledge associated with genetic and species diversity significantly enhances the resilience of livelihood systems among indigenous peoples and local communities in the face of climate change. For instance, the diversity of plant genetic resources, coupled with relevant traditional knowledge, enables these communities to select germplasm from various crops to adapt to environmental changes induced by climate change, including alterations in soil quality, temperature, slope, orientation, and sunlight exposure. This traditional knowledge related to plant genetic resources diversifies agricultural livelihoods and mitigates the risks posed by climate change to agriculture, thereby facilitating more effective adaptation for indigenous peoples and local communities.

The diversity of livestock breeds and their related traditional knowledge can help indigenous peoples and local communities adapt to the impact of climate change on animal husbandry. In order to cope with the drought, herdsmen will choose to raise more drought-tolerant livestock varieties. The diversity of forage resources and their related traditional knowledge can help herdsmen to adjust the forage varieties for feeding livestock in time and minimize the risk of climate change on livestock livelihood. At the same time, traditional knowledge related to medicinal plant resources can also reduce the health risks brought by climate change. Diversified medicinal plants and their related traditional knowledge can maintain the health and medical care of indigenous peoples and local communities.

Traditional knowledge associated with ecosystem diversity can significantly enhance the resilience of both ecological and social systems among indigenous peoples and local communities in response to climate change. For instance, to address the impacts of climate change on farmland environments, farmers can adapt their agricultural practices and irrigation schedules based on traditional knowledge related to the farmland ecosystem, thereby strengthening agricultural resilience. Similarly, to mitigate the effects of climate change on pasture environments, herders can adjust their grazing schedules and rotational grazing systems informed by traditional knowledge of grassland ecosystems, enhancing the resilience of pastoralism. In the context of basin environments, farmers can utilize traditional knowledge to anticipate disasters such as floods, landslides, and debris flows, allowing them to implement preventive measures and effective responses post-disaster, thereby bolstering community resilience against extreme weather events and climate-related disasters. Furthermore, through traditional knowledge and cultural practices associated with forest ecosystems, indigenous peoples and local communities can enhance the resilience of forest ecosystems and provide vital ecosystem services to their local societies.

6 Traditional Knowledge and Climate Change in Southwest China

Climate change is increasingly impacting global ecosystems and human societies. Traditional communities, residing in specific regions for generations, have developed unique adaptation mechanisms through ecological knowledge accumulated over long-term interactions with nature.

In China, ethnic minorities possess a wealth of knowledge and traditions, which can be defined as traditional genetic resources, traditional utilization of medicinal biological resources, traditional technology, production and lifestyle practices, and traditional cultures related to the protection and use of biological resources, including traditional geographical indication products (Xue, 2011). This traditional knowledge is closely related to climate change.

Firstly, traditional knowledge is significantly affected by climate change. Ethnic minorities are widely distributed across China, living in environments that are complex and diverse, with varying climate conditions and changes. Consequently, traditional knowledge is closely impacted by climate change. On the one hand, traditional knowledge has evolved in response to climate change. For instance, the alpine climate of the plateau influences the living environment of Tibetan people, who have developed extensive traditional knowledge in agriculture and animal husbandry for their livelihood. The subtropical warm-wet climate has fostered the paddy field culture of the Dai people. The arid climate of deserts has shaped the lifestyle of Uyghur people, who have developed advanced green agriculture and irrigation technology using wells connected by underground channels. On the other hand, climate change impacts the development of traditional knowledge. For example, Mongolian traditional knowledge of using grasses is negatively affected by global warming, which causes grasses to grow or mature earlier. Drought negatively affects Kazak people's grazing traditions, necessitating changes in their rotational grazing system. Jino people's knowledge of disaster risk warning becomes less effective due to mudslides caused by heavy rainfall.

Secondly, traditional knowledge is an effective approach for adapting to climate change. Confronted with the risks and challenges posed by climate change, ethnic minorities do not respond passively. Instead, they actively develop and innovate their traditional knowledge to adapt effectively (Yin, 2015). There are two primary approaches to climate change adaptation: mitigation and adaptation.

On the one hand, traditional knowledge can mitigate the negative impacts of climate change at the local level. For instance, the Dai people's reverence for Dragon Mountain helps protect forests, which partially alleviates the risk of local climate warming. In regions with complex ecosystems, the Hani people employ effective terrace cultivation techniques to mitigate the threat of drought on agriculture.

On the other hand, traditional knowledge aids local communities in adapting to climate change. For example, the Dong people's traditional knowledge of managing water resources allows them to adapt effectively to changes in precipitation. Similarly, the Tibetan people's understanding of classifying grasses and protective planting techniques helps them manage changes in grazing land and livestock resulting from climate warming.

In southwest of China, particularly Yunnan and the Tibetan Plateau, characterized by complex geography and diverse ethnic groups, provides rich and vivid case studies for exploring the relationship between traditional knowledge and climate change.

6.1 Perception of Climate Change and Influencing Factors

6.1.1 Perception of Climate Change

Changes in temperature, particularly warming or cooling, are the most readily perceived aspects of climate change in daily life. However, perceptions vary significantly based on microclimatic differences across altitudes. In Deqin County, Yunnan, Tibetan villagers in low-altitude areas widely reported significant warming in recent years (Byg & Salick, 2009a, 2009b; Yin, 2013; Lurongzhuoma, 2013), while those at higher altitudes primarily noted increased temperature instability (Yin, 2013; Lurongzhuoma, 2013). Miao respondents in the Wumeng Mountains generally observed temperature changes, with the most prominent being a decrease in cold weather (Yang, et al., 2017). In Northwest Yunnan (Lijiang, Nujiang, Diqing), most respondents detected abnormal climate patterns, with warming being the most significant trend. Among Yi and Tibetan groups in this region, perceptions of drought were particularly strong (Yang et al., 2022). On the Tibetan Plateau, herders in Northern Tibet clearly perceived warming (Aquzhen, 2023); most respondents in high-altitude communities (4600m) in the northern Himalayas also reported warming, alongside an increased perception of drought intensity with rising altitude (Huang et al., 2019); villagers along the Yarlung Tsangpo River basin also reported rising temperatures based on observations over the past 30 years (Li et al., 2013).

Perceptions of precipitation changes were also significant, primarily characterized by an overall decrease, including reduced snowfall (Byg & Salick, 2009a, 2009b; Aquzhen, 2023; Huang et al., 2019; Yin, 2011c). Beyond this, local residents observed numerous other changes: a. Altered rainy seasons (Yin, 2013; Lurongzhuoma, 2013; Yin, 2011a), such as erratic start and end times; b. Glacier retreat (Byg & Salick, 2009a, 2009b; Yin, 2013; Lurongzhuoma, 2013; Aquzhen, 2023; Yin, 2011b; Xu and Yin, 2024); c. Changes in avalanches (Yin, 2013; Yin, 2011b; Xu and Yin, 2024), including abnormal frequency and scale; d. Fluctuating water levels and even shrinkage of plateau lakes (Yin, 2013; Lurongzhuoma, 2013; Aquzhen, 2023; Yin, 2011c; Xu and Yin, 2024); e. Increased seasonal fluctuations in rivers (Yin, 2013; Lurongzhuoma, 2013; Aquzhen, 2023; Yin, 2011a; Xu and Yin, 2024); f. Shifts in the distribution ranges of animals and plants (Aquzhen, 2023; Huang et al., 2019; Yin, 2011b; Xu and Yin, 2024); g. Disruptions in seasonal timing (Yin, 2013; Lurongzhuoma, 2013; Aquzhen, 2023; Li et al., 2013), such as erratic transitions between seasons.

6.1.2 Factors Influencing Climate Change Perception

Perception of climate change is inherently subjective and influenced by multiple factors. Key factors include age (Yin, 2013; Lurongzhuoma, 2013; Yang et al., 2022; Aquzhen, 2023; Li et al., 2013; Yin, 2011c), gender (Yin, 2013; Lurongzhuoma, 2013; Yang et al., 2022; Aquzhen, 2023; Li et al., 2013; Yin, 2011a), occupation (Yin, 2013; Lurongzhuoma, 2013; Yin, 2011b) and geographical location (Byg & Salick, 2009a, 2009b; Yin, 2013; Lurongzhuoma, 2013; Yang et al., 2022; Huang et al., 2019; Yin, 2011c). For instance, older farmers and herders tend to be more sensitive to changes; perceptions also differ between low-altitude and high-altitude residents.

6.2 Impacts of Climate Change

Climate change is currently affecting ecosystems and human livelihoods in multifaceted and profound ways, particularly impacting groups reliant on natural environments and traditional knowledge.

6.2.1 Disturbance to Ecosystems

Climate change impacts ecosystems in cascading ways. Regarding hydrology and water resources, warming causes glacier retreat, snowline rise, and increased instability in river flows (Yin, 2011a). Reduced snowfall and frequent droughts lead to insufficient lake replenishment, causing shrinkage and reduced river volumes (Aquzhen, 2023), even drying up some water sources (Huang et al., 2019). Extreme droughts threaten terrace irrigation systems, while heavy rains trigger soil erosion and landslides, damaging terrace structures.

Climate change also significantly alters the distribution patterns of flora and fauna. Some plant species, unable to adapt, wither or disappear regionally (e.g., small-leaf rhododendron, shrubs, dying juniper trees) (Aquzhen, 2023). Forests expand into higher altitude pastures due to warming, squeezing pastureland, while low-altitude animals migrate upwards (Yin, 2011b). Shrub communities on pastures expand due to warming and reduced rainfall, crowding out grasses, leading to decreased species diversity and yield (Lurongzhuoma, 2013). Increased soil moisture evaporation further reduces grass biomass and disrupts growth cycles (e.g., delayed greening, earlier withering), significantly reducing the proportion of high-quality forage (Yin, 2013). Ecosystem stability declines further with erratic precipitation patterns; extreme droughts can even cause terrace ecosystem collapse, directly threatening regional food production and biodiversity (Xu and Yin, 2024). Warming also introduces new species, such as new mosquito types (Aquzhen, 2023), new earthworm species (Huang et al., 2019), new pests (Li et al., 2013; Yin, 2011c), causing further ecological disturbance.

6.2.2 Impacts on Agriculture and Animal Husbandry

(1) Impeded Agricultural Production

Climate change impacts agriculture through intertwined effects on yield, pests/diseases, and cropping systems. Regarding yield and quality, extreme weather (snowstorms, heavy rains, droughts) directly damages crops, reducing output (Yang et al., 2017). Reduced rainfall forces greater reliance on river irrigation, but limited water often leads to unfilled grains and lower nutritional value (Yin, 2011a; Xu and Yin, 2024). Persistent precipitation decrease also limits the growth of traditional crops like barley and wheat, reducing yields (Shen and Chen, 2020). In Dai rice paddies, rice varieties intolerant to high temperatures are being phased out due to persistent warming (Yin et al., 2020).

Pest and disease problems worsen with warming. Milder winters significantly increase pest overwintering survival rates, leading to outbreaks of new pests and traditional diseases (Li et al., 2013; Yin, 2011b; Xu and Yin, 2024). New harmful insects spread to higher altitudes (Huang et al., 2019). Proliferation of new earthworm species in fields damages soil structure, reducing hull-less barley yields (Huang et al., 2019).

Cropping systems and calendars are becoming ineffective due to climate disruption. Warming accelerates crop growth cycles, disconnecting traditional agricultural calendars from actual seasonal changes (Yin, 2011c; Xu and Yin, 2024; Yin and Xue 2012). The Bai lunar calendar and seasonal markers no longer accurately guide farming due to mismatched phenology caused by warming (Yin et al., 2020). The Jinuo shifting cultivation calendar has also lost its guiding function due to climate instability (Yin et al., 2020). Extreme weather (floods, mudslides) directly destroys farmland (Yin, 2011a); Wa shifting cultivation fields are often ruined by heavy rains and floods (Yin et al., 2020). Drought increases irrigation demand, forcing farmers to buy fertilizers to compensate for soil fertility loss (Lurongzhuoma, 2013).

(2) Impeded Animal Husbandry Development

Animal husbandry faces significant challenges, from forage resources to livestock health and traditional techniques. Regarding forage resources, warming causes grasses to wither earlier, leaving insufficient fodder for livestock overwintering, forcing winter pastures to shift to higher altitudes (Yin, 2011b; Xu and Yin, 2024). Persistent dry, hot weather directly results in "less grass on the mountains, thinner cattle, less milk," affecting herders' income (Shen and Chen, 2020). Increased soil evaporation exacerbates forage shortages (Yin, 2013).

Livestock survival and production face multiple pressures. Reduced summer rainfall limits grass growth, reducing food sources (Aquzhen, 2023); less snow and delayed snow seasons make winter survival difficult (Aquzhen, 2023). Extreme snow disasters cause weight loss and reduced survival rates due to cold and food scarcity (Yao et al., 2015). Warming increases livestock disease incidence and lowers reproduction rates (Lurongzhuoma, 2013); new diseases (e.g., foot-and-mouth disease) spread under suitable conditions created by warming (Yin, 2011c), ultimately reducing yak body condition and milk production (Yin, 2011a; Shen and Chen, 2020).

Traditional herding techniques are being forcibly adjusted. Warming alters grass growth rhythms, requiring changes to grazing cycles and significantly advancing seasonal migration times (Yin, 2013; Yin, 2011b; Xu and Yin, 2024). Traditional rotational grazing cycles lose effectiveness due to frequent climate fluctuations (Lurongzhuoma, 2013).

(3) Crisis in the Transmission of Traditional Knowledge and Culture

Climate change directly impacts traditional knowledge formed through natural rhythms. Regarding livelihood knowledge, traditional agricultural calendars lose their guiding function due to warming, unstable precipitation, and mismatched seasons (Yin, 2011c; Xu and Yin, 2024; Yin et al., 2020; Yin and Xue 2012). Tibetan rotational grazing cycles are forced to adjust due to altered grass growth patterns (Lurongzhuoma, 2013; Yin, 2011a).

Resource utilization knowledge is also affected. The distribution altitude of medicinal plants rises with warming, and altered growing environments reduce yield and quality. Traditional collection products like caterpillar fungus and matsutake mushrooms experience disrupted growth cycles due to climate fluctuations, leading to unstable quality and yield, often with declines (Yin, 2013).

Transmission of beliefs and rituals is impaired. The Tibetan "mountain deity worship" ritual (Tibetan calendar, 5th month) has been canceled about 40% of the time in the last decade due to unstable snowfall, reducing traditional activities (Yin, 2013). Traditional Tibetan taboos like "grain for people, grass for cattle" are being broken as grain becomes more abundant, challenging traditional resource allocation concepts (Lurongzhuoma, 2013). These reflect the deep cultural impact of climate change.

(4) Multiple Pressures on Livelihoods and Health

Climate change directly threatens livelihood stability and health. Regarding livelihoods, weather changes (snow disasters, drought, erratic rainfall) severely impact traditional Tibetan agro-pastoralism and crucial income sources like caterpillar fungus collection, pushing many herders into unstable living conditions (Aquzhen, 2023; Shen and Chen, 2020).

Regarding health, persistent summer heat at lower altitudes causes weakness, food spoilage leading to stomach ailments, and surges in flies and mosquitoes (Byg & Salick, 2009a, 2009b). Residents in high-altitude villages face reduced harvests and resource shortages, leading to insufficient nutritional intake and increasing health problems (Byg & Salick, 2009a, 2009b).

In summary, climate change poses a comprehensive challenge to human societies by disturbing ecosystems, impacting agriculture and pastoralism, eroding traditional knowledge systems, and threatening human health, highlighting the particular vulnerability of traditional societies reliant on natural rhythms.

6.3 Practices and Value of Traditional Knowledge in Responding to Climate Change

Challenges like extreme weather and resource fluctuations prompt a re-evaluation of the unique value of traditional knowledge in climate adaptation. Research shows that traditional knowledge, accumulated by ethnic minorities and local communities through long-term practice, forms a composite system for coping with climate uncertainty through diverse strategies. These practices encompass technical adjustments, resource management, livelihood diversification, and spiritual support, offering important insights for contemporary adaptation research.

6.3.1 Adaptive Techniques and Species Selection in Agriculture

The application of traditional knowledge in agriculture is most direct, focusing on reducing climate risks and ensuring stability through species selection and cultivation innovations. In species selection, different groups have preserved resilient traditional varieties adapted to local climates: Tibetans prioritize cold/drought-tolerant barley; Dai cultivate heat-resistant rice; Wa grow drought-tolerant upland rice; Hani and Dong use cold-resistant rice in high-altitude terraces (Yin et al., 2020); Miao in Wumeng Mountains cultivate locally adapted traditional maize and potato varieties, significantly more resilient than introduced varieties (Yang et al., 2017). Combining introduction and selection techniques enhances adaptability. For example, choosing drought-tolerant crops based on annual forecasts, adjusting planting times, or integrating modern techniques like greenhouses for cash crops (Yin and Xue 2012).

In cultivation techniques, diversification and precision effectively spread risk. Miao practice intercropping (maize, potato, legumes) for efficient resource use; their extensive farming model, with lower marginal costs, is a collective adaptation choice (Yang et al., 2017). Wa and Jinuo use mixed cropping (multiple crops/varieties per plot), leveraging different climate tolerances to avoid total crop failure (Yin et al., 2020). Tibetans and Hani employ staggered sowing based on microclimates (altitude, slope aspect) to reduce drought/flood risk (Yin et al., 2020). Bulang rotate gramineous and solanaceous crops (e.g., cotton and upland rice) to increase soil organic matter and improve resilience to waterlogging or drought (Yin et al., 2020). These demonstrate traditional wisdom in balancing cost and risk.

6.3.2 Community Collaboration and Systems Thinking in Resource Management

Traditional resource management emphasizes "system balance," achieving adaptation through community collaboration and rules. Water management is exemplary: Cross-ethnic groups follow customary law for equitable irrigation water sharing during droughts. Within groups (e.g., Tibetan villages), community meetings coordinate rotational irrigation between upstream/downstream, ensuring fairness and avoiding conflict (Yin et al., 2020).

Forest and biological resource management also reflect systems thinking. Hani enhance environmental self-regulation through integrated management of land, forests, water, and biological resources (Jiao et al., 2012). For instance, local communities strictly protect mountaintop forests for water conservation and terrace irrigation, safeguard wild food sources and traditional crops, and maintain ecological balance through diversification. Ethnic groups such as the Tibetan people avoid ecological damage caused by overlogging by rationally demarcating forest logging areas and carrying out afforestation, etc. (Xu and Yin, 2024; Huang et al., 2025), and promote the sustainable use of resources through traditional community organizational mechanisms (Yin, 2013). In addition, the protection and utilization of biological resources are closely integrated. For example, herdsmen have found through long-term practices that Anisodus tanguticus can cope with cold and food shortages; the rational management of forage grass and Tibetan medicine resources by the Tibetan people all embodies the wisdom of traditional knowledge in the protective development of biological resources (Yin, 2013; Yao et al., 2015).

6.3.3 Diversification and Dynamic Adjustment of Livelihood Strategies

To reduce dependence on single resources and mitigate climate risk, traditional communities develop diversified livelihood strategies. Labor migration is key, providing stable cash income and diversifying sources, indirectly enhancing household resilience (Yang et al., 2017; Yang et al., 2022). Some communities pursue industrial substitution (e.g., viticulture replacing traditional farming/herding), integrating traditional knowledge with modern techniques during transition (Shen and Chen, 2020).

Dynamic adjustment is central to agro-pastoral planning. Communities move beyond rigid calendars, flexibly changing planting/migration times, optimizing livestock/crop diversity based on annual conditions (Huang et al., 2019; Yin, 2011b). In pastoralism, adjusting rotational grazing—timely pasture shifts based on grass growth, processing straw for winter fodder—improves resource efficiency and alleviates shortages (Lurongzhuoma, 2013). These strategies reduce dependence on climate stability and enhance resilience.

6.3.4 The Implicit Role of Spiritual Support and Experiential Transmission

Beliefs and experiential systems within traditional knowledge provide vital spiritual and cognitive support for facing climate disasters. Spiritually, religious rituals and beliefs play crucial roles: Dragon king worship, rain-seeking ceremonies, and ritualized festivals address climate threats (Yang, 2009); herders use rituals to alleviate fear/anxiety during disasters and motivate livelihood adjustments (Aquzhen, 2023); beliefs like sacred mountains promote ecological practices (e.g., afforestation), indirectly improving microclimates (Yin, 2013; Xu and Yin, 2024; Yin and Xue 2012; Zhou and Yin, 2024).

Experiential prediction and transmission guide production decisions. Communities use inherited proverbs and phenological observations (e.g., animal/plant behavior) to forecast weather and guide farming/herding. This knowledge, based on long-term observation, enables proactive adjustments respecting climate patterns, integrating water management and ritual practices to enhance active risk management (Huang et al., 2025).

6.3.5 Contemporary Value and Policy Implications of Traditional Knowledge

In summary, the traditional knowledge of ethnic minorities is not static "cultural heritage" but a dynamic "living system" for climate adaptation. Its core value lies in forming resilience against climate uncertainty through the synergy of species adaptation, technical innovation, community collaboration, livelihood diversification, and spiritual support. This system offers significant policy insights: government adaptation strategies should recognize local differences in traditional ecological knowledge, promote livelihood diversification to reduce resource dependence (Yang et al., 2022), and facilitate the integration of traditional knowledge with modern technologies (e.g., solar, biogas to reduce deforestation) (Yin and Xue 2012; Yang, 2009).

Ultimately, traditional knowledge provides multi-level experience, from micro-techniques to macro-governance. Its wisdom of "adapting measures to local conditions" is crucial for building more resilient climate adaptation systems.

Traditional communities in Southwest China, through long-term interactions with nature, have developed an acute perception of climate change and unique adaptation systems. In the context where global climate governance increasingly emphasizes "local wisdom," protecting and activating such traditional knowledge is not only an inherent need for safeguarding biodiversity and cultural diversity, but also a key path for building climate adaptation systems with regional resilience. Therefore, in future research and practice, it is necessary to further explore the potential for modern transformation of traditional knowledge in responding to climate change and build regionally adaptive climate adaptation systems.

7 Methodology and Area of Study

7.1 The Method of Research

In the study of climate change and traditional knowledge, interdisciplinary research and survey methods represent a prevailing trend. This study, after consulting relevant literature, books, and other materials, employs a combination of field survey methods, including semi-structured interviews, key informant interviews, questionnaire surveys, and sampling techniques. It reviews the progress and achievements in research on climate change and traditional knowledge, with a particular focus on Tibetan traditional knowledge and its interaction with climate change. Additionally, the study analyzes conventions and policies developed and endorsed by United Nations agencies, intergovernmental organizations, regional organizations, and countries concerning climate change and indigenous peoples.

In this study, the traditional knowledge of the Tibetan people in Deqin has been systematically classified. The research categorizes traditional knowledge related to climate change adaptation into eight distinct categories:

I. *Traditional Knowledge in Water Resource Management for Climate Change*

Adaptation

II. *Traditional Knowledge in Livelihood and Lifestyle Practices for Climate Change*

Adaptation

III. *Traditional Knowledge on Medicinal Plant Species and Their Response to*

Climate Change

IV. *Traditional Knowledge in Forest Resource Management for Climate Change*

Adaptation

V. *Traditional Knowledge on the Use of Plant Species and Genetic Resources for*

Climate Change Adaptation

VI. *Gender Perspectives on Traditional Knowledge in Irrigation Management and*

Climate ChangeAdaptation

VII. *Traditional Knowledge in Disaster Risk Management for Climate Change*

Adaptation

VIII. *Traditional Beliefs and Sacred Mountains in Climate Change Adaptation*

The fieldwork primarily focuses on three themes: 1) the local Tibetan traditional knowl-

Fig. 1 Tibetan village in the Eastern Himalayas, Yunnan Province, China (photoed by Yin Lun)

edge related to climate; 2) the impact of climate change on traditional knowledge related to biodiversity; and 3) the relationship between traditional knowledge and adaptation to climate change.

In this study, Tibetan villages in Deqin County and their alpine pastures were selected for field investigation. These villages are situated at altitudes ranging from 2,100 to 3,300 m above sea level (Fig. 1). Due to varying altitudes, the climatic characteristics of each village differ accordingly. The villages are classified into three types based on altitude: (1) Villages in the river valleys, at approximately 2,000 m, experience a dry and warm climate; (2) Villages in the semi-hillside areas, at around 2,500 m, have a slightly humid and cool climate; and (3) Villages located above 3,000 m above sea level are characterized by a humid and cold climate (Yin, 2015).

The quantitative analysis and research methods for traditional knowledge represent another innovation of this study. Traditional knowledge, which is largely based on relative and localized experience, often lacks scientific benchmarks and measurement scales. This has posed a long-standing academic challenge in realizing quantitative analysis and research, leading to skepticism about the credibility and value of traditional knowledge within the scientific community.

This study addresses this issue by developing indices and data for traditional knowledge and exploring methods for its quantitative analysis and research. The exploration of these quantitative methods has enhanced the credibility and value of traditional knowledge.

7.2 Research Area

Deqin County, located in the Eastern Tibetan Himalayas of Northwest Yunnan Province, China, is administered by the Diqing Tibetan Autonomous Prefecture. It spans latitudes from

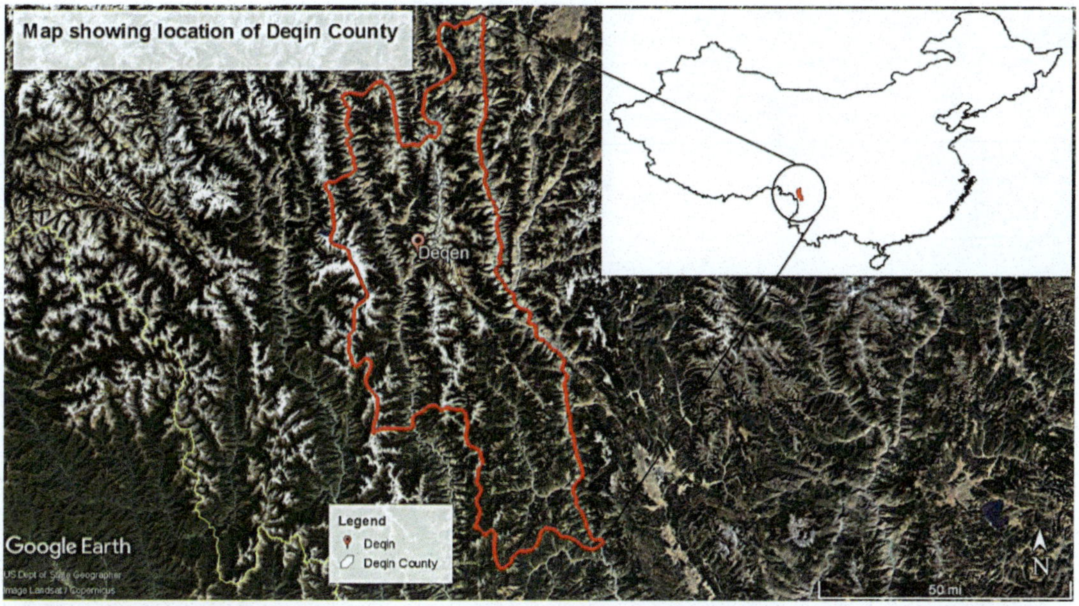

Fig. 2 Geographical location of Deqin County in China. The red boundary indicates the area of Deqin County, with an inset map of China shown in the upper right corner. Disclaimer: The geographical designations used do not imply any opinion on the part of the involved agencies regarding the legal status of territories or the delimitation of boundaries. Users who notice errors are advised to consult the source. (This map was created by Misiani Zachary on May 18, 2018)

27°33' to 29°15' N and longitudes from 98°36' to 99°33' E, covering an area of 7,596 km^2. The county borders the Tibet Autonomous Region to the northwest and Sichuan to the northeast (see Fig. 2).

Deqin County, situated at altitudes ranging from 2,000 to 6,740 m, is located at the transition between a subtropical highland climate and a humid continental climate, which is notable given its latitude. Although mean maximum temperatures remain above freezing throughout the year, minima fall below freezing from November to March. January temperatures average − 2.1°C, while July temperatures average 12.7°C, with an annual mean temperature of 5.65°C. Rainfall is concentrated between June and September, accounting for nearly 60% of the annual total of 622 mm. Snowfall is rare but can cause significant transportation issues during winter. The county receives 1,989 h of bright sunshine annually, with monthly percent possible sunshine ranging from 29% in July to 62% in December. Autumn and winter are sunnier compared to spring and summer.

Deqin is situated in the Eastern Himalayas and the central part of the Hengduan Mountains (Fig. 3), encompassing the valleys of the Salween, Mekong, and Yangtze Rivers. The region features two prominent mountain ranges: Meili Snow Mountain and Baima Snow Mountain. The Salween River bounds Deqin to the west, while the Mekong River borders it to the east. The crest of these ranges rises to over 6,000 m above sea level, creating a striking contrast with the river valleys to the east and west, which are situated between 1,500 and 1,900 m in elevation. The highest peak is Kawagebo, which reaches 6,740 m and is considered sacred by Tibetan Buddhists (Yin et al., 2013).

In Deqin, 80% of the 55,000 inhabitants are Tibetan. The Tibetans in this area reside at various elevations, ranging from relatively low-lying, warm, and dry valleys (around 2,000 m) to high, cool, and moist mountain areas (above 3,000 m). Most Tibetan people in Deqin utilize a range of ecological and climatic zones distributed along this elevation gradient (Salick et al., 2007; Yin and Xue 2012). Agro-pastoralism remains the

Fig. 3 Eastern Himalayas Hengduan Mountains, Yunnan Province, China (photoed by Yin Lun)

Fig. 4 Traditional agro-pastoralism in the Eastern Himalayas, Yunnan Province, China (photoed by Cili Zhaba)

primary and vital traditional livelihood for the local Tibetan population (Fig. 4).

This study selected Deqin as the research area for the following reasons:

(a) *Deqin County is characterized by a plateau cold temperate mountain monsoon climate. Due to its low latitude and significant vertical height difference, the climate exhibits notable vertical variations and three-dimensional characteristics. Consequently, climate change in Deqin is more complex and diverse compared to other regions.*
(b) *Due to its fragile geological conditions and ecosystem, Deqin is particularly vulnerable to climate-related disasters, such as debris flows, floods, and landslides.*
(c) *The diverse climate conditions and ecosystems of Deqin have led to the development of an agro-pastoral system among the local Tibetans that is adapted to the local climate and environment. Over time, this has resulted in the accumulation of extensive traditional knowledge about climate and phenology, which aids local villagers in predicting, preventing, and managing the impacts of climate change and related disasters.*
(d) *Local Tibetan traditional culture in Deqin encompasses traditional knowledge related to biodiversity, which has continuously interacted with changes in the natural environment and social development. This knowledge is a basis for local Tibetans to adapt to climate change and is a process of ongoing innovation.*

7.3 The Objects of Research

7.3.1 Traditional Knowledge of Climate and Climate Change

Although climate models can provide a broad view of climate change and estimate the various trends and consequences of human development in the future, they often lack precision for regional climate change. Recently, there has been increasing recognition of the traditional knowledge about climate and climate change held by indigenous peoples and local communities. This knowledge, based on the complex interplay of ecology and environment, offers valuable insights into regional climate change that instruments alone cannot capture (Berkes, 2012a, 2012b). Currently, scientific explanations of climate change primarily focus on greenhouse gas emissions. In contrast, traditional knowledge is more diverse and encompasses a broader range of factors, providing more localized and practical observations and understandings of climate change. Traditional knowledge suggests that climate change is not merely a natural phenomenon but also influenced by environmental and social factors, which in turn shape people's understanding of climate change (Reidlinger, 2001; Leonard, 2013).

The traditional knowledge of climate change among Tibetan people in Deqin encompasses various aspects, including temperature fluctuations, changes in rainfall and snowfall, seasonal variations, wind patterns, annual climate trends, as well as modifications in glaciers, snow cover, ice, rivers, and lakes. It also includes predictions and responses to extreme climate disasters. This body of traditional knowledge is rich and diverse, providing valuable insights that enhance the understanding of climate change (Xue et al., 2009).

7.3.2 The Impact of Climate Change on Traditional Knowledge

The Tibetan people in Deqin have developed resource utilization and livelihood practices that are well adapted to the local natural environment and ecosystem. They have accumulated extensive traditional knowledge related to the protection and sustainable use of biological resources. This knowledge includes traditional practices for utilizing forage resources, Tibetan medicine and medicinal plants, and traditional animal husbandry methods. However, this traditional knowledge is increasingly impacted by climate change and extreme disasters, with some practices becoming less effective while others evolve and adapt.

7.3.3 Traditional Knowledge and Adaptation to Climate Change

Currently, global discussions on climate change are increasingly focusing on adaptation as a priority for research and policy development. Perspectives on climate change vary among different groups, leading to diverse approaches for mitigating its negative effects and adapting to its impacts. In Deqin, the Tibetan people are not only observers and affected by climate change but also actively engage in adapting to changes in the local climate environment. Their adaptation measures include: diversifying and innovating cash crop cultivation, adjusting production schedules, modifying production technologies, changing resource use and lifestyles, and integrating traditional knowledge with modern science and technology.

7.3.4 The Questions of Research

As research on climate change continues to deepen and expand, interdisciplinary approaches are expected to become increasingly important. This study aims to establish a research framework that integrates natural and social sciences, based on the classification of traditional knowledge.

Traditional knowledge forms the foundation for indigenous peoples and local communities to understand their living environment, and it can be theoretically measured and evaluated. Consequently, a significant scientific challenge is to develop indices and data for traditional knowledge and establish methods for its quantitative analysis and research.

8 Traditional Knowledge for Coping with Climate Change

Deqin is situated in the eastern Himalayas, on the southern edge of the Qinghai-Tibet Plateau, and along the upper reaches of the Mekong River Basin. The region encompasses both plateau mountain areas and valley areas, resulting in a climate that combines characteristics of both cold temperate mountain monsoons and low-temperature plateau monsoons. The area exhibits a notable three-dimensional climate, with temperature variations spanning from mountainous north subtropical zones to temperate zones, cold temperate zones, and modern glacier zones.

Fig. 5 Tibetan villager in the Eastern Himalayas, Yunnan Province, China (photoed by Yin Lun)

In Deqin County, the considerable altitudinal variation across its geographical range has given rise to a remarkable diversity of biodiversity, climate types, ecosystems, and livelihood strategies. Building upon this ecological and sociocultural diversity, local Tibetan villagers (Fig. 5) have, through long-term practices of production and daily life, developed a rich and sophisticated system of traditional knowledge.

In the context of contemporary global climate change, this body of knowledge can play an important role in coping with environmental transformations, thereby constituting a traditional knowledge system for addressing climate change. More specifically, this system comprises five interrelated components: traditional phenological information for recognizing climate change; traditional natural resource utilization techniques for mitigating climate change; traditional livelihood strategies for adapting to climate change; traditional ecosystem management mechanisms for reducing the risks of climate-related disasters; and traditional cultural beliefs and practices for interpreting climate change.

Local villagers categorize the climate into three distinct types. The first type, known as *Qiarong*in Tibetan, refers to the cold highland with perennial snow, specifically the area at an altitude of about 4,000 m up to the snow-capped peaks. This zone remains covered by snow year-round and is classified climatologically as a cold temperate zone, cold zone, and modern glacier zone. The second type, called *Sineng*in Tibetan, translates to "the place where snow will melt" and refers to the hillside at an altitude of 2,600 m. This area experiences heavy snowfall in winter and melting ice in summer, corresponding to the temperate zone and cold temperate zone in climatological terms. The third type, known as *Jiashui*in Tibetan, means "the place where snow cannot reach," and pertains to the valley area at an altitude of 2,000 to 2,600 m. This zone does not receive snowfall throughout the year and is classified as a mountainous north subtropical and temperate zone climatologically.

The traditional knowledge about climate change among the Tibetan people in Deqin can be categorized into four types based on various factors, including meteorological elements, natural environment, living environment, and animals and plants.

First, changes in meteorological elements. Local villagers use the proverb: "Red clouds in the evening, sunny in the morning; dark clouds in the evening, heavy rain in the morning." This reflects their ability to forecast weather based on cloud patterns and other meteorological phenomena. Specifically, if the evening sky is adorned with red clouds, the following days are expected to be clear. Conversely, dense dark clouds in the evening predict rainstorms in the near future. Villagers also observe cloud movements; for instance, if clouds move back and forth from east to west within a day, heavy rain is anticipated in the subsequent days. Temperature fluctuations are another predictive factor. For example, if a sudden drop in temperature occurs after an extremely hot summer day, it signals the likelihood of heavy rain. Similarly, if a warm spell occurs suddenly during a cold winter, it also suggests that heavy rain is expected in the following days.

Second, changes in the natural environment. For instance, if the river suddenly diminishes or even dries up during summer, heavy rain is expected in the following days. A rainbow appearing at the river's source at sunset signals upcoming heavy rain. At night, if the stars and moon appear hazy with a surrounding moisture, it indicates that heavy rain is imminent. A lack of thunder during rain suggests that it will persist for a long time. Conversely, if rain stops abruptly and clouds settle over the village, it will rain for several days. When thunder is muffled, it indicates heavy rain; however, if thunder is loud, the rain will be lighter. If the wind precedes the rain, it typically will not be heavy, but if the wind follows the rain, heavy rain is likely. Strong winds may, however, lead to short-lived rainfall. Dark clouds at sunset predict heavy rain the next day. Low clouds on mountain peaks suggest continuous heavy rain. Intermittent rain indicates a heavy downpour in the upcoming week. Clouds surrounding the sun predict a rainstorm within a day, while clouds around the moon forecast rainstorms within a month. Rain at the beginning of the month signals that families should prepare for extended heavy rainfall. Unusual loud noises from the mountain foretell mudslides and landslides, while a sudden increase in river noise over several nights suggests floods and mudslides.

Third, changes in the living environment. For example, if people suddenly become listless and drowsy upon sitting down, it predicts heavy rainfall in the coming days. If smoke from cooking does not exit through the chimney but instead lingers inside the house, heavy rain is expected soon. When wooden doors become unusually difficult to open or close, heavy rain is likely within two days. Additionally, if the bottoms of house pillars suddenly become wet, it indicates prolonged rainfall.

Fourth, changes in animal behavior. For example, if yaks on an alpine pasture start running around and their mouths puff up as if sniffing something, heavy rain is likely in the next few days. When pigs in a pen become irritable, run around, and carry hay to their sleeping areas, heavy rain is expected over the next two or three days. If a high-altitude cuckoo suddenly flies to a lower altitude and makes a particularly loud call, it indicates that heavy rain or snowfall is imminent. In summer, if a snake drinks water by the river or sleeps in a dish without waking up, heavy rain is expected soon. If a pheasant from the mountain forest suddenly flies to the village, heavy rain and snowfall are likely within the next month. Additionally, the presence of crows and frogs calling, ants moving, fish jumping out of the water, and chickens hiding in their coops are also signs of continuous heavy rainfall. Conversely, if cattle are reluctant to return to their pens, dogs whine and groan, and pigs run wildly on a day of continuous heavy rain, it suggests the possibility of mudslides, landslides, and floods.

8.1 Traditional Climatic and Phenological Indicators for Understanding Climate Change

The traditional knowledge of Tibetan villagers in Deqin County concerning the recognition of climate change can be divided into two parts: climate information and phenological information. Traditional climate information is primarily based on observable climatic and meteorological phenomena, such as clouds, wind, air currents, sunrises and sunsets, rain, and fog, which are used to anticipate future climatic conditions. In contrast, traditional phenological information relies mainly on the behaviors of animals and plants to predict forthcoming climatic changes.

Regardless of whether the information pertains to climate or phenology, it is largely preserved in the form of local proverbs. To document this knowledge, the author conducted detailed field investigations, recording, translation, classification, and compilation of traditional proverbs in Deqin County. Since most of these proverbs are expressed in Tibetan, each was translated into Chinese. From the extensive corpus of proverbs, those directly related to climate change were selected. These were then categorized into two principal groups—climate information and phenological information—and further subdivided according to specific climatic and meteorological phenomena on the one hand, and animal and plant behaviors on the other.

8.1.1 Traditional Climatic Indicators

(1) Clouds

In the traditional climate knowledge of Tibetan villagers in Deqin County, a large proportion of observations pertain to cloud formations. For instance, the proverb "red clouds at dusk, fair weather in the morning; dark clouds at dusk, rain in the morning" illustrates how villagers interpret variations in clouds and sunsets to forecast weather. It conveys that after rainfall, if red clouds appear in the sky before sunset, clear days are expected to follow, whereas the emergence of dark clouds suggests impending rainfall.

Other sayings include: "white clouds drifting east and west indicate heavy rain in the coming days"; "an accumulation of dark clouds at sunset foretells rain by midnight"; "low-lying clouds herald rainfall, while higher clouds presage clear conditions"; "cotton-like clouds in the morning are followed by afternoon thunderstorms"; "clouds shaped like sheep signify light rain"; "layered clouds upon one another imply prolonged precipitation"; "spherical cloud forms are associated with thunder and rain"; "clouds concentrated in the east rarely bring rain, whereas those in the west often do"; "clouds resembling a white pagoda in the morning imply afternoon winds and rainfall"; "hook-shaped clouds are taken as a sign of imminent rain"; "fish-scale clouds in the morning anticipate wind and rain later in the day"; "scattered cloud clusters on the distant horizon suggest no rain but strong sunlight in the following days"; "clouds shaped like a shuttle foretell persistent sunshine"; and "tile-like cloud patterns imply hot, sunny weather in the days ahead."

(2) Wind, Thunder, and Lightning

In addition to clouds, traditional climate knowledge among Tibetan villagers in Deqin County also encompasses observations of wind, thunder, and lightning. With regard to wind, local sayings include: "when the wind is weak yet the air feels stifling, heavy rainfall will follow"; "several days and nights of continuous strong wind are usually succeeded by drought"; "persistent southerly winds are interpreted as a sign of impending drought"; and "the onset of westerly winds is associated with frost."

In relation to thunder, villagers observe that "explosive thunder is linked to light rainfall, whereas muffled thunder presages heavy rain," and that "if thunder occurs before rainfall begins, the subsequent precipitation will be limited."

As for lightning, proverbs hold that "horizontal flashes of lightning foretell heavy rainfall, whereas vertical flashes indicate lighter rain," and that "if lightning is visible but thunder is inaudible, no heavy rainfall will ensue."

(3) Sun, Moon, and Stars

In the traditional climate knowledge of Tibetan villagers in Deqin County, there also exists a wealth of observations concerning the sun, moon, and stars. Common sayings include: "a halo around the sun signals rain within the day, while a halo around the moon indicates rainfall within the month"; "bright stars foretell fair weather on the following day"; "if the sun sets into dark clouds, rain is expected the next day"; "a red sun suggests forthcoming wind and rain"; and "a yellow sun indicates heavy rainfall on the following day."

With regard to the moon, villagers observe that "a red halo around the moon signifies heavy rainfall"; "a yellow halo indicates lighter rain"; "a blue halo is associated with strong winds"; and "a dark shadow encircling the moon predicts clear skies the next day." Similarly, it is believed that "a solar halo presages evening rain, whereas a lunar halo suggests wind on the following day"; "a large ring around the sun corresponds to clear weather, while a smaller ring implies rain"; and "an excessively scorching sun in the morning is followed by heavy afternoon rain."

Other sayings focus on sunlight and stars, such as: "when sunbeams radiate in streaks from behind clouds, rain is imminent"; "a luminous halo surrounding stars is taken as a sign of approaching rain"; "flickering or twinkling stars are likewise interpreted as an indication of impending rainfall"; and "when stars and the moon appear dim at night, as though veiled by mist, it foreshadows rain on the following day."

(4) Fog, Rainbow, Haze, Dew, and Frost

Fog, rainbow, haze, dew, and frost also constitute important components of the traditional climate knowledge of Tibetan villagers in Deqin County. First, regarding fog, it is said that "dense morning fog is followed by scorching sunlight in the afternoon," whereas "light morning fog indicates clear weather throughout the day."

Second, in relation to rainbows, villagers note that "a rainbow appearing in the west predicts rainfall, while one in the east signals fair weather," and that "if a rainbow emerges at the source of a river in the evening, several consecutive days of rain will follow."

Third, with respect to haze and colored clouds, proverbs include: "red clouds in the morning foretell rain in the evening, whereas red clouds at sunset predict a clear day ahead," and "when reddish-brown clouds appear both at dawn and dusk, drought is expected in the near future."

Fourth, in terms of dew, the sayings hold that "when morning dew is abundant and lingers for a long period, several consecutive days of clear weather will ensue"; "if dew is heavy but evaporates quickly, overcast conditions are likely in the coming days"; and "the presence of both dew and clouds suggests that rainfall is imminent."

Finally, with regard to frost, villagers remark that "a single day of frost is followed by several days of cloudy weather, whereas multiple days of frost foretell several days of sunshine."

(5) Temperature

Perceptions of temperature also constitute an integral part of traditional climate knowledge. Tibetan villagers in Deqin County interpret sudden shifts in thermal sensation across different seasons as indicators of forthcoming climatic change. For example, the saying "sudden cold in summer brings rain, sudden warmth in winter brings rain" conveys that when a hot summer day is unexpectedly interrupted by a cold airflow that makes people feel chilled, or when an intensely cold winter day is suddenly accompanied by a warm airflow that produces a sensation of heat, both conditions are understood to presage heavy rainfall.

The most direct way villagers perceive climate change is through temperature variations. Changes in temperature are the most apparent and easily felt manifestations of climate change. Most local villagers believe that temperatures are rising compared to the past, while a significant number feel that temperatures are becoming increasingly unstable. However, perceptions of temperature change vary among villagers at different altitudes. Those in middle and low-altitude villages generally experience a more noticeable rise in temperature, whereas high-altitude villagers often perceive greater temperature instability. This discrepancy is linked to recent climate phenomena, such as snow disasters and temperature drops in high-altitude areas. Additionally, villagers believe that temperature changes are affecting seasons, the natural environment, and patterns of rainfall and snowfall.

(6) Rain and Rainfall

Rain itself also constitutes an essential component of traditional climate knowledge among Tibetan villagers in Deqin County. Common sayings include: "a sudden downpour quickly clears, while slow rain persists"; "if gusty winds arise during rainfall, the rain will soon cease"; and "wind before rainfall means the rain will not last long, whereas wind after rainfall indicates it will continue."

Other observations note that "alternating showers and clear spells signal rainfall lasting for several days and nights"; "if clouds accumulate over mountain peaks during rainfall, the precipitation will endure for many days"; and "drizzle following heavy rain suggests continued overcast conditions, whereas showers after heavy rain signify the return of clear skies."

Seasonal patterns are also emphasized: "an easterly wind in the dry season does not bring rain, while in the rainy season it does"; "if strong southerly winds suddenly arise during continuous rainfall, clearing weather is imminent"; and "when prolonged light rain is suddenly accompanied by strong winds, sunshine will soon follow." Finally, it is observed that "rainfall preceded by wind is of short duration, but if no wind occurs during rainfall, the precipitation tends to last much longer."

According to surveys of villagers, changes in rainfall over the past 20 years include a decrease and increased instability in rainfall patterns, a delayed onset of the rainy season, a shortened duration of the rainy season, an earlier end to the rainy season, and a reduction in intermittent rainfall during the dry season.

Firstly, annual rainfall has been decreasing over the years. Although there are instances of extreme heavy rainfall, the overall trend is a decline in precipitation. Secondly, the onset of the rainy season has been delayed from late May to late June. Thirdly, the duration of the rainy season has been reduced from three months to two months. Fourthly, the end of the rainy season has shifted from late August and early September to early August. Lastly, the frequency of intermittent rainfall during the dry season has decreased. Previously, intermittent rainfall occurred every month during the dry season, but now continuous droughts can persist for several months without any rain.

Villagers generally perceive a significant reduction in rainfall and a shortening of the rainy season. This perception is particularly pronounced in low- and medium-altitude villages. In contrast, villagers in high-altitude areas, where sudden heavy rainfalls have increased, observe that both the amount and duration of rainfall are becoming more variable. Additionally, villagers' perceptions of the onset and conclusion of the rainy season are also inconsistent. Villagers in low-altitude regions have noted a delay in the start of rainfall and an earlier end to the rainy season.

(7) Glacier, Avalanche, and Snow

Tibetans are known as the people of the snowy plateau, with snow being a significant cultural symbol in Tibetan tradition, representing holiness. Tibetans hold deep reverence for snow-capped mountains throughout the year, with many of these mountains considered sacred and protective of the Tibetan people.

Villagers living at different altitudes have experienced varying impacts from changes in glaciers, avalanches, and snow. Meili Snow Mountain and Baima Snow Mountain traverse Deqin from north to south. These two snow-capped mountains remain snow-covered year-round and host glaciers of various sizes. Many villagers believe that glaciers have been thinning, disappearing, and shrinking over the past 50 years. This observation is supported by comparisons of photographs taken by American explorers in the 1920s with those of the same locations taken today. Villagers also perceive a gradual reduction in snow coverage, noting that many snow-capped peaks now retain snow only during the winter months.

According to surveys of villagers, particularly herdsmen who frequently graze in alpine pastures, avalanches traditionally occurred during the hottest summer months, from July to August. However, in recent years, avalanches have become more frequent and their timing has become less predictable. Large-scale avalanches now occur in other months as well, often accompanied by black boulders that roll down onto the alpine meadows.

Villagers in middle- and high-altitude areas are more aware of the disappearance of glaciers and snow, as well as the increasing frequency of avalanches, due to their direct impact on their living environment and livelihood. In contrast, villagers in low-altitude areas, situated in dry and hot river valleys, are less perceptive of changes in glaciers and snow cover. As a result, some of these villagers believe there have been no significant changes. Additionally, as fewer villagers graze on alpine pastures—particularly those in low-altitude regions—some are unaware of the occurrence of avalanches.

According to surveys of villagers, changes in snowfall over the past decade include delays in the onset of snowfall, a rise in the snow line, a reduction in the duration of snow cover, and an earlier snow melt. Firstly, the onset of snowfall has been delayed from late October to December and sometimes extends into February of the following year, with occasional years experiencing no snowfall at all. Secondly, the snow line has risen; previously, snow was generally found around 2,500 m above sea level, but it is now rarely seen below 3,000 m. Thirdly, the duration of snow cover has decreased from nearly six months (October to March) to just two months. Lastly, the timing of snowmelt has advanced from April to March.

Villagers generally perceive a decrease in snowfall and a shorter duration compared to the past. This feeling is particularly strong in low- and medium-altitude villages due to drought conditions. In high-altitude villages, however, an increase in snow-related disasters has led some locals to believe that snowfall has become more frequent and its duration more variable. Regarding the timing of snowfall, villagers in middle- and high-altitude areas mainly report instability, while those in low-altitude regions primarily observe that snowfall starts later and ends earlier than before.

(8) Water Resource

Villagers also perceive climate change through alterations in local water resources. In Deqin, water resources are primarily found in large and small alpine lakes, as well as rivers and streams. Recent changes in these water bodies have led villagers to believe that these shifts are directly or indirectly related to climate change. For instance, many villagers observe that the area and water level of alpine lakes have been shrinking over the past 30 years. The lakes' areas can only be partially restored during the rainy season, which brings the highest rainfall.

Similarly, rivers have also undergone changes. Variations in precipitation, snow cover, and glaciers have led to unstable river flows. Villagers in high-altitude areas recall that, more than a decade ago, rivers were generally clear in the morning with low flow, becoming turbid with increased flow in the afternoon. This pattern has become more pronounced in recent years, with rivers experiencing sharp increases in flow, especially in the afternoon. In low-altitude villages, river water used to remain stable throughout winter and surge during summer. However, in the past decade, river flow has diminished significantly in winter, necessitating careful coordination for irrigation. Additionally, river water now dries up for longer periods, exceeding 20 days compared to over 10 days a decade ago.

(9) Season

In the local division of seasons, villagers use three distinct categorizations. The first classification divides the year into four seasons: *Re* (summer), *Gen* (winter), *Dun* (autumn), and *Xi* (spring) in Tibetan. *Re* spans from April to June, *Gen* from October to December, *Dun* from June to September, and *Xi* from January to March. This division differs from the conventional spring, summer, autumn, and winter order. In this system, summer and winter are grouped together based on the length of daylight, while autumn and spring are categorized according to the blooming and fading of flowers. Additionally, summer is associated with storage and winter with consumption, leading to their combined classification. Conversely, autumn is linked to sowing and spring to germination, resulting in their joint categorization.

The second classification is the three-season system, which divides the year into three seasons: *Reka* (summer), *Dunka* (autumn), and *Genka* (winter) in Tibetan. According to the lunar calendar, *Reka* spans from March to July, *Dunka* from August to October, and *Genka* from November to February of the following year.

The third classification divides the year into two seasons: *Reka* (summer) and *Genka* (winter) in Tibetan. *Reka* spans from March to August according to the lunar calendar, while *Genka* covers September through February of the following year.

Some villagers believe that summer lasts longer and winter shorter due to increased temperatures, reduced snowfall, and decreased rainfall compared to the past. Villagers in high-altitude areas are more sensitive to seasonal changes than those in lower and medium-altitude regions, indicating that temperature rises, snowfall, and rainfall changes are more pronounced in high-altitude areas.

8.1.2 Traditional Phenological Indicators

Traditional phenological knowledge can be divided into two major categories: one concerns animal-related phenological indicators, and the other pertains to plant-related phenological indicators.

(1) Animal Phenological Indicators

Tibetan villagers in Deqin County rely on behavioral changes in animals to anticipate variations in weather and climate, thereby forming animal-based phenological knowledge. These animals include primarily domesticated mammals such as cattle, pigs, and dogs, as well as birds, amphibians, reptiles, and various insects.

I. Cattle

In Deqin County, Tibetan villagers primarily herd yaks and yak–cattle hybrids (dzo) in high-altitude pastures, while in lower-altitude villages dzo and cattle are more commonly raised. Over generations of herding and managing these animals, behavioral changes in yaks, dzo, and cattle have become important indicators for predicting weather variation. For example, in summer pastures at high elevations, when yaks and dzo suddenly stampede across the grassland and raise their heads as if sniffing the surrounding air, herders interpret this as a sign of heavy rainfall in the coming days. At lower altitudes, villagers routinely drive dzo and cattle from cowsheds to mid-altitude spring–autumn pastures or to the lower-altitude winter pastures around villages. The mid-altitude spring–autumn pastures, in particular, are especially favored by dzo and cattle due to their relatively high elevation and abundant forage. However, when these animals display unusual behavior—such as refusing to remain in the pastures during the day and showing eagerness to return to cowsheds—villagers recognize this as an omen of impending torrential rain. Similarly, when cattle repeatedly bellow and appear restless or uncomfortable, it is taken as a signal that heavy rainfall is imminent.

II. Pigs

In addition to cattle, Tibetan villagers in Deqin County also raise Tibetan pigs, a local indigenous breed known for its sensitivity to changes in weather and climate. These pigs are generally free-ranged on the grassy hillsides near villages. When villagers observe Tibetan pigs running restlessly back and forth on the grassland and carrying branches into their pens to line their sleeping places, this is understood as a sign of impending rainfall. During summer, if pigs are seen bathing in the Jiabi River, villagers likewise interpret this behavior as an omen of forthcoming rain.

Tibetan pigs are also highly responsive to cold waves. Prior to their arrival, pigs collect branches or grasses to reinforce their nests; when the weather turns colder, they extend their snouts into the bedding, and in even colder conditions, they burrow entirely into it for warmth. Sows exhibit these behaviors with particular intensity. Such instinctive cold-resisting habits have become a reliable phenological indicator for villagers in Jiabi to anticipate the onset of cold waves.

III. Dogs

In summer, when dogs are observed soaking in small ponds or streams along the Jiabi River, villagers interpret this behavior as an indication of imminent rainfall.

IV. Birds

Observations of bird behavior also form an important component of phenological knowledge among Tibetan villagers in Deqin County, serving as a means to forecast weather and climate change. One example concerns a type of cuckoo living at elevations of 2,700–2,800 m, locally referred to as Niuzha. When this bird suddenly descends from higher altitudes to villages at lower elevations and emits cheerful calls, villagers interpret this as a sign that rainfall or snowfall will occur within the following days. Another bird inhabiting high-altitude areas is the blood pheasant (Ithaginis cruentus), known locally as Xiagua. When flocks of this species migrate from higher to lower elevations and remain there for more than a week, villagers regard this as an indicator of approximately one month of rainy or snowy conditions.

Swallows also provide key signals. When swallows fly high, it is taken as evidence of prolonged sunny weather, whereas low flight is interpreted as an omen of rain. Villagers explain that before rainfall the air becomes stifling, forcing insects that normally live several inches below the soil to surface, which in turn leads swallows to fly low to catch them. The structure of swallow nests is also believed to convey meteorological information: if rainfall is expected to be abundant, swallows add more grass to reinforce the nest for breeding, whereas sparse rainfall is associated with thinner nest linings. Thus, villagers in Jiabi use the thickness of grass within swallow nests to anticipate annual rainfall and to plan agricultural activities. In winter and spring, if swallows remain in their nests without leaving for several consecutive days, this is considered a precursor of heavy snow or frost.

Other bird species likewise provide weather cues. Magpies and doves, when suddenly calling noisily, are seen as announcing rainfall in the coming days. A local proverb states, "when crows caw restlessly, rain will follow." Villagers report that when crows circle in flocks and caw repeatedly in the sky, precipitation is imminent. In winter, sparrows gathering together for prolonged periods to feed are regarded as storing food in preparation for snowfall in the following days. If sparrows gather to bathe in the Jiabi River, rain is expected soon. Their collective calls also indicate changing weather: when they chirp together on a sunny day, overcast conditions are anticipated; when they call in overcast weather, clearing skies are expected; and during extended rainy periods, sudden sparrow calls signal that the rain will soon end.

Eagles are another important indicator species. If an eagle calls in the morning, villagers expect cloudy weather for the entire day, whereas an evening call signals clear skies on the following day. Extremely high flight is taken as a sign of approaching torrential rain, while low, soaring flight suggests that the weather will soon clear. When an eagle perches on a treetop and calls, rain is regarded as imminent. Owls are interpreted in much the same way: morning calls predict cloudy days, evening calls predict clear days, and midnight calls foretell rain.

Even domestic poultry are included in such knowledge. Chickens raised in household courtyards, when suddenly returning to their coops by themselves, are believed to signal forthcoming rain. Conversely, if chickens ascend to high ground and crow, villagers interpret this as a sign that rainfall will cease and the weather will turn fine.

V. Snakes

In addition to birds, reptiles also serve as phenological indicators through which villagers observe weather and climate change. For example, during summer, when snakes are seen crawling to the riverside to drink, coiling and sleeping along the riverbank, resting at cave entrances, or lying across mountain paths, such behaviors are interpreted as signs of impending rainfall. Villagers further believe that these phenomena also signal the likelihood of flooding in the Jiabi River following heavy rain.

VI. Toads

Along riverbanks and drainage ditches on mountain slopes, when toads gather in large numbers, croaking loudly or moving onto mountain paths, villagers interpret this as a sign of rainfall on the following day. If, during the daytime, toads are observed climbing from the Jiabi River to the surrounding slopes, this is regarded as a precursor of torrential rain at night, often accompanied by flooding of the Jiabi River. Seasonal variation is also emphasized: in the dry season, sudden collective croaking of toads is taken to indicate impending rain, whereas in the rainy season, the same phenomenon is believed to foretell clear weather on the following day.

VII. Fish

When fish are observed leaping out of the Lancang River, it is interpreted by local villagers as a sign of impending torrential rain.

VIII. Insects

Ants moving their colonies from lower to higher ground, or piling soil at the entrance of their nests—even sealing the openings entirely—are interpreted as signs of imminent heavy rainfall. A sudden increase in moths and mosquitoes, especially when they fly indoors and cluster around light sources, is likewise regarded as an omen of approaching rain. Spiders provide similar cues: when they retract their webs, villagers take this as an indication of forthcoming rain, whereas the expansion of webs is seen as a sign of clearing weather.

Bees are also considered sensitive indicators. When bees are observed busily collecting nectar and returning to their hives unusually late into the evening, this is understood as foretelling heavy rainfall. Conversely, if bees forage during light showers, villagers interpret this behavior as signaling an imminent shift to clear weather. Dragonflies, when seen swarming and circling in the air, are believed to indicate that rainfall will occur over the next two to three days.

IX. Earthworms

When earthworms emerge from the soil and writhe on the ground, or when they dig burrows, expel soil, and seal the openings with mud, villagers interpret these behaviors as signs of imminent heavy rainfall or even torrential rain. The timing of their emergence is also considered meaningful: if earthworms appear early in the day, overcast conditions are expected, whereas late emergence indicates fair weather. Moreover, during extended periods of sunny weather, the sudden mass emergence of earthworms is regarded as a precursor of heavy rainfall; conversely, during prolonged rainy spells, a sudden increase in earthworm activity on the surface foretells an impending shift to clear skies.

(2) Plant Phenological Indicators

All kinds of plants possess adaptive capacities to environmental change, and their processes of growth, germination, flowering, and fruiting are closely linked to weather and climatic conditions. Consequently, villagers in Jiabi have developed plant-based phenological knowledge, whereby they observe transformations in plant life cycles to predict and assess variations in weather and climate.

I. Crops

Cultivated crops are also regarded as important indicators for assessing weather and climate variation. Local Tibetan villagers believe that when crops and vegetables in the fields lose their color and appear wilted, this serves as a sign that rainfall will occur within two days. In the sixth lunar month, when maize leaves lose their luster and curl upward, this is likewise interpreted as an omen of impending rain.

II. Trees

In addition to cultivated crops, trees also constitute an important component of phenological indicators. For example, in August, when villagers observe the leaves of Pyrus xerophila on the mountains or the leaves of walnut trees near their houses curling upward, this is taken as a sign that summer has arrived and that four to five days of heavy rainfall will follow. The flowering of walnut trees is likewise regarded as a key signal: it marks the arrival of spring and the end of snowy weather. If spring warming occurs earlier, the flowering period of walnut trees advances; if warming is delayed, flowering is likewise postponed. Hence, local Tibetan villagers frequently rely on the flowering time of walnut trees to judge the onset of spring warming and to schedule spring sowing and plowing.

Other trees also provide useful indicators. If pomegranate trees retain their leaves or shed only a few after the onset of winter, this is interpreted as a sign of a warm winter. Conversely, when peach trees shed their leaves early and in large numbers after the arrival of winter, it indicates that cold conditions will begin early; this also foreshadows an early spring warming in the following year, often accompanied by a phenomenon known as a "false spring chill" (daochunhan).

8.1.3 Analysis of Traditional Climatic and Phenological Information Related to Climate Change

From the foregoing discussion of traditional climatic and phenological knowledge among Tibetan villagers in Deqin County, it becomes evident that whether predicting indirectly through other climatic phenomena or directly through rainfall itself, villagers place particular emphasis on changes in precipitation. Indeed, much of their understanding of other meteorological events ultimately converges on the prediction of rain.

Rain holds a special place in local knowledge, reflected both in its classification and in the terminology associated with it. In Tibetan, the word for rain is written as "—," while in the local dialect of Jiabi Village it is referred to as jia, and the act of raining is expressed as jiabu. Tibetan villagers in Deqin County classify rain into nine distinct categories (see Table 1).

In summary, Tibetan villagers in Deqin County have developed a rich body of traditional climatic and phenological knowledge concerning both the prediction of rainfall and the classification of rain itself. This is partly attributable to their residence and livelihoods within a vertically stratified climatic and geographic environment, which gives rise to diverse types of precipitation. It is also closely related to their semi-agricultural and semi-pastoral subsistence system, in which agriculture depends heavily on rainfall. At the same time, the local geographic setting makes the community particularly vulnerable to landslides, debris

Table 1 Classification of rain by Tibetan Villagers in Jiabi Village and associated knowledge

No	English name	Tibetan name	Knowledge characteristics
1	Heavy rain	(jiazha)	Rainfall is very large in volume
2	Light rain	(jiagaga)	Rainfall is very small in volume
3	Torrential rain	(cuzha)	Extremely heavy rain; this type of rainfall often triggers mudslides and poses disasters for farmland, crops, and local livelihoods
4	Sunshower	(nianshajiabu)	Rain occurs simultaneously with sunshine; changeable and irregular, usually in summer
5	Rainbow rain	(raqie)	Rain occurs at the same time as a rainbow appearing at the water source
6	Highland rain	(yabu)	Rain or snow falls in the high mountains but has not yet reached the village; usually in winter, indicating that the weather is turning colder
7	Thunder rain	(zhuoqie)	Rainfall begins after thunder, though the thunderclap is not very loud; light rain follows the sound of thunder
8	Moistening rain	(jianiao)	Gentle and nourishing rainfall that may last for several days; it does not cause disasters and benefits both crops and mountain vegetation
9	Early-month rain	(zenggujia)	Rainfall at the beginning of the lunar month; usually lasts for more than seven days and can cause flooding disasters
10	Late-month rain	(maiwengjia)	Rainfall occurring after the 15th day of the lunar month; it usually does not last long

flows, and other geohazards triggered by intense rainfall, thereby reinforcing the centrality of rain in local knowledge and practice.

8.2 Traditional Techniques of Natural Resource Use and Conservation for Climate Change Mitigation

Terrestrial ecosystems and natural resources, with forests as the most representative, function as important carbon sinks. In Deqin County, terrestrial ecosystems capable of absorbing and storing substantial amounts of carbon dioxide include forest ecosystems, meadow ecosystems, watershed ecosystems, and lake–wetland ecosystems.

Tibetan villagers in Deqin County maintain a close relationship with surrounding terrestrial ecosystems and natural resources. On the one hand, these ecosystems provide the material foundation for local livelihoods and subsistence. On the other hand, through long-term resource use, villagers have developed traditional techniques that gradually evolved into sustainable practices of resource utilization and management. Such traditional practices not only ensure the sustainable use of local natural resources but also enhance the capacity of terrestrial ecosystems to absorb carbon dioxide and increase carbon sink functions, thereby contributing to climate change mitigation.

8.2.1 Traditional Practices of Natural Resource Use and Management in Forest Ecosystems

The forest ecosystem in Deqin County is referred to in the local Tibetan language as Lure, meaning "a place with forests". This ecological zone extends from approximately 2,300 to 4,000 m in elevation and encompasses alpine forest areas where villagers collect valuable non-timber products such as matsutake mushrooms and caterpillar fungus, as well as harvest timber resources.

The daily subsistence and livelihood of Tibetan villagers in Deqin are closely tied to the forest ecosystem. First, it provides essential timber resources for house construction and fuelwood. Second, it serves as an important source of food and fodder. Third, the collection of non-timber forest products—especially matsutake mushrooms—constitutes a primary source of household cash income. Through long-term reliance on forest resources, villagers have gradually developed traditional practices of use and management aimed at ensuring the sustainability of the forest ecosystem and its natural resources.

(1) Traditional Practices of Natural Resource Use in Forest Ecosystems

Traditional natural resource-use practices within forest ecosystems encompass not only the utilization of timber resources but also the management and harvesting of non-timber resources.

I. Traditional Techniques of Timber Resource Utilization

In Deqin County, traditional practices for the use of timber resources mainly consist of three components: rotational logging, species selection, and moderate harvesting. Jiabi Village provides a concrete case study.

Rotational Logging. Traditionally, villagers in Jiabi divided the collective forestland permitted for logging into six sections. Each year, timber and fuelwood for house construction and daily use were harvested in one section in sequential order, thus completing a six-year cycle that was continuously repeated. At the end of each cycle, logging resumed again from the first section. This practice ensured that no single area was subjected to excessive harvesting, thereby preventing overexploitation of forest resources and degradation of forest ecosystems.

The rotational system was managed by the village's traditional male organization, the Arrow Association (Jianhui). To ensure effective and sustainable use of forests, the Arrow Association divided the surrounding mountains into southern and northern sectors and formulated detailed customary rules. In the northern mountains, the area from Jiabi Gully to Mujugongniang (local place name) was designated as a conservation zone where logging was strictly prohibited. From Mujugongniang to the boundary with Junda Village was classified as production forest, while the southern mountains were designated mainly for construction timber, and areas along the road were reserved for fuelwood cutting.

To prevent violations in the conservation zone, the Arrow Association invited the Living Buddha Zhabajican of Hongpo Monastery to conduct a ritual called Ri gua (literally "mountain sealing"), which prohibited the cutting of trees and vegetation under customary law. It was widely believed that transgressions would bring disaster. In the production forest, villagers were allowed to freely cut fuelwood, gather leaves, and collect materials for fertilizer. By contrast, the southern mountain forests were subject to a mixed-use regime: primarily designated for construction timber, but also used for collecting pine needles, leaves, and other materials. When timber in the northern production forests became visibly depleted, villagers turned to the southern forests for limited cutting and collection, with fuelwood harvesting restricted to the two months before the Lunar New Year. After an interval of two or three years, when the northern forests had recovered, the Arrow Association would deliberate and decide to terminate or reduce southern forest use—except for construction timber—and shift harvesting activities back to the northern production forests (Fig. 6).

Next is the selection of tree species. In Jiabi Village, the choice of tree species for constructing traditional Tibetan-style houses is complex and diverse. The species used for construction materials are quite similar due to the similarity in vegetation distribution, with coniferous trees such as pine, fir, and cypress being the primary types. These include *Taxus wallichiana*, *Cupressus duclouxiana*, *Platycladus orientalis*, *Juniperus squamata*, *Tsuga dumosa*, *Pseudotsuga forrestii*, *Pinus armandi*, *Pinus densata*, *Pinus yunnanensis*, *Picea likiangensis*, *Picea brachytyla*, *Larix potaninii*, *Abies ernestii*, *Abies georgei*, *Abies delavayi*, and *Abies forrestii*, totaling 16 species. Additionally, a smaller quantity of species from families such as *Lauraceae*, *Betulaceae*, *Aceraceae*, *Anacardiaceae*, *Juglandaceae*, and *Araliaceae* are used for special purposes.

Taxus wallichiana, *Tsuga chinensis*, *Larix potaninii*, and cypress species are known for their tough, durable, and strong timber, and were

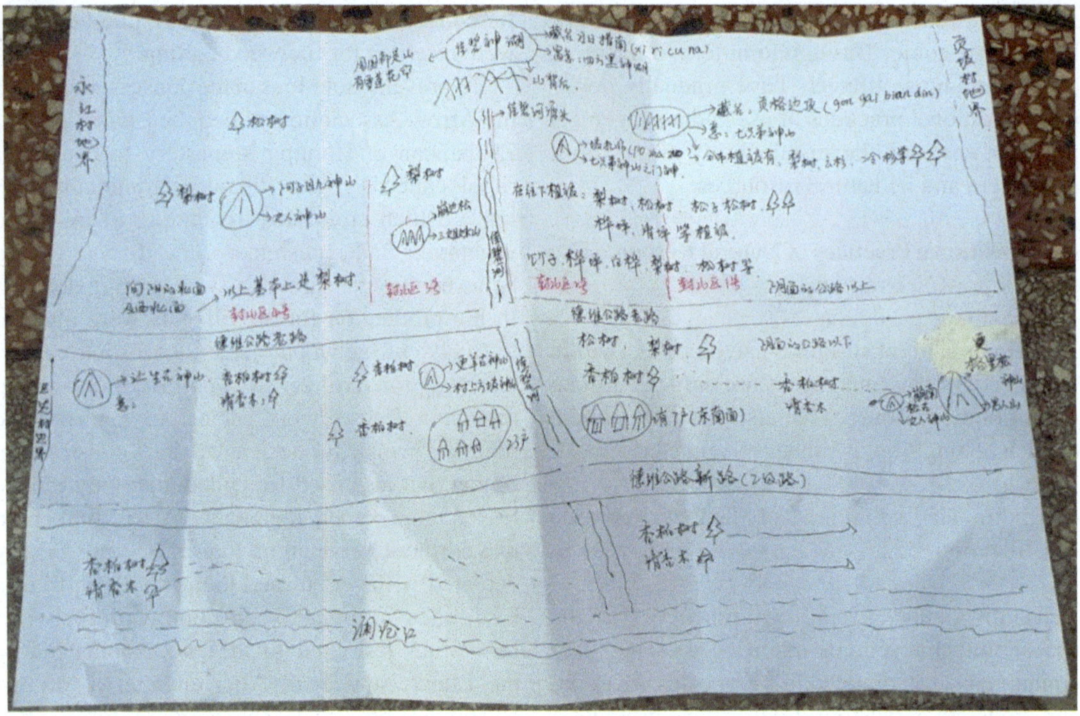

Fig. 6 Rotational logging system in Jiabi Village forests. (*Source* Illustration by Li Qingpingzu)

traditionally used as base pillars. However, they are now rarely used, first because there are no longer large-diameter logs suitable for base pillars, and second because they are protected species. *Abies ernestii* is the preferred species for wall panels in rammed-earth walls due to its large diameter, soft and lightwood, straight grain, and ease of processing. *Pinus yunnanensis* and *Pinus densata* are the first-choice materials for wall poles in rammed-earth walls, as they require uniform thickness, minimal difference in diameter between the ends, and a length of about 7 m; after the wall construction, they are reused as small corner square beams. Spruce, with its tough texture, fine structure, and straight grain, is the preferred material for flooring. *Pinus armandi*, with its fine texture, straight grain, soft quality, and ease of processing, is the first-choice building material for carving doors and windows. Species from the Betulaceae family, including *Betula albosinensis*, *Betula utilis*, and *Betula delavayi*, are the designated materials for traditional Tibetan fireplaces, a custom still followed today. In the main hall on the second floor, there is a tradition of using species with edible seeds—such as *Toxicodendron vernicifluum*, *Juglans regia*, and *Camphora glandulifera*—for the central pillar. For other pillars on each floor, including both long and short ones, there are no strict species requirements: straight, seedless trunks of pine, fir, or cypress can be used. Other building materials, such as beams, corner square beams, and clapboards, have no special restrictions; as long as the size requirements are met, various firs, spruces, and pines can be used.

It should be noted that, as modern wood-processing tools have become increasingly advanced, miscellaneous hardwoods that were previously used only in small quantities—such as *Toxicodendron vernicifluum*, *Juglans regia*, *Camphora glandulifera*, and *Gamblea ciliata* from the Araliaceae family—are now applied to a wider range of uses, with their consumption gradually increasing. In contrast, species of the genus *Populus* in the Salicaceae family, such as *Populus davidiana* and *Populus yunnanensis*, are considered taboo for use as building materials—because their Tibetan name (*mugen*) carries the same meaning as "greedy" in Chinese—and are rarely used even for firewood.

Based on the diversity of construction timber species in Jiabi Village identified through a Rapid Forest Assessment (RFA), a paired matrix ranking survey was conducted to evaluate local preferences for different timber species. Respondents compared each species in terms of wood quality, ease of collection, frequency of use, quantity used, and other criteria (see Table 2). The results show that the most preferred construction timber species are firs, represented by *Abies ernestii*, ranked 1–4. Second are spruces, represented by *Picea likiangensis*, which have a wide range of uses and are easy to collect and process, ranked 5–7. Third are pines, represented by *Pinus armandi*, which are also versatile and easy to process but have become scarce due to overexploitation in lower-altitude areas, ranked 8–10. Fourth are miscellaneous hardwoods, represented by *Gamblea ciliata*, whose utilization has been steadily increasing with advances in processing technology, ranked 11–13. Fifth are species with excellent durability but scarce resources due to their biological characteristics, such as *Pseudotsuga forrestii* (ranked 14), cypress species (ranks 15–17), and *Tsuga dumosa* (ranked 18). Sixth are species with special uses but small quantities, such as *Juglans regia*, *Camphora glandulifera*, and birches (ranks 19–21). Finally, rare species with inherently scarce resources are almost never used.

Without considering the quantity of each timber species used, a survey was conducted to record the utilization rates of construction timber species in Jiabi Village, as shown in Table 3. This helps us gain a deeper understanding of the frequency of use, usage patterns, trunk diameter, tree age, and other related aspects, thereby obtaining more information on available timber resources. The results indicate that the most frequently used and most versatile construction timber species in Jiabi Village are Pinus yunnanensis, Pinus densata, Pinus armandi, and spruces, a pattern related to the abundance of these species in the area as well as their wood quality, ease of collection, ease of processing, and traditional customs. Next are Pseudotsuga forrestii and firs, followed by cypress species and Tsuga dumosa, whose limited use is due to resource scarcity rather than tradition. Lastly are miscellaneous hardwoods for special purposes, whose usage is strongly influenced by processing technology, resource availability, and traditional customs.

Picea asperata, known in Tibetan as *nuge*, is one of the primary materials used in local house construction. It is employed for wall clamps in rammed-earth construction, pillars, and large beams, and when cut into planks, it is used for interior partitions between pillars. The wood is relatively hard, resistant to decay, has strong load-bearing capacity, and does not easily bend. *Abies fabri*, known locally as *nanbu*, is mainly used for the same purposes as spruce. The tree known as Qingping, called *nuyong* in Tibetan, has softer wood that is easy to carve, making it the preferred material for crafting household altars. Pine (*duoxin* in Tibetan) is also used in house construction. Its wood is relatively soft and has weak load-bearing strength, but it is suitable for carving, although it decays easily. The pine species producing edible seeds, known as *nongza* in Tibetan, shares similar characteristics with ordinary pine. Traditionally, the thickest central pillar in a Tibetan house must be made from *nongza*, symbolizing family happiness, health, and a good destiny, as the saying goes: "When the fruit ripens, the seed falls". The tree known as Muhuaping, called *chuwa* in Tibetan, has a long growth cycle and beautiful grain, making it mainly used for furniture making. Pear tree, also called pear firewood (*buxin* in Tibetan), is one of the main firewood sources. Its wood is hard and burns for a long time. The leaves can be cut into sections, tied into rope loops, and fermented to make farmyard fertilizer. Matsutake mushrooms often grow beneath pear trees. White birch (*ruwa* in Tibetan) has lightweight wood that does not easily crack. Locals use it in house building to make tools for ramming earth walls and for wall panels. Incense cedar (*Shuoga* in Tibetan) has aromatic leaves used for burning incense. Its fragrant wood, once dried, is one of the main fuels for incense burning. The tree known as Qingxiang (*yu* in Tibetan) bears

Table 2 Survey of village residents' preferences for various building material tree species

Species name	A	B	C	D	E	F	G	H	I	J	K	L	M	N	O	P	Q	R	S	T	U	V	Scoring	Sorting
Taxus wallichiana		B	C	D	E	F	G	H	I	J	K	L	M	N	O	P	Q	R	S	T	U	V	0	22
Cupressus duclouxiana	B		B	B	B	F	G	H	I	J	K	L	M	N	O	P	B	B	S	T	B	V	14	15
Platycladus orientalis	C	B		C	C	F	G	H	I	J	K	L	M	N	O	P	C	C	S	T	C	V	12	16
Juniperus squamata	D	B	C		D	F	G	H	I	J	K	L	M	N	O	P	D	D	S	T	D	V	10	17
Tsuga Dumosa	E	B	C	D		F	G	H	I	J	K	L	M	N	O	P	E	E	S	T	E	V	8	18
Pseudotsuga forrestii	F	F	F	F	F		G	H	I	J	K	L	M	N	O	P	F	F	S	T	F	V	16	14
Pinus armandi	G	G	G	G	G	G		G	G	J	K	L	M	N	O	P	G	G	G	G	G	G	28	8
Pinus densata	H	H	H	H	H	H	G		I	J	K	L	M	N	O	P	H	H	H	H	H	H	24	10
Pinus yunnanensis	I	I	I	I	I	I	G	I		J	K	L	M	N	O	P	I	I	I	I	I	I	26	9
Picea likiangensis	J	J	J	J	J	J	J	J	J		J	J	M	N	O	P	J	J	J	J	J	J	34	5
Picea brachytyla	K	K	K	K	K	K	K	K	k	J		K	M	N	O	P	K	K	K	K	K	K	32	6
Larix potaninii	L	L	L	L	L	L	L	L	L	J	K		M	N	O	P	L	L	L	L	L	L	30	7
Abies ernestii	M	M	M	M	M	M	M	M	M	M	M	M		M	M	M	M	M	M	M	M	M	42	1
Abies georgei	N	N	N	N	N	N	N	N	N	N	N	N	M		N	N	N	N	N	N	N	N	40	2
Abies delavayi	O	O	O	O	O	O	O	O	O	O	O	O	M	N		O	O	O	O	O	O	O	38	3
Abies forrestii	P	P	P	P	P	P	P	P	P	P	P	P	M	N	O		P	P	P	P	P	P	36	4
Camphora glandulifera	Q	B	C	D	E	F	G	H	I	J	K	L	M	N	O	P		R	S	T	U	V	2	21
Betulaceae	R	B	C	D	E	F	G	H	I	J	K	L	M	N	O	P	R		S	T	U	V	4	20
Acer miyabei	S	S	S	S	S	S	G	H	I	J	K	L	M	N	O	P	S	S		S	S	V	20	12
Toxicodendron vernicifluum	T	T	T	T	T	T	G	H	I	J	K	L	M	N	O	P	T	T	S		T	V	18	13
Juglans regia	U	B	C	D	E	F	G	H	I	J	K	L	M	N	O	P	U	U	S	T		V	6	19
Gamblea ciliata	V	V	V	V	V	V	G	H	I	J	K	L	M	N	O	P	V	V	V	V	V		22	11

Table 3 Uses of various building material tree species in Jiabi Village houses

Building material Species name	Column	Sill pillar	Wallboard	Floor	Ceiling	Wall rod	Big corner square beam	Small corner square beam	Big beam	Small beam	Door and window	Fire pit rack	Carved objects	Scoring
Cupressus duclouxiana		★						★		★	★	★		6
Platycladus orientalis		★						★		★	★	★		6
Juniperus squamata		★						★		★	★	★		6
Tsuga dumosa		★		★							★	★		5
Pseudotsuga forrestii	★	★		★	★		★	★	★	★	★			11
Pinus armandi		★			★			★		★	★		★	8
Pinus densata	★	★		★	★	★	★	★	★	★	★			12
Pinus yunnanensis	★	★	★	★	★	★	★	★	★	★	★			13
Picea likiangensis	★	★	★	★	★		★	★	★	★	★			12
Picea brachytyla	★	★	★	★	★		★	★	★	★	★			12
Larix potaninii	★	★	★	★	★		★	★	★	★	★			12
Abies ernestii	★	★	★		★		★		★	★	★			10
Abies georgei	★	★	★		★		★		★	★	★			10
Abies delavayi	★	★	★		★		★		★	★	★			10
Abies forrestii	★	★	★		★		★		★	★	★			10

(continued)

Table 3 (continued)

Building material Species name	Column	Sill pillar	Wallboard	Floor	Ceiling	Wall rod	Big corner square beam	Small corner square beam	Big beam	Small beam	Door and window	Fire pit rack	Carved objects	Scoring
Camphora glandulifera												★	★	2
Betulaceae											★	★	★	3
Acer miyabei											★	★	★	4
Toxicodendron vernicifluum											★		★	3
Juglans regia											★	★	★	4
Gamblea ciliata											★		★	3

fruit from May to June, which can be eaten by humans or fed to pigs. Its leaves remain evergreen. The wood is hard and suitable for burning, and the leaves can also be used as farmyard fertilizer.

The third is moderate logging. This mainly refers to the logging of firewood. The villagers of Jia Bi Village have specific standards for moderate logging, including tree species, altitude, slope, and specific parts of the trees. For the tree species, the villagers of Jiabi Village will first choose the tree species that are easy to grow locally for logging, such as the hawthorn tree, and avoid logging the slow-growing tree species like pine trees, cypress trees, fir trees, and walnut trees. In terms of altitude, they prioritize harvesting in forests located at mid-to-high elevations above 2,600 m, while avoiding low-altitude sparse woodlands where vegetation is already scarce, to prevent further degradation. In terms of slope orientation, villagers choose to harvest on the shady slopes of mountains, since trees from shady slopes burn longer as firewood compared with those from sunny slopes, thereby reducing the number of trees needed. In terms of parts of the tree, they preferentially cut branches and limbs rather than felling the main trunk.

II. Traditional Non-timber Resource Utilization Techniques

In Deqin County, the sale of matsutake mushrooms (*Tricholoma matsutake*) has become a major source of household income, so non-timber forest resources mainly refer to matsutake, which can directly generate cash income. Through long-term harvesting practices, local villagers have developed a set of traditional techniques for collecting and utilizing matsutake based on its weight and color. These techniques primarily involve four aspects: selection of forest vegetation, altitude, mountain slope orientation, and weather conditions.

First, matsutake mushrooms are collected according to the forest vegetation dominated by different main tree species. In coniferous forests dominated by pine trees, matsutake mushrooms that grow under the pine needles are lighter in weight and more susceptible to insect damage, although they appear whiter in color. In contrast, in mixed forests dominated by alpine rhododendron and chestnut trees, matsutake mushrooms that grow in the soil beneath these trees are heavier and have a redder color. Next, matsutake mushrooms are collected based on altitude. Mushrooms grown in high-altitude forests tend to be heavier, while those from low-altitude forests are lighter. Third, matsutake mushrooms are harvested according to slope orientation. Mushrooms grown on shady slopes are lighter, while those from sunny slopes are heavier. Fourth, matsutake mushrooms are collected based on weather patterns. After continuous rainy or thunderstorm days, mushrooms grown in the forest are heavier and of better quality. In contrast, after prolonged dry weather, the mushrooms are lighter, more prone to insect damage, and often have wormholes, resulting in hollow mushrooms. These mushrooms are less desirable in the market and cannot fetch a high price.

(2) Traditional Techniques for Maintaining Forest Ecosystem Natural Resources

Traditional techniques for maintaining natural resources in forest ecosystems not only involve the preservation of forest resources but also include the maintenance of non-timber forest product resources.

I. Traditional Forest Resource Maintenance Techniques

In Deqin County, Tibetan villagers have developed traditional techniques for maintaining forest resources and have also carried out forest vegetation restoration efforts. Taking Jiabi Village in Deqin County as a specific research case:

Firstly, the maintenance of existing forest resources involves several techniques. The villagers divide the forest into three sections based on altitude: low, middle, and high elevations. They begin maintenance at the low-altitude forests and work their way upwards to the higher elevations. Secondly, based on local customs, the surrounding mountains in Jiabi Village are categorized by gender: the mountain for women is called

"Bunanzong", and the mountain for men is called "Gelezong". Therefore, typically, female villagers maintain the forests on the women's mountain, while male villagers manage the forests on the men's mountain. Thirdly, in Jiabi Village, there are traditional female and male organizations, the "Sisters' Association" and the "Arrow Association". Traditionally, these two organizations coordinate and organize forest maintenance activities. Today, with the presence of women's groups, the village party branch, and village subgroups, the traditional organizations often collaborate with these current groups to jointly maintain the forest. Fourth, in every forest maintenance activity, villagers of different age groups—elderly, middle-aged, and young—participate. The older villagers do not directly engage in labor but sit by the side, seemingly chatting, while in fact, they are teaching and guiding the younger generations in traditional forest resource maintenance techniques. This includes knowledge on which areas of the forest need maintenance, how to carry out the maintenance, and so on. Fifth, when maintaining the forest, villagers in Jiabi Village carry out maintenance for each tree from the bottom to the top. Generally, the first step is to clear the weeds around the roots of each tree. The second step is to cut off any side branches growing from the root area of each tree. The third step is to remove any unnecessary branches and leaves from the main trunk of the tree. These techniques are adjusted according to the slope of the forested mountain. On shady slopes, all three steps are carried out in sequence, while on sunny slopes, typically only the second and third steps are performed. Villagers believe this process maximizes the water and nutrient flow to the main trunk and top of the tree, thus promoting the growth of the entire tree and ensuring the healthy growth of the forest as a whole.

Although these maintenance techniques may seem simple, they are highly effective, especially in mitigating climate change and dealing with extreme weather conditions. For example, during the drought from 2021 to 2023, which affected the entire Northwestern Yunnan region, including Deqin County, the forests on the surrounding mountains of Jiabi Village were severely damaged or even died in patches. However, the forests that had been maintained using traditional techniques continued to thrive. From a distance, it was clear that only the forests maintained using traditional methods remained green, while other areas without such maintenance had turned dry yellow. This stark color difference clearly demonstrated the effectiveness of the traditional techniques.

Next is forest vegetation restoration techniques. The villagers in Jiabi Village began restoring forest vegetation in areas that had been previously logged. However, they believe that forest vegetation restoration is not simply about planting trees, but about restoring the entire forest ecosystem. This straightforward understanding has, in practice, played a significant role in restoring forest biodiversity and the ecosystem. Firstly, before starting the restoration of forest vegetation, the villagers in Jiabi Village conduct surveys in nearby mountains that still have natural forest cover to investigate the local tree species. The surveys are conducted according to three criteria: The first criterion is to survey the tree species at three different elevations—high, medium, and low. The species distribution and growth at each elevation are investigated separately. The second criterion is to survey the slopes at these three different elevations, investigating the species distribution and growth on both the shady (north-facing) and sunny (south-facing) slopes at each elevation. The third criterion is to further survey the dry and wet areas at each of these three elevations and slopes. The species distribution and growth in the dry and wet areas of shady and sunny slopes at each elevation are recorded separately. The fourth criterion builds on the previous three, where, based on the height at which different tree species can grow, the species surveyed are classified into three height categories—high, medium, and low.

After the survey, the villagers in Jiabi Village record the distribution of tree species at different elevations, slopes, dry/wet areas, and height levels through drawing maps. Second, after recording the tree species, the villagers return to the village and consult with the older generations to inquire about the forest vegetation distribution of the mountain to be restored before it was logged. They follow the same four criteria—elevation, slope, dry/wet areas, and tree height—to ask the elders and compare the answers with the results of the field survey.

Thirdly, the villagers will present the results of the investigation and inquiries to the local experts who are familiar with the forest in the village. These experts will select six tree species, namely fragrant wood, walnut wood, Tibetan pear, cypress wood, wild mountain peach, and fragrant tree, and lead the villagers to the surveyed and naturally forest-covered mountains to collect seedlings of these six tree species. Then, they will bring the seedlings back to the village for cultivation. Fourth, after two years of nurturing the seedlings, the villagers take the six species to the areas designated for forest restoration. Tree planting is carried out following the same standards of elevation, slope, dry/wet areas, and tree height distribution. Fifth, after planting the trees, the villagers surround the newly planted seedlings with thorny branches and bushes to prevent cattle and sheep from grazing on them. They also establish village rules, customary laws, and agreements to protect the reforested area. The most effective protection measure is to invite monks from the local monastery to perform ceremonies and build white stupas to give the restored mountain the status of a "sacred mountain" and "sacred forest". According to local customs, once an area becomes a sacred mountain and sacred forest, people are prohibited from damaging the forest.

II. Traditional Non-timber Forest Product Resource Maintenance Techniques

Matsutake mushrooms are the primary non-timber forest product in the region and a major source of income for the Tibetan villagers in Deqin County. Therefore, the non-timber forest product resource maintenance techniques primarily focus on the preservation of matsutake mushroom resources. Firstly, the habitat of matsutake mushrooms, known as "fungus nests," must be protected. After harvesting the mushrooms, local villagers refill the area with leaves and soil to protect the humus layer and the fungal mycelium in the soil, ensuring that matsutake mushrooms will grow in the same place the following year. Secondly, villagers generally avoid collecting matsutake mushrooms that are smaller than five centimeters in size to ensure that the fungus nests are not damaged. Thirdly, they do not collect mushrooms that have opened, allowing these mushrooms to naturally decompose and return to the soil, which benefits the growth of matsutake mushrooms the next year. Fourthly, when collecting matsutake mushrooms, villagers use wooden sticks made from rhododendron branches instead of bamboo or metal rods. Using bamboo or metal can damage the fungus nest habitat, preventing mushrooms from growing there the following year. Among the local villagers, mastering these techniques is often referred to as possessing a "bank card", because these methods allow them to continuously harvest matsutake mushrooms, thereby generating a steady source of income. Additionally, these simple techniques protect the organic matter in the soil, enhancing the soil's ability to absorb carbon dioxide, contributing to the mitigation of climate change.

8.2.2 Traditional Natural Resource Utilization and Maintenance Techniques in the Meadow Ecosystem

In addition to forest ecosystems, meadow ecosystems are also important carbon sinks, playing a significant role in absorbing carbon dioxide and mitigating climate change. At the same time, the meadow ecosystem is an important natural resource in Jiabi Village, as villagers use it for grazing livestock. In the Tibetan language of Deqin County, the meadow ecosystem is referred to as "Bare", meaning "grazing area". This zone extends from an altitude of approximately 2,300 m to 4,500 m.

(1) Traditional Natural Resource Utilization Techniques in the Meadow Ecosystem
I. Seasonal Migration

Due to Deqin County's location in the Lancang River Gorge, which experiences a dry-hot valley climate, the local pastoral livelihood system is closely linked to the mountain environment. This system is mainly reflected in the indigenous classification of grazing lands and the annual cyclical migration of cattle and herders between the river valley grasslands and the snow-mountain meadows. Taking Jiabi Village in Deqin County as a specific research case:

To maximize the advantages of the local pasturelands and the vertical distribution of forage resources, the local herders have developed a traditional system of migration between pastures at different elevations to adapt to seasonal changes, temperature fluctuations, and variations in forage availability. Generally, the migration begins in mid-May each year when the weather starts to warm, and the forage in the transitional pastures begins to sprout. The herders drive their cattle from the winter river valley pastures, first grazing for about a month in the spring transitional pastures at an altitude of around 2,500 m. When the weather becomes warmer and the forage in the high mountain pastures starts to green, the herders move to the summer high mountain pastures above 3,000 m, where they remain until August. In late August, as temperatures drop and the forage in the high mountain pastures begins to yellow and wither, the herders and their cattle migrate to the autumn transitional pastures, where they graze for about a month. Once temperatures decrease further and the forage in the transitional pastures turns yellow, they begin the winter migration back to the river valley pastures in October, staying there until the next spring migration begins in May. This cycle repeats annually. Through this seasonal migration, livestock farming effectively adapts to the seasonal variations in climate and plant growth, ensuring the sustainability of pasture resources.

At the end of April each year, according to the lunar calendar, the villagers begin preparing for the migration of their livestock. They first send representatives to the monastery to ask the lama or monks to determine the date of departure. Once the date is set, the villagers start discussing and organizing the grazing arrangements. On the day of departure, the herders wake up around 5:00 AM. After drinking buttered tea and eating Tibetan bread, they gather the cattle. Around 7:00 AM, the herders officially leave the river valley winter pastures and head toward the high-altitude spring and autumn pastures. After about twelve hours of travel, the herders and cattle arrive at the halfway point, the autumn pasture, around 6:00 or 7:00 PM. Here, a wooden hut, known locally in Tibetan as "Gui," is built for them to rest for the night. The next morning, the herders wake up around 5:00 AM, gather the cattle again, and depart around 7:00 AM. After crossing the forested areas and traveling for three days, they finally reach the snow-capped mountain summer pastures. At the high-altitude pastures, the herders build wooden huts or tents, called "Ra" in Tibetan, where they and their cattle will stay and graze until the seventh month of the lunar calendar.

At the end of the seventh lunar month, the herders residing in the snow-capped mountain summer pastures of Jiabi Village begin preparing for the upcoming migration. The day before departure, they gather the cattle, and on the morning of departure, around 5:00 AM, the herders wake up. By around 7:00 AM, they set off, leaving the snow-capped mountain summer pastures where they have lived for three months. After three days of travel, they reach the high-altitude autumn pastures, where they and the cattle will stay and graze for about two months. At the end of the ninth lunar month, the herders and their cattle prepare to leave the high-altitude autumn pastures. Once again, they depart around 7:00 AM and, after half a day's journey, they arrive at the river valley winter pastures by 7:00 PM. The bulls are left to graze on the pasture, while the herders return to the village with the cows and calves.

By the time the tenth lunar month arrives, the entire livestock migration journey from the river valley grasslands to the snow-capped mountain meadows is complete. The herders and their cattle will stay in the river valley winter pastures near Jiabi Village for about seven months, before migrating again in the following year's fourth lunar month.

II. Grazing Resource Utilization

Taking Guonian Village in Deqin County as a specific research case, the herders in Guonian Village have developed indigenous classification criteria for local forage plants, forming a system with ten different categories:

① Classification Based on Plant Size: Tall plants (arbors)—da gui, gu hu (the Chinese names are phonetically transcribed based on Tibetan pronunciation); medium-sized plants (shrubs)—gang de shin; short plants (herbaceous plants)—s zha.
② Classification Based on Livestock Preferences: Highly palatable: gin, gu lü lü, mi lu; moderately palatable: shi lu pa nu (crow's pipa meat); very poor palatability (not eaten by livestock): a duo ri duo, jio nong na, li jia pong.
③ Classification Based on Nutritional Content: Most nutritious: di ba qiong jie, cha bu lu, jiu ba na ju.
④ Classification Based on Feeding Effects: For weight gain: ba gin ri ma, chuo bo ga; for medicinal purposes: gü duo.
⑤ For Plants with Similar Morphology: Use the terms "main" and "secondary" for differentiation. For example, jio na (main), jio gai (secondary); ru nuo (main), ru nuo ru gai (secondary).
⑥ Classification based on method of forage collection, it is classified into two types: excavation and cutting. Excavation primarily involves the utilization of subterranean tubers or roots, with common forage species including Waboma, Wajiuma, Panazua, Gianjiama, Dasimu, Azongxuzong, and Dongximu. Forage species intended for cutting are primarily tall-stemmed types, such as Hu, Huohua, and Pazua.
⑦ Classification based on habitat: Forage grasses are classified into five types according to their growing locations: grassland meadows, field margins, cultivated fields, mountain slopes, and snow mountains. For example:
   a. Forage in grassland meadows: Panazua, Gianjiama, Pazua.
   b. Forage along field margins: Waboma, Wajiuma, Yagema, Azongxuzong.
   c. Forage in cultivated fields: Huohua, Hu.
   d. Forage on mountain slopes: Jiahula, Rongmuronggong, Shuoshu.
   e. Forage in snow mountains: Yezi, Lumuyasha, Dongzha, Nahua.
⑧ Classification Based on Altitude: According to differences in elevation, forage grasses are divided into two types: those growing above 3,500 m and those growing below 3,500 m. For example:
   a. Forage species above 3,500 m: Yezi, Lumuyasha, Dongzha, Nahua.
   b. Forage species below 3,500 m: Waboma, Wajiuma, Yagema, Azongxuzong.
⑨ Classification Based on Growth Environment: Forage grasses can be categorized into two types: xerophytic (adapted to arid conditions) and hygrophilous (adapted to moist environments). For example:
   a. Xerophytic forage species: Rongqu, Gianjiama, Shuoshu.
   b. Hygrophilous forage species: Pazua, Panazua.
⑩ Classification Based on Propagation Method: Forage grasses are divided into two categories: seed propagation and vegetative propagation. For example:
   a. Species primarily propagated by seeds: Yezi, Huohua, Yagema, Lumuyasha.
   b. Species primarily propagated vegetatively (without flowering and seed production): Nawo, Rongba, Gamugeishi.

In this survey, fifty types of local ecological knowledge regarding native forage grasses were documented among herders in Guonian Village. The distinct characteristics and local evaluations of these fifty forage species were recorded, including emergence time, growing environment, livestock primarily consuming them, management and feeding practices, as well as evaluations of the "effects" after consumption by livestock. Regarding the criteria for identifying high-quality forage, multiple interviews with local experts revealed the following standards: For cattle, evaluation is based on the amount of milk produced and the hardness of the butter made from it. For pigs, assessment depends on the smoothness of their coat and the thickness of their meat.

During the survey, four Tibetan local experts from Guonian Village—Zhongwang, Zhaba,

Lurong Yixi, and Qiliti—ranked the local forage species based on their feeding effects:

① Zhongwang's Ranking: 1. Panazua; 2. Kianjiama; 3. Pazua
② Zhaba's Ranking: Among grasses for cutting: 1. Hu; 2. Huohua; 3. Pazua. Among grasses for excavation: 1. Waboma; 2. Wajiuma; 3. Panazua; 4. Kianjiama; 5. Dasimu; 6. Azaxuza; 7. Dongximu.
③ Lurong Yixi's Ranking: 1. Huohua; 2. Hu; 3. Panazua; 4. Jiahula; 5. Waboma; 6. Wajiuma; 7. Kianjiama.
④ Qiliti's Ranking: 1. Yezi; 2. Lumuyasha; 3. Rongqu (above 4,000 m); 4. Gongzua (above 3,000 m); 5. Dongzha; 6. Nahua; 7. Mutong; 8. Siren; 9. Mogong.

Among these four local experts, Zhongwang, Zhaba, and Lurong Yixi engage in year-round grazing on winter pastures located on flatlands at around 3,000 m above sea level. Therefore, most of the forage grasses they are familiar with grow around the village and on flat terrains.

In contrast, Qiliti herds year-round on summer alpine pastures at approximately 4,000 m in elevation, giving him greater familiarity with forage grasses found in snow-mountain areas. Local herders have a comparative standard for evaluating forage from these two regions: Forage above 3,500 m is characterized by rounded leaves, higher nutritional value, slower digestibility, and promotes weight gain. Forage below 3,500 m has flatter leaves, lower nutritional content, and is easier to digest. As a result, when cows consume forage from above 3,500 m, their milk yield increases, and the butter produced from this milk is harder.

(2) Traditional Natural Resource Maintenance Techniques in the Meadow Ecosystem

Traditional natural resource maintenance techniques for meadow ecosystems encompass the stewardship of both the entire rangeland resources and the specific forage resources.

I. Rotational Grazing

Taking Hongpo Village of Deqin County as a specific case study, rotational grazing represents the most effective knowledge, mechanism, and technique employed by local herders for the holistic management of pasture resources. This practice helps prevent damage to rangeland resources caused by overgrazing, thereby contributing to the maintenance of the meadow ecosystem functioning as grazing land. It enhances the carbon dioxide absorption capacity of the meadow ecosystem, thus playing a role in mitigating climate change. Herders in Hongpo Village practice rotational grazing across all three types of pastures.

First is the summer snow-mountain pasture. Located on the eastern bank of the Lancang River Valley, at an altitude of around 4,000 m in the Baima Snow Mountain alpine meadows of the Yunling Range, the summer snow-mountain pasture in Hongpo Village is divided by the villagers into two sections. The first is called "Jiabazhura", which means "a place where robbers haunt." It is said that before liberation, due to poor transportation and treacherous terrain, bandits often inhabited this area. It covers an area of approximately 500 mu (about 33 hectares). The second section has the Tibetan name "Diere", meaning "land of four families." Legend has it that this was the pasture of four former slave-owning families in the history of Hongpo Village, covering about 400 mu (around 27 hectares). In total, the summer snow-mountain pasture spans approximately 600 to 700 mu (about 40 to 47 hectares). Based on local herders' indigenous knowledge, the forage here is considered highly nutritious and is deeply valued by the community.

Second is the spring–autumn high mountain pasture. Located on the eastern bank of the Lancang River Valley, at an altitude of around 3,000 m in meadow and sloping terrain, the spring–autumn high mountain pasture in Hongpo Village comprises four sections. The first is named *"Muqugunian"* in Tibetan, meaning "land on the slope," covering approximately

900 mu (about 60 hectares). The second is called *"Bazhonggu"*, of unknown meaning, with an area of over 50 mu (around 3.3 hectares). The fourth section is named *"Gedaomu"*, also of unknown meaning, covering about 70 mu (approximately 4.7 hectares). According to the indigenous knowledge of local herders, the forage in this area is considered to have relatively high nutritional value.

Third is the winter valley pasture. Also situated on the eastern bank of the Lancang River Valley, at an altitude of around 2,000 m, the winter valley pasture in Hongpo Village consists of three sections located on the slopes surrounding the village and on the terraces along the river. The first section is named "Gamudon" in Tibetan, with an unknown meaning, covering approximately 10 mu (about 0.67 hectares). The second is called "Chishapo", meaning "a place with walnut trees," and spans over 100 mu (around 6.67 hectares). The third section, named "Tongduibura", translates to "land by the river." It is located on a terrace beside the Lancang River and covers an area of more than 60 mu (approximately 4 hectares). According to the indigenous knowledge of local herders, some forage grasses in this area are considered to be highly nutritious.

Herders in Hongpo Village subdivide each of the aforementioned pastures into 4–5 zones. When grazing on the summer snow-mountain pasture, they practice rotational grazing by moving livestock sequentially between these zones. The decision to rotate is based on the growth condition of the forage and the extent of grazing by the cattle. If the forage in one zone is consumed rapidly, the herders move to the next zone, thus forming a rotational grazing mechanism. The specific duration of rotation varies depending on the type of pasture.

On the summer snow-mountain pastures, where forage is abundant, highly nutritious, and the area is extensive, with a large number of yaks, each zone can be grazed for approximately 20 days per rotation. On the spring–autumn high mountain pastures, where forage quality is moderate and the pastures—mostly small clearings in forested areas—are limited in size, and where pian cattle (hybrids of yaks and cattle) predominate, each zone supports about 12 days of grazing per rotation. On the winter valley pastures, although the area is somewhat larger than that of the spring–autumn pastures, forage production is limited, and the livestock consist mainly of pian cattle and yellow cattle; each zone can be grazed for around 8 days per rotation.

The subdivision of pastures into multiple rotational zones is designed to allow cattle to graze on a diversity of forage species within the same pasture in sequence, thereby maximizing the utilization of forage resources. This approach not only improves grazing efficiency but also effectively protects pasture resources by preventing degradation in any single area due to overgrazing. Although the primary purpose of rotational grazing is the efficient use of natural pasture resources, it objectively contributes to the maintenance of the meadow ecosystem.

II. Forage Management

In Hongpo Village, forage maintenance primarily refers to the conservation-oriented planting of high-quality forage species. Based on traditional knowledge, local herders have surveyed and identified superior forage resources, selecting nine excellent species such as *Phlomis atropurpurea*, *Juncus allioides*, and *Cyananthus flavus var. flavus*. Cultivated pastures have been established at different altitudes, where these selected forage species are actively planted and protected. This practice of conservation-oriented planting of superior forage effectively functions as an ecological restoration experiment for the meadow ecosystem. To some extent, it enhances the carbon sink capacity of the areas within the meadows where these conservation measures are implemented (Table 4).

8.2.3 Traditional Natural Resource Utilization and Maintenance Techniques in the River Ecosystem

Taking Hongpo Village in Deqin County as a specific case study, the Jabi River, which flows through Jabi Village, is a tributary of the Lancang River (Fig. 7). It originates from the sacred Mount Buzazuo at an altitude of 4,500 m and converges

Table 4 Nine excellent forage species and basic information

No	Chinese name	Scientific name	Family	Tibetan name	Growth period	Growing site	Effect
1	Huicai	*Chenopodium album*	Amaranthaceae	Hui	Grows when rainfall is scarce in April	cultivated fields	After consuming it, livestock can efficiently improve their body condition
2	Yuntai	*Brassica campestris*	Brassicaceae	Huo hua	Grows in April	cultivated fields	This forage is highly nutritious. After consuming it, livestock can efficiently improve their body condition
3	Pugongying	*Taraxacum mongolicum*	Asteraceae	Wang bao mu	Grows in March	Grassland	This forage is highly nutritious. After consuming it, livestock can efficiently improve their body condition and milk yield
4	Tianyeqianliguang	*Senecio oryzetorum*	Asteraceae	Wang ju mu	Grows in April	Forest	This forage is highly nutritious. After consuming it, livestock can efficiently improve their body condition and milk yield
5	Pingcheqian	*Plantago depressa*	Plantaginaceae	Gai jia mu	Grows in April	Grassland	This forage is highly nutritious. After consuming it, livestock can efficiently improve their body condition and milk yield. It is also used to treat bovine diarrhea
6	Qiaomai	*Fagopyrum esculentum*	Polygonaceae	Xie ye ma	Grows in April	Forest	The grass seeds are highly nutritious. After consuming it, livestock can efficiently improve their body condition

(continued)

Table 4 (continued)

No	Chinese name	Scientific name	Family	Tibetan name	Growth period	Growing site	Effect
7	Shenzicaosu	*Phlomis atropurpurea*	Lamiaceae	Pa na zhua	Grows in May	Grassland Meadow	This forage is highly nutritious. After consuming it, livestock can efficiently improve their body condition
8	Congzhuangdengxincao	*Juncus allioides*	Juncaceae	Gong zhua	Grows in April	moist areas in alpine meadow	This forage is highly nutritious. After consuming it, livestock can efficiently improve their body condition and milk yield
9	Huangzhonghua	*Cyananthus flavus var flavus*	Campanulaceae	Yi zi	Grows in May	Mountainside and alpine scree slopes	The grass seeds are highly nutritious. After consuming it, they can efficiently improve their body condition and milk yield; a key drawback is their scarcity

Fig. 7 Lancang Mekong River, Yunnan Province, China (photoed by Yin Lun)

into the Lancang River at the Jabi River estuary, around 2,200 m above sea level, after passing through Jabi Village.

Along with the Gongpo River, which also originates from the same mountain, the Jabi River is personified in local Tibetan belief as two children of sacred Mount Buzazuo—the Jabi River being the elder brother and the Gongpo River the younger sister. According to local legend, because villagers drink its water, more boys are born in Jabi Village along the Jabi River, whereas more girls are born in Gongpo Village along the Gongpo River.

Besides bringing male offspring to Jabi Village, the Jabi River is also regarded as having a volatile nature. Every year, it causes floods and mudslides that destroy houses and farmland. Therefore, during the rainy season, villagers invite the Living Buddha from Ganden Sumtseling Monastery to perform rituals by the river. The Living Buddha recites scriptures and casts barley grains into the water while also preparing consecrated medicinal pellets. These pellets, along with barley, are placed in earthenware pots by the villagers. The pots are then situated at locations prone to mudslides, symbolically meant to pacify the spirit of the Jabi River.

The daily life and production activities of Jabi Village residents rely heavily on the river ecosystem: the Jabi River not only provides a source of drinking water for the local community but also supplies essential water resources for agricultural irrigation. Therefore, through their interaction with and use of the river ecosystem and its natural resources, the villagers have gradually developed traditional utilization and conservation practices to ensure the sustainability of both the river ecosystem and its natural resources.

(1) Traditional Natural Resource Utilization Techniques in the River Ecosystem

Over the long course of their production and daily lives, the villagers of Jiabi have developed a set of traditional systems, knowledge, and technologies for utilizing the water resources of the Jiabi River.

In terms of management systems, to ensure the rational and fair use of water resources, the Jiabi villagers divide the water into two categories: drinking water and irrigation water. For drinking water, based on the village's topographical structure, the village is divided into three sections—left, middle, and right. Three main water diversion canals have been built to channel river water to these three sections respectively. Each household then connects a branch diversion canal to the main canal to bring water into their homes. Every month, villagers take turns cleaning the canals regularly in groups of three households. For irrigation water, in line with the village's cultural practices, the village is divided into two parts: Upper Village and Lower Village. During the dry season when water is scarce, farmers in Upper Village are regulated to irrigate their farmland in the morning, while those in Lower Village irrigate from afternoon to evening.

Based on the village's topographical structure, the drinking water and irrigation water in Jiabi Village are used both separately and interchangeably. Households close to the river directly set up wooden troughs or water pipes leading to the river, channeling water outside their houses before storing it in water vats or pouring it into vats. According to local traditions, every household has a water vat placed in the main hall, beside a specific area designated for Tibetan women. This location, being relatively shaded, helps keep the water cool effectively. Households far from the river generally rely on canals for water supply, where drinking water and irrigation water are used interchangeably and adjusted according to various circumstances. For instance, the amount of drinking water required is small, but it is essential to ensure a steady, minimal flow. When irrigation is needed, farmers upstream along the canal may temporarily block the drinking water supply to downstream households to ensure sufficient water volume for irrigation, and downstream households must not complain. During irrigation, if two or three households need to water their fields simultaneously, the household that first diverted water must be allowed to finish irrigating first, and the order of subsequent irrigation is determined by the sequence of water diversion. If the canal capacity allows two or three households to use water at the same time, the water demand of the first household to start irrigating must be guaranteed; subsequent households needing water must divert additional water directly from the river. However, at the water distribution points of the ditches, it is mandatory to ensure that the water supply for the first irrigating household is not affected. These traditional ethics and detailed water-use rules have effectively regulated the functions of drinking water and irrigation water during periods of drought and water scarcity, maintaining harmony among villagers. There is a Tibetan saying: "Those who drink dirty water downstream suffer no illness or moral decay, while those who pollute the water upstream will surely face retribution." Along the Jiabi River, three time-honored water mills (which have existed for many years) operate day and night to grind flour and feed, corresponding to the village's residential communities. These mills follow a traditional rotation system for usage, but adjustments can be made based on the urgency of need.

(2) Traditional Natural Resource Maintenance Techniques in the River Ecosystem

The water resources of the Jiabi River serve as the foundation for the livelihoods and production of Jiabi Village's residents. It can be said that without the Jiabi River, there would be no Jiabi Village—this river is the "mother river" of Jiabi Village, a symbolic term referring to a river that nurtures and sustains a community. Therefore, while utilizing the water resources of the Jiabi River Basin, local residents have also developed traditional techniques for water resource conservation. Objectively, this set of techniques has enhanced the capacity of the Jiabi River Basin's ecosystem to absorb carbon dioxide.

First is the protection of the ecosystem in the upper reaches of the Jiabi River Basin. Gongga Baideng Mountain, situated at an altitude of over 4,000 m, is the source and upper reaches of the Jiabi River. The water sources in the upper reaches of the Jiabi River consist of two parts: the main part is a stream gushing out from the mountain, and the other part is streams formed by melting snow on Gongga Baideng Mountain. To protect the primary water sources of the Jiabi River, the villagers of Jiabi Village maintain the ecosystem in the river basin around the upper reaches and source area. Their efforts include two key aspects: First, formulating customary laws for protecting the river basin ecosystem, such as prohibiting water pollution, banning the felling of forest vegetation near water sources, forbidding the destruction of soil and soil layers in water source areas, and restricting grazing in water source regions. Second, villagers implement measures to maintain the river basin ecosystem, such as building stone enclosures around water sources to protect water outlets, planting trees near water sources, and conserving alpine meadows and soils in the vicinity of water sources.

Second is the protection of the ecosystem in the middle reaches of the Jiabi River Basin. When the Jiabi River flows through Jiabi Village—located at an altitude of approximately 2,400 to 2,600 m—this section constitutes the middle reaches of the river. The ecosystem in the middle reaches of the Jiabi River is fragile with low vegetation coverage, which can affect the stability of water supply for the villagers' drinking and irrigation needs. Therefore, the villagers of Jiabi Village maintain the river basin ecosystem in this middle section. Similarly, their efforts involve two key aspects: First, formulating customary laws for protecting the river basin ecosystem, such as prohibiting pollution of tributaries designated specifically for drinking water, banning the felling of forest vegetation near water sources, forbidding the destruction of soil and soil layers along both banks of the river channel, and restricting grazing in the areas surrounding tributaries used for drinking water. Second, implementing measures to maintain the river basin ecosystem, such as constructing stone embankments along both banks of the river to protect the river channel, planting trees near the riverbanks, and conserving the herbaceous plants and other vegetation along the river channel.

Third is the protection of the ecosystem in the lower reaches of the Jiabi River Basin. After flowing through Jiabi Village, the Jiabi River moves toward farmlands located at an altitude of 2,200 m, and then merges into the Lancang River—this section constitutes the lower reaches of the Jiabi River. The lower reaches of the Jiabi River have a relatively large water volume, which exerts significant erosion on the land along both banks of the river dikes. Notably, this area is precisely the critical farmland for Jiabi Village. Therefore, to reduce the river's erosion of farmlands, the villagers of Jiabi Village maintain the river basin ecosystem in the lower reaches. Their efforts follow the same two-pronged approach: First, formulating customary laws for protecting the river basin ecosystem, such as prohibiting sand and soil extraction along both banks of the lower-reach river channel, banning the destruction of vegetation along the river channel, and forbidding the destruction of soil and soil layers near water outlets. Second, implementing measures to maintain the river basin ecosystem, such as constructing stone embankments along the river channel to prevent river water from eroding farmlands, and conserving the soil, sediment, and vegetation along both banks of the river channel.

While seemingly simple, these customary laws and measures—targeting the protection of upstream water sources, middle-reach river channels, and lower-reach water outlets—are the most direct and effective techniques for maintaining the Jiabi River Basin ecosystem.

8.3 Traditional Livelihood Practices for Climate Change Adaptation

Traditional livelihoods are dynamic and experience-based; they have continuously adapted to ongoing environmental and social changes, and even today, they actively respond to the impacts of climate change. For a long time, Tibetan villagers in Deqin County have maintained a semi-agricultural and semi-pastoral traditional livelihood, encompassing both agricultural and pastoral activities.

Tibetan villagers in Deqin County share an intimate connection with this semi-agricultural and semi-pastoral livelihood. On the one hand, this livelihood provides local villagers with food supplies. On the other hand, in the long-term practice of this livelihood, villagers have developed traditional knowledge—a body of wisdom that enables them to determine and adjust the timing of grazing, planting, crop varieties, livestock breeds, and related techniques, as well as select genetic resources of different crops. These adaptations are targeted at mitigating climate change impacts, including changes in soil quality, temperature, slope gradient, aspect, and sunlight. Meanwhile, this traditional knowledge related to plant genetic resources also diversifies agricultural and pastoral livelihoods, thereby helping villagers adapt to climate change more effectively.

8.3.1 Traditional Knowledge of Agricultural Livelihood Practices

(1) Traditional Knowledge of Agricultural Land Classification

Tibetan villagers in Deqin County classify agricultural lands into three major categories based on the criterion of whether they can be irrigated by rivers: Ganren (dry lands), Reren (fallow-rotational lands), and Qinwang (irrigated lands).

Ganren (dry lands), which means "farmland on mountains" in Tibetan, are characterized by high terrain. They are distributed on slopes at an altitude of over 2,500 m or in open spaces within alpine forests. These lands cannot be directly irrigated by river water and require the construction of canals for irrigation, while also relying on rainfall—thus, they are marked by extensive cultivation with low yields. Ganren support one crop per year, primarily cold-resistant and drought-tolerant species such as highland barley, corn, potatoes, and nettle. Sowing takes place in mid-March of the lunar calendar, and harvesting occurs from August to September of the following year in the lunar calendar. Local villagers note that Ganren sustain a greater variety of crops; due to their higher altitude and proximity to forests, the grains grown here possess unique texture, flavor, and nutritional value.

Reren (fallow-rotational lands), meaning "fallow-rotational farmland" in Tibetan, are located at an altitude between 2,300 and 2,500 m. They lie on slopes far from rivers, cannot be irrigated by river water at all, and depend entirely on rainfall. Cultivation of Reren originates from the traditional slash-and-burn livelihood practice, and these lands are defined by fallow rotation: they support one crop per year (e.g., wheat, buckwheat, turnips), with sowing in mid-September of the lunar calendar and harvesting in April of the following year in the lunar calendar. After harvesting, the land is left fallow for one year before being cultivated again.

Qinwang (irrigated lands), which translates to "farmland that can be watered" in Tibetan, are situated around villages. At an altitude typically between 2,000 and 2,300 m, they feature relatively flat terrain and are characterized by intensive cultivation with high per-mu yields. These lands can be directly irrigated by diverting river water and support two crops per year, mainly wheat, corn, highland barley, various legumes, and vegetables. Most farming activities in Qinwang are undertaken by women, except for tasks like plowing, sowing, and harvesting—these are either completed or joined by men. A local proverb reflects this division of labor: "Irrigated land is like an old man who has been ill for a long time; young girls are his eternal attendants."

Up until more than 20 years ago, the cultivation of dry lands and fallow-rotational lands still occupied a crucial position in the agriculture of Jiabi Village. The crops grown on these lands mainly included corn, highland barley, buckwheat, bitter buckwheat, potatoes, nettles, and turnips. At that time, the ecological environment was excellent: forest coverage was high, which significantly regulated the microclimate, and rainfall was abundant. The dry lands and fallow-rotational lands were fertile, supporting vigorous crop growth—their per-mu yield almost matched that of irrigated lands (the average per-mu yield of irrigated lands was 750 jin, approximately 375 kg).

Dry lands spanned a large altitude difference, resulting in variations in suitable crop species, as well as staggered sowing and harvesting schedules. Fallow-rotational lands, due to their rotational fallow system, allowed villagers to reasonably arrange their work schedules and farming methods in daily life. When fallow-rotational lands were left uncultivated, shrubs and grasses grew vigorously, turning these areas into ideal grazing grounds. Generally, dry lands and fallow-rotational lands were distributed in depressions and hillsides with gentle slopes and sufficient sunlight—regions suitable for growing a variety of crops. For instance, nettle was not only edible but also provided bark from its stems, which could be used as material for weaving ropes. Another example is turnip: primarily used as livestock feed, it could also be pickled for human consumption. During the peak flowering season in summer, dry lands and fallow-rotational lands became covered with abundant blossoms, forming a "sea of flowers" that created favorable conditions for raising domestic bees. During that period, wild bees commonly built nests in cliff crevices and caves. The stems of crops grown on dry lands and fallow-rotational lands, along with the vine plants growing naturally in these fields, served as the main feed for pigs, cattle, and sheep. Nowadays, due to inconvenient transportation in the hilly areas where dry lands and fallow-rotational lands are located, coupled with the national Grain for Green Policy (a program converting farmland to forest), local villagers have gradually abandoned the cultivation of these lands.

The main crops grown on irrigated lands include wheat, corn, highland barley, and various legumes. Currently, a small portion of French grapes has been introduced and cultivated. Irrigated lands support two crop seasons per year: corn is planted from June to October, followed by wheat or highland barley from November to June of the following year, following this rotational farming pattern. Typically managed through intensive cultivation, irrigated lands receive watering, fertilization, soil loosening, and seedling thinning at fixed intervals. In traditional agricultural practices, black millet and a small amount of Chinese cabbage are usually intercropped in corn fields. Black millet can be used both as food grains and pig feed, while legumes are planted at the edges and corners of the fields. The crop cultivation methods in Jiabi Village are closely linked to the transhumance of livestock. For example, after the autumn corn harvest, the fields are generally left fallow for one month to graze the cattle and sheep returning from high-altitude pastures. This grazing continues until the livestock have completely consumed the corn leaves and weeds, leaving the fields as dry as sand dunes with dust rising—only then are the livestock removed from the irrigated lands, which then migrate back to the mountain pastures. This thorough removal of weeds in the fields facilitates the growth of wheat and highland barley in the subsequent crop season. To ensure grain production on irrigated lands, economic forest trees such as walnuts and fruit trees are planted slightly away from the farmlands. This prevents the trees from blocking sunlight to the fields (which would otherwise affect grain yields), thus achieving maximum land utilization.

(2) Farming Techniques

Agricultural cultivation in Deqin County is divided into two seasons: the Spring Crop and the Winter Crop. The Spring Crop primarily consists of corn, known as Dama in Tibetan. Sowing begins in late May of the lunar calendar and harvesting takes place in September of the same lunar year. For Spring Crop cultivation, before sowing corn, farmers first divert water to irrigate the farmland once, followed by plowing and sowing seeds—with a dense planting density of over 3,000 plants per mu (1 mu $\approx$ 0.067 hectares). During the entire growth period, the farmland is irrigated twice by diverting water. The Winter Crop mainly includes wheat (called Nian in Tibetan) and a small portion of highland barley (called Gere in Tibetan). Sowing starts in October of the lunar calendar and harvesting occurs in May of the following lunar year. For Winter Crop cultivation, before sowing wheat, farmers plow the land twice and apply farmyard manure, then broadcast the seeds. The farmland is irrigated five times throughout the growth period. For double-cropping (two harvests per year) on irrigated lands, cultivation is also divided into Spring Crop and Winter Crop seasons: the Spring Crop focuses on wheat and highland barley, while the Winter Crop centers on corn.

Spring Crop Cultivation Techniques. Due to the varying altitudes across villages in Deqin County, the sowing time for the Spring Crop differs among villagers of different villages. However, sowing generally begins between late April and early May of the lunar calendar. Several days before or after the wheat harvest, villagers divert water to irrigate the land, followed by plowing once and broadcasting corn seeds—with a dense planting density of over 3,000 plants per mu (1 mu $\approx$ 0.067 hectares). During the entire growth period, the fields are hoed once and irrigated twice by diverting water.

Winter Crop Cultivation Techniques. Villagers across villages in Deqin County begin sowing the Winter Crop between late September and early October of the lunar calendar. Before harvesting corn, the fields are hoed; after corn harvest, livestock are grazed in the fields for approximately one month. Then, farmyard manure is applied—about 100 baskets per mu (1 mu $\approx$ 0.067 hectares), with each basket weighing roughly 40 kg (totaling approximately 4,000 kg per mu). After plowing the land once or twice, wheat or highland barley seeds are broadcast. Villages at higher altitudes generally grow highland barley, while those at lower altitudes grow more wheat. Throughout the entire growth period of wheat and highland barley, the fields need to be irrigated five times, top-dressed with farmyard manure once or twice, and hoed once or twice.

Over the past decade or so, significant changes have occurred in both Spring Crop and Winter Crop cultivation. For instance, highland barley, once a major Winter Crop, has gradually been replaced by wheat—especially in villages at lower altitudes. Buckwheat, which was once widely cultivated on dry lands, is barely grown at all now. Even for the two main crops (wheat and corn), changes in cultivation have emerged, primarily reflected in the fact that local traditional varieties have been largely replaced by exotic new varieties sold by seed companies. Local villagers note that while traditional varieties have a slower growth cycle, their taste is superior to that of the new varieties.

(3) Knowledge of Crop Pests and Diseases

The most common crop diseases in Deqin County include Sani (wheat rust) and Gena (sterile ear disease, affecting wheat and corn ears). In recent years, wheat rust has occurred with increasing frequency, and the area of affected farmlands has expanded. Through long-term observation and summarization of crop pests and diseases, local villagers attribute the widespread outbreak of wheat rust to four main factors: First, climate change—particularly rising temperatures—has led to a higher disease risk in farmlands where sowing was advanced beyond the traditional schedule. Second, farmlands regularly treated with chemical fertilizers such as urea and compound fertilizers are highly susceptible to the disease. Third, farmlands planted with introduced new grain varieties are more prone to wheat rust. Fourth, beyond wheat, wheat rust also easily spreads to vegetables such as chili peppers and tomatoes. To address wheat rust, in addition to using chemical pesticides, villagers also employ traditional prevention and control methods: they insert stakes of a tree species called Yewang around affected farmlands, and spray lime and kitchen ash on the fields to prevent the further aggravation or spread of wheat rust. Furthermore, villagers use phenological knowledge to predict wheat rust outbreaks: a local bird species called sara—a golden-feathered bird larger than a pigeon—often appears in farmlands simultaneously with wheat rust. Therefore, if sara birds are spotted when wheat or highland barley is in the mid-maturity stage, villagers spread the word: "Sara is here—wheat rust is coming." They then assist each other to collectively prevent and control the disease. The symptom of Gena (sterile ear disease) is that the ears of wheat or corn turn black and fail to produce grains. Villagers believe that preventing this disease must start with seeds: for example, grain seeds should be dried thoroughly in the sun, and villagers should regularly exchange grain seeds with those from nearby villages.

To address crop pest infestations, villagers invite Living Buddhas or monks from local monasteries to preside over a ritual called Duduo—a ceremony intended to appease the desires of the deceased. Local villagers believe that crop pests occur because the unfulfilled desires of the deceased cause them to be reborn as pests, which then damage crops. Therefore, villagers request Living Buddhas or monks to conduct the Duduo ritual. During the ceremony, over a hundred types of grasses and various kinds of soil are collected, mixed with butter or walnut oil, and then burned into scorched earth. After Living Buddhas or monks chant sutras to bless the scorched earth, villagers spread it over farmlands with pest infestations—this practice is believed to drive away the pests.

8.3.2 Adaptation of Traditional Agricultural Livelihood Practices to Climate Change

The adaptation of Traditional Agricultural Livelihood Practices to climate change is primarily reflected in two aspects: the selection of genetic resources of local traditional highland barley varieties and the adjustment of farming practices.

(1) Genetic Resources of Local Traditional Highland Barley Landraces

Historically, highland barley (Hordeum vulgare L. var. nudum) served as the staple crop in the river valley villages of Deqin County, accounting for over 60% of cultivated land before 1958. However, policy adjustments and market demands shifted agricultural focus to maize and wheat, leading to a decline in barley cultivation. In recent years, climate change has subjected low-altitude maize crops to heat stress and drought, prompting villagers to rediscover the climate resilience of highland barley.

Located in the Lancang River Basin on the southeastern edge of the Qinghai-Tibet Plateau, Deqin County features pronounced vertical climatic variations across altitudes ranging from 1,800 to 6,750 m. Its complex terrain and diverse microclimates create ideal conditions for highland barley cultivation. With a millennia-old cultivation history, this crop has been a dietary cornerstone for the local Tibetan communities. Through centuries of natural selection and farmer-led breeding, over 30 distinct local traditional landraces have evolved.

In recent years, with the intensification of global climate change, climate change has exerted a significant impact on highland barley cultivation in the Lancang River Basin of Deqin County. Based on interviews with villagers in Jiabi Village, these impacts are primarily manifested in three aspects:

First is the rise in temperature. Higher temperatures have shortened the growth cycle of highland barley, advancing its heading stage and maturity stage. While this has increased highland barley yields to a certain extent, it has also led to a decline in grain quality. Meanwhile, high temperatures have adverse effects on the growth and development of highland barley—for instance, causing heat stress and insufficient grain filling. As Jiabi Village is in a high-altitude area, rising temperatures have left some traditional highland barley landraces unable to adapt to the new climatic conditions, resulting in reduced yields or decreased quality. Additionally, high temperatures have intensified the occurrence and spread of pests and diseases, posing a greater threat to highland barley production.

Second is the change in precipitation patterns. Deqin County features complex terrain and diverse climates, so changes in precipitation patterns have an especially pronounced impact on local highland barley production. These changes have altered both the total rainfall and the temporal distribution of rainfall during the highland barley growing season. In some parts of the village, decreased precipitation and uneven temporal distribution of rainfall have exposed highland barley to drought stress; in other parts, increased precipitation and concentrated rainfall periods have led to flooding disasters. The frequent occurrence of both droughts and floods has caused reductions in highland barley yields or quality, posing a threat to the food security of local Tibetan villagers.

The third issue is the frequent occurrence of extreme weather. Deqin County is at a high altitude with a highly variable climate, and extreme weather events are common. These events expose highland barley to hazards such as hail, frost, and strong winds. Such disasters not only disrupt normal growth and development but also seriously affect yield and quality. They can lead to reduced harvests or total crop failure, posing serious threats to the food supply and livelihoods of local Tibetan villagers.

In Deqin County, villagers commonly grow and store seeds of 11 traditional local barley varieties: Chaguo barley, Long Black barley, Yellow barley, White barley, Short White barley, Red-awned barley, Green barley, Red barley, Short-awned barley, Double-brow barley, and Purple barley. These diverse varieties have developed unique growth traits and climate adaptability over a long history. Drawing on generations of experience, villagers in Jiabi Village have selected varieties suited to different elevations and climates, creating an integrated system of "altitude–vertical climate zone–variety matching" to adapt to climate change (Table 5).

These traditional barley varieties have been grown in Deqin County for generations and were developed by local Tibetan villagers through long-term farming practice. Today, facing climate change, they draw on this accumulated knowledge to adapt by selecting from these diverse genetic resources, ensuring stable yield and quality.

First, using traditional climate and phenological knowledge. Villagers use local calendars and environmental cues to guide variety selection. They observe the snowline on surrounding mountains: the speed of spring snowmelt predicts water availability. If melting is early, they plant early-maturing varieties to avoid drought. They also watch bird migration; for example, if cuckoos call earlier, it signals rapid warming, so they delay sowing cold-tolerant varieties to prevent frost damage. These local varieties are highly climate-adapted. During severe droughts around 2010, introduced varieties lost yield, while local ones remained stable.

Second, choosing cold-tolerant varieties for low temperatures. Winters in Deqin are cold and spring warms slowly. To cope, villagers plant hardy varieties such as Erdaomei barley, which grows well at high altitudes and low temperatures. With proper field management and irrigation, these varieties develop normally in cooler conditions, keeping yield and quality steady.

Third, choosing drought-tolerant varieties to cope with dry conditions. In Deqin's low-altitude, hot, and dry river valleys, rainfall is scarce and evaporation is high, making drought common. To reduce its impact on barley growth, villagers plant drought-tolerant varieties. For example, red barley can grow well under dry conditions. With proper irrigation and soil management, these varieties maintain good growth, limiting yield loss and quality decline caused by drought.

Fourth, choosing disease-resistant varieties to handle pests and diseases. With climate change, pests and diseases are increasing in Deqin's barley fields. To address this, villagers plant disease-resistant varieties. For example, Duanbian barley has good resistance to common barley diseases such as rust and powdery mildew. Combined with sound field management and pest control, these varieties help reduce yield loss and quality decline caused by pests and diseases.

Fifth, matching varieties to local climate and soils. Deqin has complex terrain, diverse climates, and varied soils. To maximize yield and quality, villagers choose barley varieties suited to each area's conditions. In high, cold areas, they plant cold-tolerant types such as Erdaomei and red barley. In drier zones with low rainfall and high evaporation, they use drought-tolerant varieties such as Duanbian barley. By aligning variety choice with local climate and soil, and using proper field management, they maintain stable production despite climate change.

In summary, by preserving diverse traditional barley varieties, practicing vertical-layer planting, and following an "altitude–climate zone–variety matching" system, Tibetan villagers in Deqin have built a traditional farming strategy well adapted to climate change. This practice shows that traditional knowledge is not static heritage but an effective tool for resilience.

Table 5 Climate adaptability of traditional local barley varieties

Name	Traditional knowledge characteristics	Planting altitude	Climate adaptation features	Adaptation effect
Chaguo barley	Strong cold tolerance, short growth period (120–130 days)	Suitable for high-altitude cold mountain areas above 3,000 m	Cold, humid	Avoids heat stress during heading in summer
Long Black barley	Strong cold tolerance, spike length 80–120 mm, grain color dark brown to black	Suitable for high-altitude cold mountain areas above 3,000 m	Cold, humid	Avoids heat stress during heading in summer
Yellow barley	Strong cold tolerance, yellow grains, growth period about 130 days, excellent frost resistance	Suitable for high-altitude cold mountain areas above 3,000 m	Cold, humid	Avoids heat stress during heading in summer
White barley	White grains, balanced stress resistance, adapts to rainfall fluctuations in mid-mountain areas, often intercropped with beans to enhance system stability	Suitable for mid-altitude semi-mountain areas at 2,500–3,000 m	Warm, semi-arid to semi-humid	Uses balanced resistance with bean intercropping to stabilize yield
Short White barley	Short growth period (about 120 days), strong drought tolerance	Suitable for low-altitude warm and dry river valleys below 2,500 m	Hot, arid	Deep root system absorbs subsoil water to cope with uneven rainfall
Red-awned barley	Red-awned husk, strong drought tolerance, deep roots, high lodging resistance, stable yield in poor soils of the warm, dry Lancang River valley	Suitable for low-altitude warm and dry river valleys below 2,500 m	Hot, arid	Suited to slopes with concentrated rainfall and soil erosion risk
Green barley	Green husk, strong drought tolerance, deep roots, stable yield even in poor soils	Suitable for low-altitude warm and dry river valleys below 2,500 m	Hot, arid	Deep root system absorbs subsoil water to cope with uneven rainfall
Red barley	Red grains, capable of growing under harsh conditions	All altitudes	Cold, arid	Strong cold and drought tolerance
Short-awned barley	Short spike type, plump grains, high yield, grows in various soils and climates	Suitable for mid-altitude semi-mountain areas at 2,500–3,000 m and low-altitude warm, dry river valleys below 2,500 m	Warm, arid	Strong disease resistance and adaptability
Double-brow barley	Distinctive double-brow spike, large grains, grows well in high-altitude and arid areas	All altitudes	Cold, arid	Strong cold and drought tolerance
Purple barley	Able to grow in mountain areas with unstable rainfall	Suitable for mid-altitude semi-mountain areas at 2,500–3,000 m and low-altitude warm, dry river valleys below 2,500 m	Warm, arid	Strong drought tolerance

(2) Farming Practices

Tibetan farmers in Deqin have developed a unique adaptive farming system shaped by complex terrain and fragile ecosystems. It integrates traditional ecological knowledge, seasonal calendars, agricultural biodiversity, crop rotation, and traditional water management to build resilient agriculture.

I. Traditional Farming Methods

Main crops include barley, wheat, maize, beans, and tubers, selected according to local climate. Warm valley areas favor wheat and maize, while high mountain areas grow cold-tolerant barley. Due to marked vertical climate differences, planting and harvest times vary by elevation.

In terms of farming systems, villagers have formed distinctive practices through long-term experience, including crop rotation, intercropping, and relay planting. Crop rotation alternates different crops on the same land to prevent nutrient depletion and pest buildup. Intercropping grows two or more crops together to improve land use and yields. Relay planting sows a new crop among an existing one before harvest to make full use of soil and sunlight.

In water management and use, Deqin County has relatively abundant water resources, but their distribution is uneven. River valleys receive less rain and face water scarcity, while higher mountain areas get more precipitation due to elevation. Over time, villagers have developed unique water management practices, including building cisterns and irrigation channels and using rain and snowmelt for irrigation.

For soil protection and improvement, Deqin has diverse soils, from alluvial soils in valleys to yellow–brown, brown, and dark-brown soils at higher elevations. Some areas show signs of degradation from long-term farming and improper use. Villagers address this by applying organic fertilizers, rotating crops, and leaving fields fallow to protect and improve soil quality.

II. Impact of Climate Change on Traditional Farming

First, rising temperatures affect crop growth. In recent years, Deqin County has seen a steady increase in temperature. This has clear effects: on one hand, higher temperatures can lengthen the growing season and improve yield and quality; on the other, they can cause overly vigorous growth, lodging, and greater pest and disease risks. For cold-tolerant crops in high mountain areas, warming disrupts growth cycles and reduces yield and quality.

Second, shifting rainfall patterns affect farming. Like the rest of the Eastern Himalayas, Deqin's rainfall has become less stable. Total rainfall fluctuates, and its timing and distribution are uneven. Reduced rainfall brings frequent droughts, cutting yields; heavy rain brings floods that damage fields and crops. Uneven timing and distribution also disrupt growth cycles and lower production.

Third, extreme weather is becoming more common. In the Lancang River basin, events such as drought, floods, hail, and frost occur more often. Drought dries soils and stresses crops; floods wash away fields; hail damages leaves and grains; frost causes cold injury and yield loss.

III. Strategies of Traditional Farming to Adapt to Climate Change

Climate change, with uneven rainfall and frequent natural disasters, has affected traditional farming livelihoods. Villagers adapt by adjusting farming methods, forming climate adaptation strategies for traditional agriculture.

First is the dynamic adjustment of traditional ecological knowledge and seasonal systems. Tibetan villagers in Deqin integrate climate change into farming decisions through traditional knowledge. 1) They use flexible seasonal divisions—four-season (spring, summer, autumn, winter), three-season (summer, autumn, winter), and two-season (summer, winter)—to adjust planting and harvest times based on phenological changes. For example, if spring is colder than usual, maize sowing is delayed to avoid frost. 2) They observe natural signs such as the speed of mountain snowmelt, river levels, and vegetation growth to predict rainfall trends. For example, if the Jiabi River flow decreases, villagers repair irrigation channels early to prepare for drought.

Second is traditional knowledge of agricultural biodiversity and crop rotation. Tibetan villagers practice mixed and intercropping, planting maize, wheat, and beans on the same land. If one crop fails due to climate stress, others can still yield, ensuring food security and serving as a buffer against climate shocks. They also use crop complementarity to increase resilience; for example, beans fix nitrogen, improving soil fertility and reducing fertilizer use. A local saying, "Three years without changing seed, and the soil loses strength," reflects this awareness. In dry years, they expand drought-tolerant crops; in wet years, they increase water-demanding crops. This flexible adjustment keeps land productivity within climate limits. During droughts and other extreme weather, fields under traditional rotation have smaller yield losses than monocropped fields.

Third is the traditional farmland management system. Tibetan villagers in Deqin manage both dryland and irrigated land. On slopes, they build dry fields to conserve soil and water, allocating crops by elevation: maize at lower altitudes and cold-tolerant barley higher up. Irrigated land follows a "large spring–small spring rotation," planting maize in the main spring season and wheat in the minor spring season to match seasonal rainfall shifts. The dryland system acts as a climate regulator in mountain farming. By stacking stones to create terraces, villagers turn steep slopes into arable land, slowing surface runoff and increasing soil organic matter. These terraces not only help mitigate climate impacts but also create a "step effect," where each level has a controlled temperature range, providing stable microclimates for crops and further supporting climate adaptation.

Fourth is the traditional water management system. Although Deqin County has relatively abundant water resources, their distribution is uneven in time and space. Facing climate change, local villagers adapt by improving water management and use: On one hand, they adjust irrigation schedules and water volumes according to local climate and rainfall. In river valleys, where rainfall is scarce and evaporation high, they focus on building and maintaining irrigation facilities. In high mountain areas, they emphasize using and storing rain and snowmelt for irrigation. On the other hand, they introduce modern technologies and management to improve efficiency and quality, such as water-saving irrigation, rainwater harvesting, and reducing waste and pollution. They also strengthen education and awareness to improve water management and protection among Tibetan villagers. Villagers maintain irrigation networks through traditional groups such as "Arrow Associations" and "Sisters' Associations." Each spring, the Arrow Association organizes men to clear river channels, while the Sisters' Association organizes women to clean field canals. For fair water use, water is divided into drinking water and irrigation water: 1) For drinking water, the village is divided into left, middle, and right sections based on topography. Three main channels bring river water to each section, and each household connects to branch channels. Every month, villagers clean the canals in rotation, with one team for every three households. 2) For irrigation water, the village is divided into upper and lower sections by custom. During dry seasons, they share the limited Jiabi River water through staggered schedules: upper-section farmers irrigate in the morning, and lower-section farmers irrigate in the afternoon and evening.

Fifth is the traditional vertical planting system. On the steep slopes of mountain valleys, Tibetan villagers in Deqin have built a three-dimensional vertical planting system using generations of traditional knowledge. At around 2,200 m in the dry, hot valley zone, they grow drought-tolerant crops such as barley, buckwheat, and wheat; at 2,600 m on mid-mountain slopes, they plant wheat, barley, and maize; at 2,800 m in high mountain areas, they grow potatoes and barley. This vertical differentiation acts as a dynamic buffer to climate variability—when one elevation zone is hit by extreme weather, crops in other zones can still provide basic yields.

Sixth is innovation in traditional farming systems. Over time, Tibetan villagers in Deqin developed distinctive farming systems, but climate change has made some practices less suitable. To adapt, they innovate: (1) Based on local climate and crop growth cycles, they adjust farming schedules. In valley areas, with rising temperatures and less rain, they now sow and harvest earlier; in high mountain areas, they maintain later sowing and harvesting to take advantage of cooler conditions and longer growth periods. (2) They also use modern technology and management to improve efficiency and quality, such as mechanized farming, precision fertilization, and integrated pest management to raise yields and quality. In addition, they strengthen agricultural cooperation and organization to scale up and improve farming practices.

8.3.3 Traditional Knowledge of Pastoral Livelihoods Practices

(1) Traditional Knowledge of Pasture Classification

Due to differences in terrain, altitude, temperature, and soil, meadows in Deqin County are divided into five main types: alpine meadow pastures, shrub-meadow pastures, forest-meadow pastures, sparse-wood pastures, and mountain shrub-grass pastures. Alpine meadow pastures are found above 3,800 m, with harsh climates, short frost-free periods, and short grass growth seasons. Vegetation is low and mat-like, dominated by perennial species such as dragon's blood, sedges, wormwood, and grasses. These are used for grazing cattle, horses, and sheep in summer and autumn. Shrub-meadow pastures occur at similar altitudes but have larger areas of broadleaf forest dominated by alpine rhododendron, making grazing less efficient; they are also used in summer and autumn. Forest-meadow pastures occur between 2,500 m and 3,800 m, characterized by dense forests and varied shapes. Below 2,800 m they serve as year-round grazing grounds; above 2,800 m they are used in summer and autumn. Sparse-wood pastures also lie between 2,500 m and 3,800 m, with scattered shrubs and remnants of logged areas, some suitable for year-round grazing. Mountain shrub-grass pastures, with mountain landform features, occur between 1,800 m and 3,200 m, with 10–30% forest cover. They are used year-round for all livestock and include larger pastures such as those in Guonian Village.

Based on these different meadow types, Tibetan villagers in Deqin classify pastures by location, elevation, and grazing season into three types: summer snow-mountain pastures, spring–autumn high mountain pastures, and winter valley pastures. The summer snow-mountain pasture is called "Rura" in Tibetan, meaning "snowy pasture," located around 4,000 m in alpine meadows; the spring–autumn high mountain pasture is called "Rumei," meaning "middle pasture," located around 3,000 m in meadows and slopes, serving as transitional grazing land; the winter valley pasture is called "Rubo," meaning "pasture near home," located around 2,000 m on slopes near villages. From the summer snow-mountain pasture to the spring–autumn high mountain pasture to the winter valley pasture, pasture species and quantity increase, but nutritional value declines, meaning that forage in the summer snow-mountain pasture is generally more nutritious than in the other two, while the winter valley pasture has the richest variety and largest quantity of forage.

(2) Traditional Knowledge of Forage Resources

Through long-term grazing practice, Tibetan herders in Deqin County have accumulated rich knowledge of forage. Surveys of herders and local experts in Jiabi Village recorded 46 forage species belonging to 28 families and 38 genera. These were cataloged and summarized as shown in Table 6.

Table 6 Survey of forage species identified by herders and local experts

No	Family	Genus	Chinese name	Scientific name	Tibetan name	Habitat	Forage quality grade[1]	Trend[2]
1	Osmundaceae	Claytosmunda	Rongziqi	*Claytosmunda claytoniana (L.) Metzgar & Rouhan*	Jia hu la	Moist valley	B	a
2	Ranunculaceae	Anemone	Caoyumei	*Anemone rivularis Buch.-Ham. ex DC*	Za dan gu song	Hillside grassland or roadside	C	b
3	Brassicaceae	Brassica	Yuntai	*Brassica rapa var. oleifera DC*	Huo hua	In cultivated fields	B	c
4	Saxifragaceae	Rodgersia	Yuyeguidengqing	*Rodgersia pinnata Franch*	Rong men rong gong	Under forest canopy at elevations of 2,500–3,500 m	B	b
5	Caryophyllaceae	Stellaria	Fanlou	*Stellaria media (L.) Vill*	Gai mu gai xia	Field margins, roadside areas, or streamside grasslands	D	b
6	Polygonaceae	Fagopyrum	Qiaomai	*Fagopyrum esculentum Moench*	Xie ye ma	Wasteland or roadside	A	b
7			Xiaoyeqiaomai	*Fagopyrum leptopodum (Diels) Hedberg*	Gong me	In fields	D	c
8		Persicaria	Niboerliao	*Persicaria nepalensis (Meisn.) H. Gross*	Dong me	Fields, watersides, and moist valley habitats	B	a
9			Yuansuiliao	*Bistorta macrophylla (D. Don) Soják*	Rong qu	Watersides and wetlands	C	a

(continued)

[1]Grades A, B, C, and D represent, respectively, excellent, good, moderate, and poor forage quality.
[2]Grades a, b, and c represent, respectively, declining, stable, and increasing population trends.

Table 6 (continued)

No	Family	Genus	Chinese name	Scientific name	Tibetan name	Habitat	Forage quality grade	Trend
10			Chibingliao	*Bistorta sinomontana (Sam.) Miyam*	Gai dou gu duo	Field margins, grasslands, and forest edges	C	a
11			Bingchuanliao	*Persicaria glacialis (Meisn.) H. Hara*	Dong me	Field margins, along ditches, and in marshy areas	B	a
12		Rumex	Chiguosuanmo	*Rumex dentatus L*	A beng ke le	In mixed forests at elevations of 2,500–4,000 m	B	a
13	Amaranthaceae	Chenopodium	Li	*Chenopodium album L*	Hui	Alpine meadows at elevations of 3,000–4,000 m	A	c
14	Geraniaceae	Geranium	Yunnanlaoguancao	*Geranium yunnanense Franch*	Dong zhua	Alpine meadows at elevations of 3,000–4,000 m	D	b
15	Balsaminaceae	Impatiens	Huafengxian	*Impatiens chinensis L*	Dong ri yue	Alpine meadows at elevations of 3,000–4,000 m	C	b
16	Rosaceae	Sorbus	Xinanhuaqiu	*Sorbus rehderiana Koehne*	Meng lou	In mixed forests at elevations of 2,500–4,000 m	B	a
17		Potentilla	Yunnanweilincai	*P.fulgens*	Da gai	Alpine meadows at elevations of 3,000–4,000 m	B	b
18	Fabaceae	Astragalus	Xiachuangqi	*Astragalus zacharensis Bunge*	Sai long mu	Hillside grasslands and field margins	D	c

(continued)

Table 6 (continued)

No	Family	Genus	Chinese name	Scientific name	Tibetan name	Habitat	Forage quality grade	Trend
19	Fagaceae	Quercus	Ciyegaoshanli	*Quercus spinosa David ex Franch*	Shu bu	Alpine meadows at elevations of 2,400–3,900 m	B	a
20	Sapindaceae	Acer	Qingzhaqi	*Acer davidii Franch*	La ba	In mixed forests at elevations of 2,000–3,500 m	B	a
21	Apiaceae	Physospermopsis	Gelvzi	*Carum carvi L*	Dong xia mu	Hillsides and field edges	C	b
22			Ailingziqin	*Physospermopsis nana (Franch.) Pimenov & Kljuykov*	Mu xia	Alpine meadows and alpine scree slopes at elevations of 3,000–4,500 m	B	a
23	Viburnaceae	Sambucus	Xuemancao	*Sambucus adnata Wall. ex DC*	Shuo shu	In under storey vegetation or shrubs along gullies	B	b
24	Caprifoliaceae	Dipsacus	Chuanduanxu	*Dipsacus asper Wall*	Ca xia mu	Grassy flats, forest edges, and field edges	C	c
25	Asteraceae	Cirsium	Miantouji	*Cirsium eriophoroides (Hook. f.) Petr*	Da si mu	Hillsides, grasslands, and shrub lands	D	c
26		Ligularia	Kuanjituowu	*Ligularia latihastata (W. W. Sm.) Hand.-Mazz*	Na guo	Grassy flats, forest edges, and hillsides	B	c
27			Chuitoutuowu	*Ligularia cremanthodioides Hand.-Mazz*	Mu tong	Hillsides and grasslands	C	b
28		Taraxacum	Pugongying	*Taraxacum mongolicum Hand.-Mazz*	Wang bao mu	Grassy flats and fields	A	b

(continued)

Table 6 (continued)

No	Family	Genus	Chinese name	Scientific name	Tibetan name	Habitat	Forage quality grade	Trend
29		Arctium	Niupang	*Arctium lappa L*	A zong xi za	Grassy flats and field edges	C	c
30		Senecio	Tianyeqianliguang	*Senecio oryzetoru*	Wang qu mu	Hillsides, wastelands, and field margins	A	c
31	Campanulaceae	Cyananthus	Lijianghuangzhonghua	*Cyananthus flavus C. Marquand*	Yi zi	Alpine meadows and scree slopes at elevations of 2,800–4,500 m	A	a
32	Boraginaceae	Microula	Weikongcao	*Microula sikkimensis (C. B. Clarke) Hemsl*	Ye ge mu	Alpine meadows or village-edge grasslands	B	b
33	Orobanchaceae	Pedicularis	Gyanhuamaxianhao	*Pedicularis siphonantha Don*	Di di zhua zhua	Alpine moist meadows	D	a
34	Plantaginaceae	Plantago	Pingcheqian	*Plantago depressa Willd*	Gai jia mu	Grasslands, gullies, and roadsides	A	b
35	Rubiaceae	Galium	Xinanlalateng	*Galium elegans Wall*	Qiong jia	Field margins, field edges, and roadsides	C	b
36	Malvaceae	Malva	Yekui	*Malva verticillata L*	Zhong ba	Grasslands near villages	B	b
37	Lamiaceae	Salvia	Gansishuweicao	*Salvia przewalskii Maxim*	Rong zui ma	Alpine meadows at elevations of 2,800–4,000 m	B	a
38		Phlomoides	Shenzicaosu	*Phlomoides atropurpurea (Dunn) Kamelin & Makhm*	Pa na zhua	Alpine meadows at elevations of 2,800–3,900 m	A	a

(continued)

Table 6 (continued)

No	Family	Genus	Chinese name	Scientific name	Tibetan name	Habitat	Forage quality grade	Trend
39	Iridaceae	Iris	Xinanyuanwei	*Iris bulleyana Dykes*	Bie bi	Hill country, meadows, and hillside grasslands	B	c
40	Juncaceae	Juncus	Congzhuangdengxincao	*Juncus allioides Franch*	Gong zhua	Alpine moist meadows	A	a
41	Cyperaceae	Carex	Xiaotaicao	*Carex parva Nees*	Ba zhua	Alpine moist meadows	B	a
42			Shishengtaicao	*Carex limosa L*	Zhua sha	Alpine moist meadows	C	a
43	Poaceae	Agrostis	Lushuijianguying	*Agrostis nervosa Nees ex Trin*	Zhua chu	Alpine meadows at elevations of 2,500–3,500 m	B	b
44		Elymus	Maipincao	*Elymus tangutorum (Nevski) Hand.-Mazz*	Rang ba	Alpine meadows at elevations of 2,500–3,500 m	B	a
45			Chuisuipijiancao	*Elymus nutans Griseb*	Rang gu	Alpine meadows at elevations of 2,500–3,700 m	C	b
46		Fargesia	Jianqiaojianzhu	*Fargesia acuticontracta T. P. Yi*	He ya	Under forest canopy at elevations of 2,000–3,500 m	B	b

(3) Grazing Practices

Cattle constitute the principal livestock herded by the Tibetan pastoralists in Deqin County, comprising primarily yaks (Bos grunniens), dzo (a hybrid of yak and domestic cattle), and Chinese yellow cattle (Bos taurus). Local villagers typically form herding groups, each consisting of 5 to 6 households and managing about 50 to 60 head of cattle. In Tibetan, such a group is called a Zaka. Each Zaka is attended by one or two herders, referred to as Nongza in Tibetan. Within a Nongza, roles are further differentiated: the elder herder, known as Zaben, is responsible for all production arrangements, whereas the younger herder, or Zaru, undertakes supportive tasks, usually involving labor-intensive duties such as herding cattle and collecting firewood. Upon arrival at their traditional village pastures, particularly the alpine pastures, Zaka groups have no clearly demarcated grazing territories. They share the entire pasture, with each group free to choose grazing sites in forests, lakeshores, or riverbanks, provided they do not infringe upon the traditional pastures of other villages.

Situated in the dry–hot valley climate of the Lancang River Grand Canyon, Deqin County has developed a livestock-based livelihood system closely adapted to the surrounding mountainous environment. This adaptation is reflected in the local ethnoclassification of pastures and in the annual cyclical migration of herds and herders between valley meadow pastures and alpine meadow pastures.

To maximize the advantages of the vertical distribution of pastureland and forage resources, local herders have developed a traditional system of transhumance between pastures at different elevations, in accordance with seasonal, temperature, and vegetation changes. Generally, the spring migration begins in mid-May on the Gregorian calendar. As temperatures rise and grass in transitional pastures begins to green, herders drive their cattle from the winter valley pastures to the spring transitional pastures at an altitude of around 2,500 m, grazing there for about one month. When warmer conditions signal that grass in the alpine summer pastures (above 3,000 m) has also begun to grow, the herders move the cattle further upslope, remaining there until August. In late August, as temperatures fall and alpine forage withers, the herders and their cattle migrate to the autumn transitional pastures, grazing there for roughly one month. With further cooling and pasture senescence, they embark on the winter migration back to the valley pastures in October. This cyclical transhumance, repeated annually, allows the pastoral system to adapt effectively to the seasonal variations of local climate and plant growth, while ensuring sustainable use of grassland resources.

By late April on the lunar calendar, villagers begin preparations for the migration. They first send representatives to the local monastery to consult a living Buddha or monk to determine an auspicious departure date. Once the date is set, villagers discuss and organize the herding arrangement. Two modes are practiced: individual household herding or collective herding of all village cattle under the care of a single household. On the departure day, herders wake at about 5:00 a.m., drink butter tea, and eat tsampa (roasted barley flour). They then gather the cattle, setting off from the winter valley pasture at around 7:00 a.m. toward the spring/autumn highland pastures. After approximately twelve hours of travel, they reach the mid-mountain autumn pasture at about 6:00–7:00 p.m., where Gui (log cabins) have been built. Herders and cattle rest here overnight.

The next morning, they again rise at 5:00 a.m., assemble the cattle, and set off at 7:00 a.m. After crossing forested zones and traveling for three days, they finally reach the alpine summer pasture. There, herders erect Ra—either log cabins or tents—and remain with their cattle until the seventh month of the lunar calendar.

By late July (lunar calendar), herders in the alpine summer pastures prepare for the return migration. On the day before departure, cattle are gathered. At around 5:00 a.m. on departure day, they set off by 7:00 a.m., leaving the alpine summer pasture where they have lived for three months. After three days of travel, they arrive at the highland autumn pasture, where they remain for about two months. In late September (lunar calendar), they again depart at about 7:00 a.m. and, after half a day's travel, arrive back at the winter valley pastures by around 7:00 p.m. Bulls graze freely on the pastureland, while cows and calves return to village homes.

By the tenth month of the lunar calendar, the annual cycle of transhumance from valley meadow pastures to alpine meadow pastures is complete. Herders and cattle then remain in the valley winter pastures near Jiabi Village for about seven months, until the next migration begins in the fourth lunar month of the following year.

8.3.4 Adaptation of Traditional Pastoral Livelihoods to Climate Change

For generations, pastoralists in Deqin County have developed distinctive herding practices and a traditional knowledge system, including a pastoral calendar and a grazing system.

The pastoral calendar serves as a time-management tool established by herders according to natural rhythms and climatic variations, guiding grazing activities and resource utilization. Its formulation relies primarily on observations of climate, vegetation, and animal behavior. For instance, herders determine the timing of grazing initiation and migration routes by monitoring indicators such as the timing of snow melt on mountain slopes, the growth cycles of vegetation, and behavioral changes in livestock. Beyond directing grazing activities, the pastoral calendar also governs critical production stages, such as livestock breeding and shearing. Herders select optimal breeding times according to the pastoral calendar, ensuring that young animals are born during seasons of resource abundance to maximize survival rates (Fig. 8).

In recent years, climate change has disrupted seasonal rhythms, leading to phenomena such as earlier springs and prolonged winters. In response, herders adjust the temporal management and resource-use strategies within the pastoral calendar to adapt to changing conditions. For example, when spring arrives early, herders may move livestock to high-altitude pastures ahead of schedule to avoid shortages in forage caused by drought at lower elevations.

Secondly, the grazing system constitutes a resource management approach developed by herders through long-term practice, aimed at achieving the sustainable utilization of pastures. In recent years, herders have observed that, due to climate change and rising temperatures, the timing of forage growth and maturation has advanced, and the species composition of pasture plants has also shifted. Consequently, their rotational grazing system has been correspondingly modified. The increase in the number of rotational grazing zones and the reduction in grazing duration within each zone are direct responses to accelerated and shortened growth cycles of forage grasses, earlier flowering and senescence, reduced plant height, overall sparser vegetation cover, and declines in both yield and species diversity—all of which are closely linked to climate change.

From the perspective of traditional knowledge, the changes in the number of grazing zones and the grazing duration per zone provide clear evidence of climate change's impact on the knowledge system governing pasture resource use. Field observations and surveys tracking rotational grazing practices in alpine pastures indicate that the number of designated grazing zones has been increasing annually, while the grazing duration within each zone has been decreasing, despite a continual decline in total cattle numbers. This shows that the expansion of grazing zones and shortening of grazing periods are not caused by an increase in herd size.

In this context, herders have adapted their grazing system to mitigate the impacts of climate change in three main ways:

(1) ① Seasonal Rotational Grazing—Herders divide pastures into summer pastures, typically located in high-altitude areas, and winter pastures, located in low-altitude areas. This practice not only prevents overuse of pastures but also reduces the adverse effects of climate change on pasture resources.

(2) ② Pasture Resource Allocation—The grazing system incorporates negotiated agreements among herders regarding pasture use rights and allocation rules, ensuring equitable access to resources.
(3) ③ Innovation in Grazing Systems—Faced with climate-related challenges, herders have introduced innovations into traditional grazing systems, the most effective of which is small-scale rotational grazing, wherein pastures are subdivided into smaller zones that are rotated more frequently to reduce grazing pressure on any single area.

Through the pastoral calendar and grazing system, herders in Deqin County demonstrate the critical role of traditional knowledge in adapting to climate change. Nevertheless, as climate change intensifies, this knowledge system also faces mounting challenges. Looking ahead, it will be important to further explore the integration of traditional knowledge with modern science—for example, by incorporating meteorological data and ecological monitoring technologies—so as to foster innovation and development in the pastoral calendar and grazing system, thereby enhancing their capacity to adapt to the impacts of climate change.

8.4 Traditional Ecosystem Management Mechanisms for Reducing Climate-Related Disaster Risks

Throughout the history of villages in Deqin County, there is a deep collective memory of climate-related disasters. When elders recount village history to younger generations, they often begin with a story of such a disaster. For example, in Jiabi Village, oral histories from the elders recount that, long ago, the Jiabi River Valley was broad and home to over eighty households, making it one of the largest villages in the area.

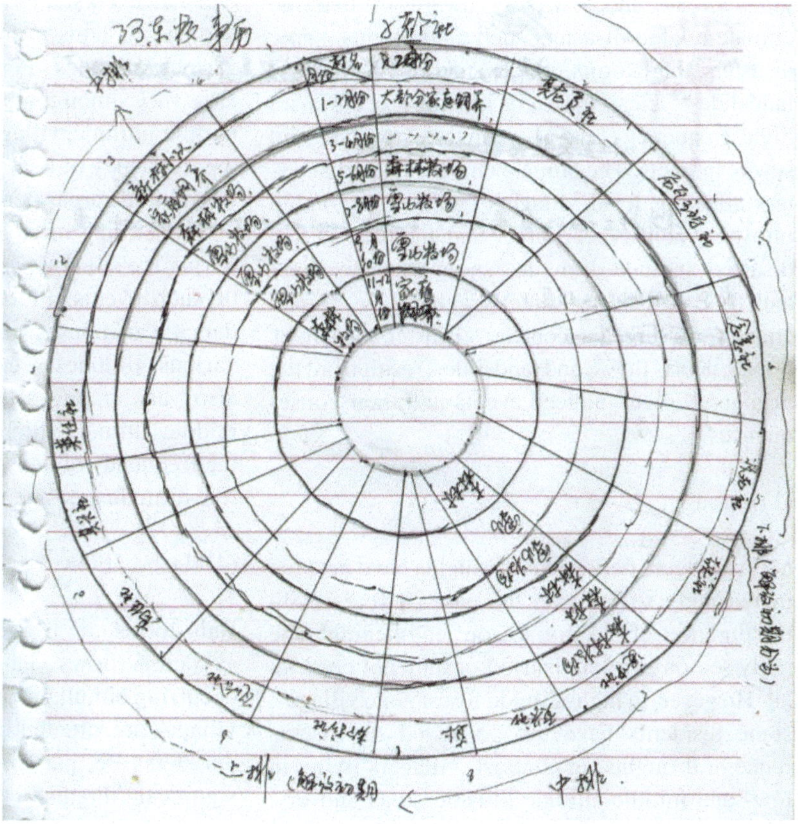

Fig. 8 Pastoral calendar of the Tibetan herdsmen in Deqin County (Source: Lichingping's drawing)

At one point, however, three consecutive days and nights of torrential rain triggered a massive debris flow. The slope of Gonggebianding Mountain beside the village collapsed, destroying and burying the entire settlement. A small hill was formed in the river valley, permanently altering its geomorphology. Only three households survived, and over many years their descendants grew into the thirty households that exist today.

This story—and many like it—has become deeply embedded in the cultural memory of Deqin's Tibetan villagers, serving as a constant reminder to respect and protect the environment. It has also motivated the formation of traditional ecosystem management mechanisms designed to reduce climate-related disaster risks.

8.4.1 Perception of Climate Disaster

In Deqin, climate disasters are frequent due to the region's pronounced topographic climate characteristics and complex geological structure. While long-term climate change trends are notable, local villagers are more acutely aware of extreme climate-related disasters such as droughts, snow disasters, hailstorms, floods, debris flows, and landslides. These disasters not only occur with high frequency but also have extensive and severe impacts. The climate change-related disasters affecting local villagers can be categorized into two main types: direct and indirect. Direct disasters include droughts, snow disasters, hail, and frost, which are directly caused by climate changes. Indirect secondary disasters, such as floods, debris flows, and landslides, result from the primary climate-induced events and their consequences.

(1) Drought

Most villagers believe that droughts have become increasingly severe over the past 20 years, with the duration of the dry season lengthening. The rainy season is often delayed or may not occur at all. However, in high-altitude forest zone villages, some residents have not perceived the occurrence of droughts. In contrast, villagers living in low- and middle-altitude hillside areas and dry, hot valleys have a stronger awareness of these changes.

(2) Snow Disaster

Snow disasters have always been a major issue in Deqin. Recently, the instability of snowfall has become more pronounced, with sudden snowfalls and snowstorms occurring over short periods. This unpredictability often leaves villagers with insufficient time to implement response measures, leading to significant losses. There are notable differences in villagers' perceptions of snow disasters: those in high- and middle-altitude villages perceive an increase in the frequency of snow disasters, while villagers in low-altitude areas feel that the incidence of snow disasters is decreasing.

(3) Rainstorm

Rainstorms frequently lead to floods, particularly during the summer, with an increasing frequency of such events. Floods have caused significant damage to houses and fields across villages at various altitudes. Perceptions of heavy rainfall also vary among villagers. Those in high- and middle-altitude villages observe an increase in the frequency of heavy rainfall, while villagers in low-altitude areas perceive no significant change.

Rainstorms often lead to floods, particularly during the summer, with an increasing frequency of such events. Floods have caused significant damage to houses and fields across villages at various altitudes. Perceptions of heavy rainfall also vary among villagers. Those in high- and middle-altitude villages observe an increase in the frequency of heavy rainfall, while villagers in low-altitude areas perceive no significant change.

(4) Debris Flow, Landslide, and Flood

Rainstorms can trigger secondary disasters such as debris flows, landslides, and floods, often occurring simultaneously. In Deqin, where many villages are situated in canyons, these secondary disasters are particularly damaging and pose significant threats to the safety of lives and property. Many villages have historical memories of being devastated by landslides and mudslides, with some even becoming part of local myths and legends, highlighting the long-term destructive impact of these disasters.

Perceptions of debris flows, landslides, and floods vary among villagers. Those in medium-altitude villages, which are predominantly located on slopes, have a heightened awareness of these disasters due to their frequent occurrence following rainstorms. In contrast, villagers in high- and low-altitude areas generally have less familiarity with such events.

Due to Deqin's steep and complex terrain, neighboring villages experience varying climatic disasters. Differences in the perception of these climate disasters are closely linked to the altitude and location of the villages, shaping villagers' understanding of these events. This variation in perception indicates that the extreme climate disasters faced by villages in different altitudes and environments are not uniform. By integrating these cognitive differences with disaster research, it is possible to reveal local disaster information that may be overlooked by broader extreme climate disaster models, providing valuable insights for disaster response policymaking.

8.4.2 Major Climate-Related Disaster Risk Analysis

Due to its high mountain–gorge terrain, complex geological conditions, and the impacts of climate change, Deqin County faces significant climate-related disaster risks, including rainstorms, droughts, hail, frosts, snow disasters, and secondary hazards such as floods, debris flows, and landslides triggered by heavy rainfall. Among these, drought and flood disasters are the most frequent and widespread. Droughts often result in reduced crop yields, pasture degradation, and shortages of drinking water for both humans and livestock. Large-scale floods can trigger debris flows and landslides, posing severe threats to the safety of lives and property. In addition, Deqin frequently suffers from hailstorms and frosts, which adversely affect agricultural production. Below are five climate-related disasters and their associated secondary hazards that have the greatest impacts on Deqin County.

First is the rainstorm disaster. In Deqin County, rainstorms are severe natural events that have profound effects on local residents' livelihoods. Triggered by sustained or intense rainfall, they cause rivers to swell rapidly, leading to floods, landslides, and debris flows. Rainstorms not only threaten human safety but also severely damage production and living infrastructure.

Second is the drought disaster. Droughts are typically characterized by prolonged periods of below-average precipitation combined with above-average temperatures, leading to falling river levels and, in some cases, complete drying up of streams. Such droughts affect agriculture by severely reducing soil moisture, causing crop yield losses and, in extreme cases, crop death, thus impacting both staple grains and other agricultural products. In pastoral areas, drought causes pasture degradation and poor forage growth, leading to feed shortages for livestock. Water scarcity can also increase livestock mortality.

Third is the flood disaster. Flooding in Deqin is usually caused by prolonged heavy rain or intense rainstorms, during which large volumes of water in a short period cause river levels to rise sharply beyond the channel's capacity. The county's rugged topography and interspersed mountain–gorge landscape accelerate flood convergence, producing peak flows that exacerbate flood severity. In recent years, as extreme weather events have become more frequent, flooding in Deqin has grown more severe and complex. Floods have particularly pronounced effects on agriculture: they wash away farmland, damage soil structure, render fields unsuitable for cultivation, and reduce yields. Floods also promote outbreaks of pests and diseases, further compounding agricultural losses. Moreover, they can destroy homes, fields, and livestock shelters located near rivers, causing significant property damage.

Fourth is the debris flow disaster. In the Lancang River Basin within Deqin County, the alternating high mountains and deep gorges, combined with steep slopes and abundant water inputs from rainfall or snowmelt, create conditions conducive to frequent debris flows. These events carry tremendous force, capable of destroying houses, bridges, roads, and other infrastructure within moments, posing serious threats to life and property. In the past, excessive land reclamation and deforestation in Deqin severely damaged forest cover, thereby increasing both the frequency and intensity of debris flows.

Fifth is the landslide disaster. The high mountain–gorge terrain of Deqin, characterized by sparse vegetation, steep slopes, fractured rock, and concentrated rainfall, is prone to landslides of various scales. Landslides are highly sudden and destructive, and they often occur together with secondary disasters such as debris flows, further threatening local communities' safety and property.

8.4.3 Analysis of Traditional Ecosystem Management Mechanisms for Reducing Climate Disaster Risks

Over the course of long-term production and daily life, Tibetan villagers in Deqin County have developed a distinctive set of traditional ecosystem management mechanisms that effectively help reduce the risks associated with climate-related disasters. First, local villagers tend to classify heavy rainfall and its consequent disasters—such as floods, debris flows, and landslides—as one category, while drought is regarded as a separate category of disaster. Second, the core of these traditional ecosystem management mechanisms focuses on two major dimensions: forests and watersheds.

(1) Traditional Ecosystem Management Mechanisms for Reducing the Risks of Heavy Rainfall, Floods, Debris Flows, and Landslides

The mechanisms aimed at mitigating the risks of heavy rainfall, floods, debris flows, and landslides encompass forest protection, ecological restoration, ecological monitoring and early warning, and disaster response and rescue.

First is mountain closure for forest regeneration. Since ancient times, Tibetan villagers in Deqin County have maintained the tradition of closing off designated mountain areas to allow forest recovery. Through village regulations and community agreements, they demarcate mountain closure zones and prohibit tree felling or the destruction of forest vegetation. This practice has effectively conserved local forest resources and promoted the restoration and stability of forest ecosystems. For example, Tibetan women in the community have voluntarily formed a traditional collective organization known as the "Sisters' Association," which implements collective forest management. They establish village rules for mountain closure, ban logging, and protect forest vegetation, starting from high-altitude zones, applying customary ecological laws to safeguard forests and trees.

Second is afforestation. In addition to mountain closure, villagers actively engage in tree planting activities, selecting indigenous tree species well adapted to the local climate and soil conditions. Afforestation is carried out in ecologically fragile areas, such as steep slopes and wastelands. This not only improves the local ecological environment but also enhances the disaster prevention and mitigation capacity of forests, thereby strengthening the resilience and stability of forest ecosystems.

Third is ecological monitoring and early warning. Through generations of practice, Tibetan villagers in Deqin County have developed unique methods of traditional forest monitoring. They assess the health and potential risks of forest ecosystems by observing phenomena such as tree growth patterns and the movement of wildlife. Although these methods are relatively simple and intuitive, they provide an effective means for the early warning of natural disasters such as forest fires and pest outbreaks.

Fourth is disaster response and rescue. Rich experience in disaster response has been accumulated through generations. Villagers organize public meetings and mutual aid activities to raise awareness of disaster risks and strengthen collective response capacity. When disasters occur, they are able to quickly mobilize rescue teams to conduct evacuation and relief operations, thereby effectively reducing losses.

Fifth is forest management decision-making. Decision-making in forest management emphasizes collective wisdom. Through village assemblies and committee meetings, the opinions and suggestions of villagers are widely solicited, and forest management plans and policy measures are formulated jointly. This approach not only fosters a strong sense of participation and belonging among villagers but also improves the scientific basis and feasibility of decisions.

By implementing mountain closure, afforestation, and other measures, Tibetan villagers in Deqin County have effectively conserved forest resources, enhanced the stability and resilience of forest ecosystems, and strengthened their disaster prevention and mitigation capacity. These practices have increased the water retention and soil conservation functions of forests, thereby reducing soil erosion and lowering both the frequency and intensity of disasters such as debris flows.

Over time, villagers have also developed a distinctive understanding of disaster risks and corresponding coping mechanisms. By observing natural phenomena and passing down historical experiences, they have accumulated substantial knowledge and skills for disaster preparedness and response. These traditional practices play a crucial role during disaster events, enabling timely warnings, rapid evacuations, and effective rescue operations, while simultaneously enhancing disaster awareness and adaptive capacity among the community.

(2) Traditional Ecosystem Management Mechanisms for Reducing the Risk of Drought Disasters

The traditional ecosystem management mechanisms employed to reduce the risk of drought disasters encompass customary ecological laws and traditional organizations, water resource management strategies, and land-use practices.

First, customary ecological laws and traditional organizations play a central role. The "Sisters' Association" and the "Archery Association" are traditional, community-based organizations established spontaneously by local villagers to enforce collective management. These organizations formulate detailed village regulations for forest closure and afforestation, strictly prohibiting tree felling and the destruction of forest vegetation. The regulations specify the scope of closed forests, the selection of tree species, elevation limits, and slope positions. According to customary ecological laws, villagers select locally adapted seedlings—such as Himalayan cypress—and carry out afforestation in ecologically fragile areas such as steep slopes and wastelands. These measures not only increase forest cover and enhance soil water retention capacity, but also effectively slow the rate of temperature rise, mitigate the threats posed by drought, and improve both the ecological environment and the community's resilience to drought disasters.

Second, water resource management and utilization demonstrate deep local ecological knowledge. Recognizing the scarcity and finiteness of water resources, villagers have developed strict rules and customs for water allocation and use. During irrigation seasons, they schedule irrigation timing and volumes according to crop growth cycles and water requirements, thereby avoiding waste. They also make use of traditional hydraulic infrastructure—such as canals and cisterns—to collect and store rainwater for use during dry periods. These facilities improve water-use efficiency and strengthen the community's capacity to withstand drought. During drought seasons, stored rainwater is used for both irrigation and domestic needs, ensuring the continuity of agricultural production and daily life. Moreover, villagers place great emphasis on the construction and maintenance of irrigation systems. Based on local topography, geomorphology, and climate conditions, they plan and build systems that ensure adequate water supply for crops, while conducting regular maintenance and repairs to guarantee proper functioning. These measures not only enhance the efficiency and stability of agricultural production but also reduce the impact of drought disasters on farming activities.

Third, land-use planning and management reflects a high degree of ecological wisdom.

Villagers allocate land for cultivation, forestry, pasture, and construction according to topography, soil conditions, and climate patterns. In drought-prone areas, they minimize land reclamation to protect native vegetation and soil structure, preventing erosion and desertification. Furthermore, through agricultural practices such as crop rotation and fallowing, they restore and enhance soil fertility, ensuring sufficient nutrients and moisture for crop growth. These measures increase the sustainability of agricultural production while reducing the negative effects of drought.

Dry land and irrigated farmland constitute two major local land-use types. Tibetan villagers plan and develop both types according to terrain and climate, ensuring effective utilization and conservation of soil and water resources. Irrigated land boosts land-use efficiency and crop yields while reducing risks of erosion and desertification. The management of dryland farming emphasizes vegetation restoration and conservation—such as planting trees and grasses—to strengthen water retention and land stability. These efforts effectively mitigate the impact of drought on agriculture and enhance production sustainability.

In agricultural production, villagers extensively apply crop rotation and fallowing. Rotation involves planting different crop species in different seasons or years to avoid excessive depletion of soil nutrients and the spread of pests and diseases. Fallowing entails temporarily suspending cultivation to allow soil to rest and recover. These practices enhance soil fertility and productivity while improving water retention and stability. During drought seasons, crop rotation and fallowing are particularly important, as they help soils retain moisture and nutrients, thereby creating favorable conditions for crop growth and reducing drought-related agricultural losses.

9 Conclusion

When this research began investigating climate change and traditional knowledge in Deqin, the term "climate change" was unfamiliar to many local villagers. Despite their remote geographical location, villagers' observations of climate change were often more direct and profound than those of scientific research.

Surveys in Deqin reveal that local villagers possess traditional knowledge for observing and measuring climate change. They use their senses—hearing, vision, touch, and smell—to gather information about weather and climate, such as cloud patterns, temperature, humidity, wind, and precipitation. Villagers typically interpret this perceptual information through cultural and experiential frameworks related to weather and climate. This traditional knowledge can be instrumental in managing livelihoods and mitigating the impacts of climate-related disasters. In this study, traditional climate knowledge is defined as the direct and indirect observation and understanding of local climate and phenology accumulated over generations, which is used to predict and address climate change and extreme climate events. This knowledge is categorized into two components: climate and phenology.

This study also reveals that, even within a small geographical area, the impacts of climate change can vary significantly due to differences in local climate, geographical environment, and the age, gender, and livelihoods of villagers. Observations of climate change by local villagers are often closely linked to their gender, age, livelihood activities, and living environment, which can result in variations in their traditional knowledge.

For instance, studies on long-term climate change indicate that older individuals generally possess more knowledge and experience compared to younger people. However, when it comes to extreme climate events, studies suggest that there is no clear correlation with age; even relatively young villagers are perceptive of phenomena such as unstable climate patterns and frequent climate disasters.

At the same time, most of the villagers' traditional knowledge related to climate change is closely tied to their living environment. Local villagers often base their weather and climate observations on the specific characteristics of their village. Due to variations in altitude, sunshine direction, temperature, precipitation, and wind patterns, even neighboring villages can experience distinct climatic conditions, leading to differing observations and knowledge of climate change among villagers.

In addition to the living environment, gender is also a significant factor influencing villagers' observations and understanding of climate change. Female villagers, who primarily reside in and engage in livelihood activities near the village, tend to be more sensitive to changes in their immediate environment. Conversely, male villagers, who are involved in animal husbandry and move between various alpine pastures, are generally more attuned to climate phenomena in high-altitude areas.

Under the influence of gender, age, livelihood activities, and living environment, individuals within the same area have developed a diverse range of traditional knowledge about climate change. This diversity is a significant advantage of climate change research based on traditional knowledge: it offers insights into local climate changes that may be overlooked by global climate models.

For Tibetan villagers in Deqin, climate change is both a familiar and unfamiliar concept. It is familiar because they have lived in a complex and variable climate environment for generations, resulting in rich and diverse traditional knowledge. However, it is also unfamiliar due to the increased instability of climate change and the heightened risk of climate-related disasters over the past two decades, which has rendered some traditional knowledge less effective and necessitated adaptations.

However, this does not diminish the value of traditional knowledge. Traditional knowledge represents not only the accumulated understanding of climate change by local people but also the process through which they have developed and refined this knowledge. Therefore, traditional knowledge possesses strong vitality and adaptability, continually renewing, innovating, and evolving over time. Moreover, traditional knowledge is not contradictory to scientific data and models; rather, they complement each other. Integrating both can provide a more comprehensive and detailed understanding of climate change and climate-related disasters, particularly for specific local regions.

References

Adesina, F. O., Siyambola, W. O., Oketola, F. O., Pelemo, D. A., Ojo, L. O., & Adegbugbe, A. O. (1999). Potentials of agroforestry for climate change mitigation in Nigeria: Some preliminary estimates. *Global Ecology and Biogeography, 8*, 163–173.

Ahmed, C. I., Hu, W. B., & Kumar, S. (2016). Indigenous knowledge about prediction in climate change. *International Journal of Humanities and Social Science, 5*, 45–62.

Ajibade, L. T., & Shokemi, O. O. (2003). Indigenous approaches to weather forecasting in Asa LGA, Kwara State, Nigeria. *Indilinga Afr J Indigenous Knowl Syst, 2*, 37–44.

IPCC. (2014). Annex II: Glossary [Mach, K.J., S. Planton and C. von Stechow (eds.)]. In: Climate Change 2014: Synthesis Report. Contribution of Working Groups I, II and III to the Fifth Assessment Report of the Intergovernmental Panel on Climate Change [Core Writing Team, R.K. Pachauri and L.A. Meyer (eds.)]. IPCC, Geneva, Switzerland, 117–130.

Aquzhen. (2023). Climate change Perception, Interpretation and Adaptation in Northern Tibet: A Case Study of Dhoni Village Nagqu City. PhD diss., Yunnan University, Kunming. (In Chinese)

Arnakak, J. (2002). Incorporation of InuitQaujimanituqangit, or Inuit traditional knowledge into the Government of Nunavut. *The Journal of Aboriginal Economic Development, 3*, 33–39.

Béné, C., Wood, R. G., Newsham, A., et al. (2013). Resilience: new utopia or new tyranny? reflection about the potentials and limits of the concept of resilience in relation to vulnerability reduction programmes. Ids Working Papers, 2012(405), 1–61.

Berkes, F. (1999). *Sacred Ecology: Traditional Ecological Knowledge and Management Systems*. Taylor and Francis.

Berkes, F. (2009). Indigenous ways of knowing and the study of environmental change. *Journal of the Royal Society of New Zealand, 39*, 151–156.

Berkes, F. (2012a). *Sacred Ecology* (3rd ed.). Routledge.

Berkes, F., & Jolley, D. (2001). Adapting to climate change: Social-ecological resilience in a Canadian western Arctic community. *Conservation Ecology, 5*(2), 18.

Michael W. Binford, Alan L. Kolata, Mark Brenner etc., (1997). Climate variation and the Rise and fall of an andean civilization. *Quaternary Research*, 47, 235–248.

Boer, G. J., Stouffer, R. J., Dix, M., Noda, A., Senior, C. A., Raper, S., Yap, K. S. (2001). Projections of future climate change. In Kim, J.-W., Stone, J. (Eds.), Climate Change 2001: The Scientific Basis. Cambridge, UK: Cambridge University Press.

Bureau of Meteorology (BoM). (2003). Indigenous weather knowledge project. www.bom.gov.au/iwk/index.shtml.

Byg, A., & Salick, J. (2009a). Local Perspectives on a global phenomenon-Climate change in Eastern Tibetan villages. *Global Environmental Change, 19*, 156–166.

Byg, A., & Salick, J. (2009b). Local Perspectives On A Global Phenomenon—Climate Change In Eastern Tibetan Villages. *Global Environmental Change, 19*, 156–166.

Chaudhary, P., & Bawa, K. (2011). Local perceptions of climate change validated by scientific evidence in the himalayas. *Biology Letters, 7*(5), 641–643.

Chaudhary, P., Rai, S., Wangdi, S., Mao, A., Rehman, N., Chettri, S., & Bawa, K. (2011). Consistency Of Local Perceptions Of Climate Change In The Kangchenjunga Himalaya Landscape. *Current Science, 101*(4), 504–513.

Colombi, B. J. (2009). Salmon Nation: Climate Change and Tribal Sovereignty. In S. A. Crate & M. Nuttall (Eds.), *Anthropology and climate change* (pp. 110–124). Left Coast Press.

Couzin, J. (2007). Opening doors to native knowledge. *Science, 315*(5818), 1518–1519.

Susan, A., Crate, (2009). Gone the Bull of Winter? Contemplating Climate Change's Cultural Implications in Northeastern Siberia, Russia. In Susan A. Crate, Mark Nuttall. Anthropology and climate change (Eds.), From Encounters to Actions. Left Coast Press, 56–63.

Crate, S. A., & Nuttall, M. (Eds.). (2009). *Anthropology and climate change, From Encounters to Actions* (pp. 9–34). Left Coast Press Inc.

Crump, J. (2008). Many strong voices: climate change and equity in the Arctic and Small Island Developing States. Indig Aff, 1–2(2008), 24–33. Available via http://www.iwgia.org/. Cited 25 Jan 2021.

Dayuan, X. (2011). The Categories and Benefit-sharing of Traditional Knowledge Associated with Biodiversity. *Journal of Resources and Ecology, 2*(1), 289–299. https://doi.org/10.3969/j.issn.1674-764x.2011.01.001

Drever, C. R., Peterson, G., Messier, C., Bergeron, Y., & Flannigan, M. D. (2006). Can forests management based on natural disturbances maintain ecological resilience? *Can. Jour. for. Res, 36*, 2285–2299.

Fikret, B., & Folke, J. C. C. (2000). Rediscovery of Traditional Ecological Knowledge as Adaptive Management. *Ecological Applications, 10*(5), 1251–1262.

Flavier, J. M. et al. (1995). The Regional Program for the Promotion of Indigenous Knowledge in Asia. In Warren, D.M., L. J. Slikkerveer and D. Brokensha eds. The Cultural Dimension of Development: Indigenous Knowledge Systems. London: Intermediate Technology Publications.

Folke, C., Carpenter, S. R., Walker, B., et al. (2010). Resilience thinking: Integrating resilience, adaptability and transformability. *Ecology & Society, 15*(4), 299–305.

Ford, J. D., Cameron, L., Rubis, J., et al. (2016). Including indigenous knowledge and experience in IPCC assessment reports. *Nature Climate Change, 6*, 349–353.

Galloway-McLean K. (2010). Advance guard: climate change impacts, adaptation, mitigation and indigenous peoples—a compendium of case studies. United Nations University-Traditional Knowledge Initiative, Darwin. Available via http://www.unutki.org/news.php?doc_id=101&news_id=92. Cited 20 Jan 2021.

Gao, Y. S., Sang, Y. Y., eds. (2004). Rise and decline of ancient loulan and its relations to climate change and disaster. *Journal of Chengdu University, 23*(3), 50–52.

Garay-Barayazarra, G., Puri, R.K. (2011). Smelling the monsoon: Senses and traditional weather forecasting knowledge among the Kenyah Badeng farmers of Sarawak, Malaysia. Indian J Trad Know 10, 21–30.

GCA (Global Commission on Adaptation). 2019.Adapt now: a global call for leadership on climate resilience. Global Commission on Adaptation, Global Center on Adaptation and World Resource Institute.

Gilchrist. G., Mallory, M., Merkel, F. (2005). Can local ecological knowledge contribute to wildlife management? Case studies of migratory birds. *Ecology and Society 10*(1), 20, 1–12.

Grenier, L. (1998). Working with Indigenous Knowledge: A Guide for Researcher. Ottawa: IDRC (International Development Research Center).

Gunderson, L. (2000). Ecological resilience: In theory and application. *Ann. Rev. Ecol. Syst, 31*, 425–439.

Shen Haimei, Chen Xiaoyi. (2020). Research on Adaptation to Aridification in the Eastern Himalayas: Gendered Traditional Water Knowledge and Viticulture in the Meili Tibetan Area of Yunnan. Journal of Southwest Minzu University (Humanities and Social Sciences Edition), 41(9),9–15. (In Chinese)

Hammer, K., & Teklu, T. (2008). Plant genetic resource: Selected issues from genetic erosion to genetic engineering. *Journal of Agriculture and Rural Development in the Tropics and Subtropics, 109*, 15–50.

Hardesty, D. L. (1977). *Ecological Anthropology*. Wiley.

Hobart, Mark, ed. (2002). An anthropological critique of development: The growth of ignorance. Routledge.

Holling, C. S. (1973). Resilience and stability of ecosystems. *Ann. Rev. Ecol. Syst, 4*, 1–23.

https://doi.org/10.1017/S0032247400017058

https://doi.org/10.3354/cr019109

https://doi.org/10.3724/SP.J.1003.2009.08256

Huang, Q., Yin, L., Kuang, Y., et al. (2025). Traditional Knowledge in Forest Resource Management for Climate Change Adaptation in a Tibetan Village in the Eastern Himalayas Region of China. Indigenous Peoples' Knowledge and the Sciences, 229.

Huang, X., Zhou, Y., Fang, J., et al. (2019). Climate change has more adverse impacts on the higher mountain communities than the lower ones: People's perception from the northern Himalayas. *Journal of Mountain Science, 16*(11), 2625–2639.

Hunn, E. S. (2008). *A Zapotec Naturai History*. University of Arizona Press.

Huntington, H. P. (2000). Using traditional ecological knowledge in science: Methods and applications. *Ecological Applications, 10*, 1270–1274.

IIRR. (1996). What is Indigenous Knowledge? In Recording and Using Indigenous Knowledge: A Manual. pp. 7 -18. Silang, Cavite: International Institute for Rural Reconstruction (IIRR).

IK&DM. (1998). Reaction. Indigenous Knowledge and Development Monitor (IK&DM). 6 (3).

IPCC. (2007). Climate Change 2007: The Scientific Basis, Contribution of Working Group I to the Fourth Assessment Report of the Intergovernmental Panel on Climate Change, Cambridge University Press, Cambridge, United Kingdom and New York, NY, USA.

IPCC. (2007a). Climate Change 2007: Impacts, Adaptation and Vulnerability. Contribution of Working Group II to the Fourth Assessment Report of the Intergovernmental Panel on Climate Change [Parry, M. L., Canziani, O. F., Palutikof, J. P., van der Linden, P. J and Hanson, C. E. (eds.)] Cambridge, UK: Cambridge University Press, 976.

IPCC. (2007).Climate Change 2007: Synthesis Report. Contribution of Working Groups I, II and III to the Fourth Assessment Report of the Intergovernmental Panel on Climate Change. Cambridge University Press: Cambridge, UK.

IPCC. (2012). Managing the risks of extreme events and disasters to advance climate change adaptation (SREX). Special Report of the Intergovernmental Panel on Climate Change (IPCC). Geneva, IPCC Secretariat.

IPCC. (2013). Climate Change 2013: The Physical Science Basis. Contribution of Working Group I to the Fifth Assessment Report of the Intergovernmental Panel on Climate Change, T.F. Stocker, D. Qin, G.-K. Plattner, M. Tignor, S.K. Allen, J. Boschung, A. Nauels, Y. Xia, V. Bex, and P.M. Midgley, editors. Cambridge University Press, Cambridge, U.K. and New York, U.S.A., 1535.

IPCC. (2014). Climate Change 2014: Impacts, Adaptation, and Vulnerability. Part A: Global and Sectoral Aspects. Contribution of Working Group II to the Fifth Assessment Report of the Intergovernmental Panel on Climate Change [Field, C. B., Barros, V. R., Dokken, D. J. et al. (eds.)]. Cambridge, UK and New York: Cambridge University Press, 1132.

Jacob, C., McDaniels, T. and Hinch, S.G. (2010). Indigenous culture and adaptation to climate change: sockeye salmon and the St'át'imc people. Mitigation and Adaptation Strategies for Global Change, 15, 859–876.

Javaid Akhter Bhat & Steve Taiwo. (2016). Plant Genetic Resources and Traditional Knowledge in Light of Recent Policy Developments. In R.K. Salgotra, B.B. Gupta (eds.), Plant Genetic Resources and Traditional Knowledge for Food Security, Springer Science+Business Media Singapore. https://doi.org/10.1007/978-981-10-0060-7_10.

Jiao, Y., Li, X., Liang, L., et al. (2012). Indigenous ecological knowledge and natural resource management in the cultural landscape of China's Hani Terraces. *Ecological Research, 27*(2), 247–263.

Johnson, L. M., and E. S. Hunn (eds.) (2010) Landscape Ethnoecology. New York and Oxford: Berghahn.

Yang Jingbiao. Adaptation Mechanism on Climate Change of Traditional Agriculture Management in Hani Rice Terraces. PhD diss., Beijing Forestry University, Beijing. (In Chinese)

Kanani, P. R., & Pastakia, P. (2001). Everything is written in the sky! Participatory meteorological assessment and prediction on traditional beliefs and indicators in Saurashtra. *Eubios Journal of Asian and International Bioethics, 9*, 170–176.

Karjalainen T, Kelloms ˇki S, Pussinen A. (1994). Role of wood-based products in absorbing atmospheric carbon. Silva Fennica, 28(2), 67–80.

King, D. N. T., Skipper, A., & Tawhai, W. B. (2008). Maori environmental knowledge of local weather and climate change in Aotearoa – New Zealand. *Climatic Change, 90*, 385–409.

Tse-ring, K., Sharma, E., Chettri, N. & Shrestha, A. (eds.) (2010). Climate change vulnerability of mountain ecosystems in the Eastern Himalayas; climate change impact an vulnerability in the Eastern Himalayas—synthesis report. Kathmandu, Nepal: ICIMOD.

Kyle Powys Whyte. 2015. The Ethics of Traditional Knowledge Exchange in Climate Change Initiatives. The Climate and Traditional KnowledgesWorkgroup. https://earthzine.org/the-ethics-of-traditional-knowledge-exchange-in-climate-change-initiatives/

Langton, M., Ma rhea, Z. & Palmer, L. (2005). Community-orientated protected areas for Indigenous peoples and local communities. Journal of Political Ecology, 12, 23–50.

Lantz, T. C., & Turner, N. J. (2003). Traditional phenological knowledge of Aboriginal peoples in British Columbia. *Journal of Ethnobiology, 23*(2), 263–286.

Heather Lazrus. 2009.The Governance of Vulnerability: Climate Change and Agency in Tuvalu, South Pacific. In Susan A. Crate, Mark Nuttall. Anthropology and climate change, From Encounters to Actions (Eds.), Left Coast Press, 46–58.

Lefale, P.F. (2003). A Samoan Perspective on the Role of Traditional Knowledge in Adapting our Understanding of Climate Change, In: Helander, E and Mustonen, T(Eds).: Snowscapes, Dreamscapes: Snow change Book on Community Voices of Change, 495–515.

Leonard, S. (2013). The role of culture and traditional knowledge in climate change adaptation: Insights from East Kimberley, Australia. Global Environ. Change. https://doi.org/10.1016/j.gloenvcha.2013.02.012

Li, C., Tang, Y., Luo, H., et al. (2013). Local farmers' perceptions of climate change and local adaptive strategies: A case study from the Middle Yarlung Zangbo River Valley, Tibet. *China. Environmental Management, 52*(4), 894–906.

Luo, S. Z. (2006). Discuss of Climate Culture in Qinghai Province, *Journal of Qinghai Meteorology,* 42–49.

Lurongzhuoma. (2013). Research on the Impact of Climate Change on Traditional Livelihood Knowledge of Tibetan People in Northwest Yunnan. PhD diss., Minzu University of China, Beijing. (In Chinese)

McGregor, D. (2004). Traditional ecological knowledge and sustainable development: towards co-existence. In In the Way of Development. Indigenous Peoples, Life Projects and Globalization (M. Blaser, H. A. Feit, and G. McRae, eds). London and New York: Zed Books, 72–91.

Memory, E. L. (2007). Evolving concepts related to achieving benefit sharing for custodians of traditional knowledge. *African Journal of Traditional Complementary and AlternativeMedicines, 4*, 443–468.

Mugabe, J., Kameri-Mbote, P., Mutta, D. (2001). Traditional knowledge, genetic resources and intellectual property protection: towards a new international regime. International Environmental Law Research Center, Geneva, Switzerland (http://www.ielrc.org/content/w0105.pdf)

Nakashima, D.J., K. Galloway McLean, H.D. Thulstrup, A. Ramos Castillo, and J.T. Rubis. 2012. Weathering Uncertainty: Traditional Knowledge for Climate Change Assessment and Adaptation.

Nakashima, D., & Roué, M. (2002). Indigenous knowledge, peoples and sustainable practice. In P. Timmerman (Ed.), *Social and economic dimensions of global environmental change* (Vol. 5, pp. 314–324). Wiley.

Nazarea, V. D. (Ed.). (1999). *Ethnoecology: Situated Knowledge/Located Lives*. University of Arizona Press.

Nelson, D. R., Adger, W. N., Brown, K. (2007). Adaptation to environmental change: contributions of a resilience framework. Social Science Electronic Publishing, 32(32). https://doi.org/10.1146/annurev.energy.32.051807.090348

Nichols, T., Berkes, F., Jolly, D., Snow, B. and the community of Sachs harbour 2004. Climate Change And Sea Ice: Local Observations From The Canadian Western Arctic. Arctic, 57(1),68–79.

NiuZhongbao and Liu Hongyu, eds. 2011. Introduction of Climate Philosophy, Beijing, China Meteorological Press, 24–26.

Nyong, A., Adesina, F., & Osman Elasha, B. (2007). The value of indigenous knowledge in climate change mitigation and adaptation strategies in the African Sahel. *Mitigation and Adaptation Strategies for Global Change, 12*(5), 787–797.

Olsson, P., & Folke, C. (2001). Local ecological knowledge and institutional dynamics for ecosystem management: A study of Lake Racken watershed, Sweden. *Ecosystems, 4*, 85–104.

Olsson, P., Folke, C., & Berkes, F. (2004). Adaptive co-management for building resilience in social– ecological systems. *Environ Manag, 34*(1), 75–90.

Orlove, B. S., Chiang, J. C. H., & Cane, M. A. (2002). Ethnoclimatology in the Andes. *American Scientist., 90*, 428–435.

Parrotta, J. A., & Agnoletti, M. (2012). Chapter 13: Traditional forest-related knowledge and climate change. In J. A. Parrotta & R. L. Trosper (Eds.), *Traditional forest-related knowledge: Sustaining communities, ecosystems and biocultural diversity* (pp. 491–534). Springer.

Parry, M. L., Canziani, O. F., Palutikof, J. P., van der Linden, P. J., & Hanson, C. E. (2007). Cross-chapter case study. In M. L. Parry, O. F. Canziani, J. P. Palutikof, P. J. van der Linden, & C. E. Hanson (Eds.), *Contribution of working group II to the fourth assessment report of the intergovernmental panel on climate change* (pp. 843–868). Cambridge University Press.

Peterson, L. C., & Haug, G. H. (2005). Climate and the Collapse of Maya Civilization– A series of multi-year droughts helped to doom an ancient culture. *American Scientist, 93*(4), 322–329.

Prno, J., Bradshaw, B., Wandel, J., et al. (2011). Community vulnerability to climate change in the context of other exposure-sensitivities in Kugluktuk, Nunavut. *Polar Research, 30*, 1–21.

Reidlinger, D., & Berkes, F. (2001). Contributions of traditional knowledge to understanding climate change in the Canadian Arctic. *Polar Record, 37*, 315–329.

Rosenzweig, C., & Hillel, D. (2008). *Climate Variability and the Global Harvest: Impacts of El Niño and Other Oscillations on Agro-Ecosystems*. Oxford University Press.

Fang Ruoyu, (2006). Comparative Study of Kazak Animal Husbandry and its Relations to Climate of Solar Terms, (4), 84–88.

Salgotra, R.K., Zargar, S.M., Sharma, M., Sood, M. (2018). Traditional Knowledge: A Therapeutic Potential in the Scenario of Climate Change for Sustainable Development. Development, 1–9.

Salick, B., & Byg, A. (2007). *Indigenous Peoples and Climate Change*. A Tyndall Centre Publication.

Schüklenk, U., & Kleinsmidt, A. (2006). North–south benefit sharing arrangements in bioprospecting and genetic research: Acritical ethical and legal analysis. *Developing World Bioethics, 6*, 122–134.

Siebenhuner, B., Dedeurwaerdere, T., & Brousseau, E. (2005). Introduction and overview to the special issue on biodiversity conservation, access and benefit-sharing and traditional knowledge. *Ecological Economics, 53*, 439–444.

Smit, B., & Wandel, J. (2006). Adaptation, adaptive capacity and vulnerability. *Global Environmental Change, 16*(3), 282–292.

SoejartoDD, F. H. H. S., & Tan, G. T. (2005). Ethno-botany/ethnopharmacology and mass bioprospecting: Issueson intellectual property and benefit-sharing. *Journal of Ethnopharmacology, 100*, 12–22.

Solomon, S., Qin, D., Manning, M., Alley, R.B., Berntsen, T., Bindoff, N.L., et al. (2007). Technical Summary. In: Climate Change 2007: The PhysicalScience Basis. Contribution of Working Group I to the Fourth AssessmentReport of the Intergovernmental Panel on Climate Change [Solomon, S., D. Qin, M. Manning, Z. Chen, M. Marquis, K.B. Averyt, M. Tignor and H.L. Miller (eds.)], New York: Cambridge University Press.

Speranza, C. I., Kiteme, B., Ambenje, P., Wiesmann, U., & Makali, S. (2010). Indigenous knowledge related to climate variability and change: Insights from droughts in semi-arid areas of former Makueni District. *Kenya. Climatic Change, 100*(2), 295–315.

Stainback, G. A., & Alavalapati, J. (2002). Economic analysis of slash pine forest carbon sequestration in the southern US. *Journal of Forest Economics, 8*, 105–117.

Strauss, S., & Orlove, B. S. (Eds.). (2003). *Weather, Climate, Culture* (pp. 3–11). Berg Publishers.

Swiderska, K. (2006). Banishing the biopirates: a new approach to protecting traditional knowledge. Gatekeeper Series No129, International Institute for Environment and Development, Environmental Economics Programme, London.

Thomas, C. D., Cameron, A., Green, R. E., Bakkenes, M., Beaumont, L. J., Collingham, Y. C., Erasmus, B. F. N., de Siqueira, M. F., Grainger, A., Hannah, L., Hughes, L., Huntley, B., van Jaarsveld, A. S., Midgley, G. F., Miles, L., Ortega-Huerta, M. A., Peterson, A. T., Phillips, O. L., & Williams, S. E. (2004). Extinction risk from climate change. *Nature, 427*, 145–148.

Thompson, I., Mackey, B., McNulty, S., et al. (2009). Forest resilience, biodiversity, and climate change. a synjournal of the biodiversity/resilience/stability relationship in forest ecosystems. *Cbd Technical, 43*, 1–67.

Toledo, V. M. (1992). What is ethnoecology? Origins, scope and implications of a rising discipline.Ethnoecological, 1(1), 5–21.

Toni Huber and Poul Pedersen. (1997). Meteorological Knowledge and Environmental Ideas in Traditional and Modern Societies: The Case of Tibet. *The Journal of the Royal Anthropological Institute, 3*(3), 577–597.

Tume, S. J. P., Kimengsi, J. N., & Fogwe, Z. N. (2019). Indigenous Knowledge and Farmer Perceptions of Climate and Ecological Changes in the Bamenda Highlands of Cameroon: Insights from the Bui Plateau. *Climate, 7*(12), 138.

Turner, N. J., & Clifton, H. (2009). "It's so different today": Climate change and indigenous lifeways in British Columbia. *Canada. Global Environmental Change, 19*(2), 180–190.

UNFCCC. (2016). Decision 1/CP.21 Adoption of the Paris Agreement. 29 January 2016.FCCC/CP/2015/10 Add. 1, Annex (Paris Agreement)

Vayda, A. P., & McCay, B. J. (1975). New directions in ecology and ecological anthropology. *Annual Review of Anthropology, 4*(1), 293–306.

Vedwan, N. (2006). Culture, Climate and the Environment: Local Knowledge and Perception of Climate Change among Apple Growers in Northwestern India. *Journal of Ecological Anthropology, 10*, 4–18.

Vedwan, N., & Rhoades, R. E. (2001). Climate change in the Western Himalayas of India: A study of local perception and response. *Clim. Res., 19*, 109–117.

Vlassova, T. K. (2006). Arctic residents' observations and human impact assessments in understanding environmental changes in boreal forests: Russian experience and circumpolar perspectives. *Mitigation and Adaptation Strategies for Global Change, 11*, 897–909.

Walker, B., & Salt. D. (2006). Resilience thinking: sustaining ecosystems and people in a changing world. Island Press, Washington, DC., USA.

Wanrui, W. (2012). Research of Origin of China Climate Culture. *Journal of Shanxi Meteorology, 5*, 50–54.

Richard A. Warrick & William E. Riebsame. (1983). Societal Response to CO2 – Induced Climate Change: Opportunities for Research. In R. Chen, E. Boulding and S. Schneider, eds. Social Science Research and Climate Change, 20–60.

Wenhui, Y. (2009). Climate Resources and Belief – Traditional Knowledge of the Bai Ethnic People in Relation to the Change of Climate. *Academic Exploration, 2*, 89–93. (In Chinese).

Wenzel, G. W. (2004). From TEK to IQ: Inuit Qaujimajatuqangit and Inuit cultural ecology. *Arctic Anthropology, 41*, 238–250.

Xu, J. C., Grumbine, R. E., Shrestha, A., et al. (2009). The Melting Himalayas: Cascading effects of climate change on water, biodiversity and livelihoods. *Conservation Biology, 23*(3), 520–530.

Xue, D. Y., & Cai, L. (2006). New hotspot of the Convention on Biological Diversity: Traditional knowledge protection. *Environmental Protection, 12B*, 72–74.

Xue, D. Y., & Guo, L. (2009). On the concept and protection of traditional knowledge. *Biodiversity Science, 17*(2), 135–142. (in Chinese).

XuWeilin, Y. L. (2024). Traditional Knowledge and the Achievement of the "Dual Carbon" Goals – A Practical Case from Guonian Village, Deqin County. *Yunnan Province. Eco-Ethonolgy Reviews, 1*, 11–25. (In Chinese).

Yang, W. H. (2010). Climate, Resources and Religion: Bai Indigenous Knowledge and Climate Change. In Yin Shaoting, Culture and Environment in China, Kunming, the People's Press of Yunnan, 199–206.

Yang, H., He, J., Su, Y., et al. (2022). Adaptation to climate change: Ethnic groups in Southwest China. *Environmental Hazards, 21*(2), 117–136.

Yang, X. Y., Wang, Y. J., Quan, W., et al. (2017). Investigation on Traditional Knowledges of Miao Community to Cope with Climate Change in Wumeng Mountains. *Journal of MUC (Natural Sciences Edition), 26*(4), 54–60. (In Chinese).

Yao, F. U., Yun, W., Yong-Ping, Y., et al. (2015). Indigenous knowledge and use of Anisodus tanguticus in pastoral communities of Eastern Tibet, China, and Its implications for local adaptation. *Plant Diversity, 37*(6), 881–890.

Yin, S. T. (2010). *Culture and Environment in China*, Kunming, the People's Press of Yunnan.

Yin, L. (2011a). Changes in Indigenous Tibetan Knowledge about Environment: A Case Study of Jiabi Village, Deqin County, Yunnan, China. In Global Change, Biodiversity and Livelihoods in Cold Desert Region of Asia edited by K.G. Saxena, Luohui Liang, Xian Xue. Bishen Singh Mahendra Pal Singh, Dehra Dun, India, 112–129.

Yin, L. (2011b). Water Knowledge, Use, and Governance: Tibetan Participatory Development Along. - the Mekong (Langcangjiang) River, in Yunnan, China. In Water, Cultural Diversity, and Global Environmental Change: Emerging Trends, Sustainable Futures? edited by Johnston, B.R.; Hiwasaki, L.; Klaver, I.J.; Ramos

Castillo, A.; Strang, V. UNESCO. Springer Dordrecht Heidelberg London New York, 185–200.

Yin, L. (2011c). Local Perspective on Climate Change of Tibetan People – Case of Guonian Village in Deqin County. *Yunnan. Think Front, 37*(4), 24–28. (in Chinese).

Yin, L. (2013). *Research on traditional knowledge and climate change of Tibetan in Deqin County*. Minzu University of China, Beijing. (In Chinese).

Yin, L. (2015). *Climate Anthropology*. Intellectual Property Press. (in Chinese).

Yin, L. & Xue, D. Y. (2012). Gender Perspective on Climate Change and Disaster -- Case of Hongpo Village in Deqin County, Yunnan. *Journal of Yunnan Normal University, 44*(5). (in Chinese)

Yin, L. & Xue, D. (2013). Tibetan Sacred Mountain Believing and Global Climate Change – Case of Hongpo village in Deqin County. *Yunnan. Journal of Yunnan Nationalities University, 30*(3), 41–45. (in Chinese).

Yin, L., & Xue, D. Y. (2013). Tibetan Sacred Mountain Religion and Global Climate Change: A Case Study of Hongpo Village of Deqin County of Yunnan Province. *Journal of Yunnan Nationality University, 30*(003), 41–45.

Yin, L. & Zheng, Y. Y. (2015). Ecological culture of Tibetan in Yunnan. *Journal of Minzu University of China (Natural Science Edition), 24*(4), 79–84.

Yin, L., Zachary, M., & Zheng, Y. Y. (2018). The Traditional Knowledge Associated to Biodiversity in an Age of Climate Change. *Earth Sciences, 7*(5), 209–215. https://doi.org/10.11648/j.earth.20180705.12.

Yin, L., Zachary, M., Zheng, Y. Y., et al. (2020). The impacts of climate change on the traditional agriculture of ethnic minority in China. *Journal of Environmental Engineering and Science, 9*, 43–55.

Zhou, H., & Yin, L. (2024). Using traditional knowledge to reduce disaster risk-A case of Tibetans in Deqen County, Yunnan Province. *International Journal of Disaster Risk Reduction, 108*, Article 104492.

Community-Based Action Research for Climate Change of Tibetan People in Eastern Himalayas, China

2

1 Background

Over the past 20 years, I have implemented a series of actions and projects about community-based solutions to improve the climate change adaptation in the eastern Himalayas of China. The Eastern Himalaya region is a global biodiversity hotspot and a vulnerable area significantly affected by climate change.

I'll always remember a time twenty years ago when I sat with an elderly Tibetan monk facing the snow-capped mountains. He told me, "Look, the glaciers and snow are shrinking, just like the mountains are wearing less and less clothing." His words deeply touched me. Since then, I've felt deeply the risks and challenges that climate change poses to the local Tibetan villagers, and addressing climate change has gradually become a key focus of my work there. I believe that to help ethnic minority communities cope with climate change, we must connect scientific and traditional knowledge, integrate natural and social sciences, and emphasize the integration of theory and practice. "Connecting scientific and traditional knowledge" means understanding the current state of research and practice on both scientific and traditional knowledge in addressing climate change, drawing on the wisdom and experience of both knowledge systems. "Integrating natural and social sciences" means emphasizing interdisciplinary collaboration and exchange, such as between ecology and anthropology, by categorizing and quantifying traditional knowledge related to climate change to influence scientific research and policymaking. "Integrating theory and practice" is paramount, meaning integrating climate change theory with the realities of local Tibetan communities and implementing practical actions to address climate change at the rural level. Therefore, I worked with local Tibetan people to develop community-based action research strategies for climate change adaptation based on their traditional knowledge.

During these 22 years, I've undergone a series of transformations: First, my work shifted from purely academic anthropological research to action research on biodiversity conservation, and later further to action research on climate change response. Second, my role shifted from anthropological scholar to community development worker, and later to community facilitator in addressing climate change. Third, my research methods shifted from anthropological participant observation to participatory methods in community work, and later to community-based action research. Interestingly, along with these changes, the way the locals addressed me also shifted in three stages: when I was conducting anthropological research, the villagers addressed me as "Doctor". When I was running community development projects, they switched to calling me "Project Director". Finally, when I was assisting villagers with climate change response projects, they addressed me as "Brother". I believe this shift not only reflects a change in address but also a shift

Y. Lun, *Climate Change and Traditional Knowledge*, Sustainable Development Goals Series,
https://doi.org/10.1007/978-3-032-12015-1_2

in my relationship with the villagers: from what anthropologists consider "the other" to what the Tibetan villagers actually consider "one of their own".

To get back to the point, I believe that in the current context of climate change, and even the climate crisis, traditional research methods such as participant observation, questionnaires, literature collection, and in-depth interviews are no longer sufficient to understand the role, significance, and value of Tibetans in responding to climate change in Deqin County, China's eastern Himalayas. Research on Tibetan responses to climate change requires breakthroughs and advancements beyond traditional anthropological research methods such as fieldwork. Community-based action research offers a new approach and direction: building on traditional anthropological fieldwork and case studies, it explores concrete practices grounded in local responses to climate change. Therefore, using the Tibetan community of Deqin County, China's eastern Himalayas, as a case study, I attempt to develop the fundamental principles of a community-based approach to climate change research. These principles must be centered on local people, emphasize the interactive relationship between climate change and social culture, and firmly embrace the crucial role of traditional knowledge.

First, I reflected on traditional ethnographic fieldwork. Ethnographic fieldwork is the most fundamental and central research method in anthropology. Fieldwork involves a slow accumulation of observation, experience, and understanding, a process that allows anthropologists to gain insights into the worldviews and cultural connotations of local people. By living and conducting research within a community for a period of time, engaging in participant observation of daily life and social relations, ethnographic fieldwork can achieve a contextual understanding of cultural realities that is difficult to obtain through methods such as simple structured interviews. Therefore, as a traditional anthropological research method, ethnographic fieldwork offers a unique opportunity to gain insights into the relationships between policy, ecology, and culture. However, with the intensification of global climate change, for traditional ethnic communities like the Tibetan community in Deqin County in China's eastern Himalayas, climate change is not simply an academic issue that can be studied dispassionately or objectively, but rather a real threat and challenge that requires diligent understanding and response. During my fieldwork, as an anthropologist, I experienced the challenges of climate change alongside the local Tibetan community, experiencing its widespread impact on local society, culture, economy, and daily life. Because local Tibetan communities live in socially and ecologically marginalized and vulnerable areas, their ability to adapt to climate change is often limited. While traditional field research methods have provided some insights into local Tibetan communities' responses to climate change, they are far from sufficient. The limitations of traditional field research methods have hindered anthropologists' understanding of how local Tibetan communities are responding to climate change.

In reflecting on traditional fieldwork methods, I've faced a series of fundamental questions directly related to the basic research and practical application of anthropology. I've also been considering: What is the responsibility of anthropologists to the communities they research in the face of climate change? What is the role of anthropologists? What findings can anthropologists gain from fieldwork? How can these findings be applied? What challenges do current fieldwork methods face? How can anthropologists understand the complex relationship between climate change and traditional ethnic societies? How can anthropologists help traditional ethnic societies cope with the impacts of climate change?

Therefore, against the backdrop of the unprecedented impact of climate change on Tibetan communities, I believe that as a fieldworker, I have a responsibility and obligation to combine academics and practice, to re-examine and reflect on the role of anthropologists in field research, and to employ a variety of research methods, including ethnography, participant observation, collaborative research, community-based research, and action research. Ultimately, I aim to work with local Tibetan communities to conduct community-based action research to address climate change. For me, this is both a method and an ideal.

Through their long-term practices, local Tibetan residents have accumulated a wealth of traditional knowledge on biodiversity and climate change. However, for various reasons, this traditional knowledge, rooted in local production and daily life, is often viewed as backward and ignorant. It is not only neglected, unused, and undistributed but is even in danger of disappearing. Top-down approaches to addressing climate change risks and disasters often fail to take into account local realities and needs, resulting in ineffective results and a lack of local engagement. Therefore, I have focused on the traditional knowledge generated and accumulated by local communities. I hope to collaborate with local people to explore and organize their traditional knowledge, actively innovate and apply it, leverage the role of traditional social organizations, enhance villagers' ability to manage and use resources, and ultimately contribute to local communities' resilience to climate change.

Therefore, through over two decades of field research and specific projects on climate change response in Tibetan communities in Deqin County, eastern China, I have gradually explored and developed a community-based action research methodology. In these projects, I have engaged local Tibetan villagers, guided by their values, perspectives, and perspectives, to assist them in exploring, organizing, innovating, and utilizing their traditional knowledge. This has leveraged the positive role of traditional organizations and customary laws, exploring climate change response pathways that align with local ecological and social values, and promoting local adaptation. The goal of this community-based approach is to maximize the participation of local Tibetan villagers in project design and decision-making, empowering them to be the primary actors in project implementation. I believe that, through this approach, these projects have contributed to strengthening local communities' capacity to respond to climate change.

Community-based research is a novel concept and a highly practical approach to exploration. Looking back on my work with the Tibetan community in Deqin County, China, in the Eastern Himalayas, I can see that it has progressed through three phases: the first phase was participant observation, the second phase was participatory research, and the third phase was active exploration of community-based action research.

I believe that the specific principles of community-based action research for climate change should include: (1) being led by the local community; (2) focusing on climate change and traditional knowledge; (3) integrating interdisciplinary research across social and natural sciences; (4) being oriented toward solving practical problems; and (5) encouraging all stakeholders, both academic and non-academic, to form knowledge partnerships in practice to jointly generate and disseminate knowledge.

This research approach has the following two key characteristics:

1. Interdisciplinary and comprehensive research centered on traditional knowledge

Traditional knowledge has demonstrated its unique value in addressing climate change. The traditional knowledge of local Tibetans is not simply a cognitive understanding of the ecological environment; it is more of an adaptive process, a process of observation, learning, and adaptation passed down from generation to generation. Therefore, the research approach to climate change and traditional knowledge requires two conceptual shifts: first, regarding the positioning of traditional knowledge. The conflict between traditional knowledge and scientific knowledge raises the crucial question of the authority of knowledge. According to the Western positivist tradition, there is only one kind of science: Western science. Any knowledge and perspectives originating outside this Western scientific system are difficult to accept and may even be rejected. However, traditional knowledge is formed within local cultural and social contexts, drawing on specific local environments, practices, and experiences—precisely lacking in scientific knowledge and technology. Therefore, research on traditional knowledge requires researchers to rethink

Western scientific definitions of knowledge and approach traditional knowledge with a serious academic attitude.

Second, regarding the relationship between traditional knowledge and scientific knowledge. Traditional knowledge and scientific knowledge are not mutually exclusive; rather, they possess a potentially complementary relationship. Traditional knowledge can broaden and enrich research on climate change across time and space. Traditional knowledge can make valuable contributions to a better understanding of climate change and provide local information often overlooked by scientific research, enabling empirical research on climate change. For example, scientific research on climate change is macroscopic and large-scale, unable to address specific localities like the Tibetan communities in Deqin County. Traditional knowledge, however, can provide local insights.

2. Community action research focused on solving practical problems

My work methods within the Deqin Tibetan community have evolved through three phases: participant observation, participatory research, and community-based action research. First, I used participant observation, a traditional method of anthropological research. However, as my understanding of local Tibetan traditional knowledge and my evolving understanding of multiculturalism and knowledge systems deepened, I began to reflect on traditional field research methods, such as participant observation. These Western-based approaches seemed to sever the connection between local communities and their own histories, environments, social relationships, and languages, as well as their ways of thinking and experiencing the world. Against this backdrop, I began to draw on and apply participatory research methodologies. Through the development of participatory research methods, the local Tibetan community members became not just my research subjects but my research collaborators. Participatory research methods include traditional methods such as participant observation, daily journaling, and in-depth interviews, as well as group discussions, participatory rural assessments, rapid rural assessments, and action research. The use of participatory research methods allows for a more objective and impartial treatment of traditional and scientific knowledge, thereby fostering a collaborative research model for me and the local Tibetan villagers. At the same time, I believe we should develop a research approach focused on helping communities address the practical challenges of climate change. Rather than using traditional survey research methods to "grab" local knowledge, we should share knowledge with local Tibetan villagers, fostering a platform for dialogue and exchange between scientific and traditional knowledge. This, in turn, should help them address climate change.

However, in participatory research, the role of local Tibetan villagers remains that of participants, often relying on external research or project implementation conducted by me. This passive participation fails to fully leverage the villagers' leadership and active role. Therefore, the extent to which this research approach truly reflects the challenges facing local Tibetan communities in addressing climate change is questionable.

Because research methods ultimately boil down to the role, status, and function of community residents, my research on the Deqin Tibetan community builds on previous research methods and explores the use of community-based research to stimulate the initiative and enthusiasm of community residents. This approach, coupled with action-oriented research focused on specific issues, aims to help local Tibetan villagers address climate change. This research approach comprises two parts:

1. Community-Based Research. While anthropological approaches to climate change focus on the relationship between climate change and culture, they have historically neglected the local villagers' own reflections and understandings on this issue, their traditional knowledge and innovations, and the adaptation measures they have adopted based on this knowledge. This neglect has resulted in anthropological research methods lacking perspectives and approaches from the perspectives of traditional ethnic societies. There-

fore, I would like to adopt an approach to climate change research based on the worldview and values of the Deqin Tibetan community. Through exploration and application within the Deqin Tibetan community, I believe that community-based research should involve local villagers in reflecting on and learning about climate change based on their own values and culture. Villagers should identify their own key challenges and threats and formulate them as research questions. Based on this process of innovation and application of traditional knowledge, they should enhance their ability to cope with and adapt to the impacts of climate change. Furthermore, community-based research is an empowering mechanism, transforming the role of villagers. Local villagers are not only actively involved in the research process, but also serve as the leaders, with me as a facilitator. Not only do I have to assist the local Tibetan community in conducting research on climate change response, but I also have to play the role of a mediator and use the research results to influence relevant stakeholders such as the government and scientific research departments.

2. Action-based research. Action-based research explores concrete practices that are oriented toward solving practical problems and are based on the integration of knowledge and practice. I believe that "objective" research cannot be conducted solely through fieldwork. When I entered the Deqin Tibetan community, I became a part of the community, facing the impacts of climate change alongside the villagers. Therefore, I have a responsibility to use action-based research to help local villagers cope with climate change. As climate change intensifies, the Deqin Tibetan community also needs to take practical measures based on local conditions. At the same time, the community-based research method has also transformed my role. I am not only a participant-observer in the Deqin Tibetan community, but also, to a certain extent, a participant in it. I have become more deeply integrated into the local community, allowing me to understand and study deeper issues that traditional research methods such as participant observation cannot reach.

Therefore, community research oriented toward solving practical problems has changed my research methods: first, the Deqin Tibetan community can identify the challenges it faces and select and determine the research issues through community-based research, rather than anthropologists coming to the community with questions and hypotheses as in traditional research methods; second, the roles of the Deqin Tibetan community and me in the research process have changed. The community villagers have transformed from being the subjects of research to being the leaders of research, and I have transformed from being a participant-observer to being a participant in the research; third, the research method that combines knowledge and practice can not only promote the practical application of research results, but also test and develop new research theories and methods in practice.

2 Review

Due to the severe environmental changes resulting from climate change, such as sea-level rise, prolonged droughts, and increased frequency of flooding in many parts of the world, adaptation has become the primary strategy for coping with climate change to mitigate its impacts on life, property, and health. There is growing recognition of the need for society to adapt to the effects of climate change (IPCC, 2012). Adaptation is a crucial component of addressing the climate crisis, especially given the current rates of emission reductions (Campbell, 2009; Petheram et al., 2010; Pielke et al., 2007).

Climate change is a complex and contentious issue, and adapting to it presents even greater challenges. Addressing the physical complexities and social controversies associated with climate change is particularly daunting for indigenous peoples and local communities. These groups, who have contributed the least to greenhouse gas emissions, are disproportionately affected by the impacts of climate change. They are more vulnerable to its effects due to their residence in geographically sensitive areas and their heavy reliance on natural resources for their livelihoods. Vulnerability to climate change is not solely determined by geography or dependence on natural resources; it also involves social, economic, and political factors that shape how different groups are impacted (Action Aid, 2005). Indigenous peoples and local communities, who are already grappling with poverty and climate-related challenges, may find themselves unable to cope or even survive as climate change intensifies. It is essential to support these communities in their adaptation efforts (Reid et al., 2009a, b). However, until recently, most adaptation efforts have focused on national planning and top-down approaches based on climate change modeling, with little attention given to communities' experiences of climate change, their coping strategies, and their long-standing management of climate variability and extremes.

In response to these challenges, community-based approaches have emerged as an effective means for indigenous peoples to adapt to climate change, while also serving as a valuable resource for policymakers, researchers, and practitioners working in the climate change adaptation field (Barnett & Campbell, 2010). Communities are not merely passive observers; they actively adapt to the new conditions brought about by changing climatic circumstances (Salick & Byg, 2007; Macchi, 2008). Indigenous peoples are coping with biodiversity loss and adapting to climate change through various strategies, including migration, irrigation, water conservation techniques, land reclamation, adjusting the timing, location, and elevation of crop cultivation, and diversifying livelihoods (Macchi et al., 2008). They have developed numerous techniques using traditional ecological knowledge and customs to "evolve fine-tuned social systems" that enable them to cope with the challenges of natural hazards (Byg & Salick, 2009; Xu & Rana, 2005).

This chapter provides an academic literature review on community-based adaptation and climate change. It begins by reviewing key concepts and discussing the relationships between community-based adaptation and traditional knowledge, as well as between community-based adaptation and ecosystems. Finally, the chapter proposes that community-based adaptation should be integrated with scientific knowledge and policy frameworks to enhance its effectiveness and broader applicability in addressing climate change.

3 Methodology

The literature review method involved three major steps, utilizing the search terms "community-based", "adaptation", and "climate change" across various websites. First, the search targeted policy and technical reports from international, multilateral, and bilateral development organizations and agencies. This included relevant United Nations (UN) agencies, the IPCC, international environmental organizations, adaptation funds, bilateral and multilateral aid agencies, and development institutes. Second, the search focused on publications within the social sciences, acknowledging the increasing contributions of social scientists to this field. Third, the search concentrated on case study reports by NGOs and community-based organizations. The search process varied depending on the layout of each website. In some cases, rather than relying solely on search terms, links were followed to identify relevant projects. Many searches also led to other portals and databases through these links.

Based on the initial literature review, a systematic process was undertaken to identify key concepts and themes related to community-based adaptation and climate change over time. The literature was categorized into five thematic areas for further analysis, with each category comprising approximately 30 articles and reports. The first category, "community-based adaptation to climate change", primarily addresses the concept of community-based adaptation and its relationship with climate change. The second category, "action research for climate change adaptation", focuses on the relationship between community-based action research and climate change adaptation. The third category, "community-based adaptation and traditional knowledge", explores the role of traditional knowledge in the adaptation process. The fourth category, "community-based adaptation and ecosystems", examines the relationship between ecosystem-based adaptation and community-based adaptation. The fifth category, "community-based adaptation and policy", discusses how to integrate community-based adaptation into policy frameworks. This literature review simultaneously shaped the framework and main content of this chapter.

4 Key Concepts

4.1 Adaptation

The general definition of climate change adaptation necessitates further elaboration, particularly within the context of conservation, as many strategies classified as adaptation closely resemble established conservation practices. From an ecological perspective, "adaptation" refers to changes in an organism's behavior, physiology, or other characteristics that enhance its survival in a new environment. From an evolutionary perspective, adaptation involves the development of novel traits and genetic changes driven by natural selection. Evidence of adaptation includes shifts in the timing of life cycle events (phenology) and changes in range or habitat usage, suggesting that some species are already responding to ongoing environmental changes. However, the current pace of climate change is extraordinarily rapid from an evolutionary standpoint, likely surpassing the capacity of many organisms to adapt in the traditional sense. Furthermore, other human-induced stressors have further diminished or even eliminated their ability to adapt. Consequently, within the context of natural systems, climate change adaptation has evolved into a management strategy that focuses on identifying, preparing for, and responding to anticipated climate changes to promote ecological resilience, maintain ecological function, and support biodiversity and sustainable ecosystem services (Glick et al., 2009).

Under the 1992 United Nations Framework Convention on Climate Change (UNFCCC), the foundational international treaty addressing global warming, both mitigation (i.e., the reduction of greenhouse gas emissions) and adaptation were initially prioritized. However, due to delays and disputes over mitigation targets and the steps required to transition to a low-carbon future, adaptation gained prominence on the international climate change agenda by the early 2000s (McNamara & Buggy, 2017). It has long been acknowledged that "the threat of global warming calls for a new paradigm of resource planning, one that enhances rather than replaces traditional planning approaches based on empirical analysis, economic efficiency, and environmental protection" (Riebsame, 1990). This perspective was reinforced by the release of the Intergovernmental Panel on Climate Change (IPCC) Fourth Assessment Report in 2007, which convincingly argued that mitigation alone would not suffice to prevent all climate change impacts (Huq et al., 2013; Ott et al., 2008).

In the climate change literature, the term "adaptation" more precisely refers to "managed adaptation to climate change" (CCSP, 2008a; Adger et al., 2007; Heinz Center, 2008). The IPCC's Fourth Assessment Report defines adaptation as "initiatives and measures designed to reduce the vulnerability of natural and human systems to actual or expected climate change effects" (IPCC, 2007). By 2014, adaptation was further defined as "the process of adjustment to actual or expected climate and its effects. In human systems, adaptation aims to moderate or avoid harm or to exploit beneficial opportunities" (IPCC, 2014). Adaptation responses can be categorized as anticipatory, occurring before climate change impacts are observed (proactive adaptation); autonomous, arising spontaneously without deliberate intent (reactive adaptation); or planned, resulting from a deliberate decision to address imminent or experienced climatic changes (IPCC, 2014). Since 2007, adaptation has been a central element in international negotiations under the UN Framework Convention on Climate Change, with the Paris Agreement elevating adaptation to a level comparable to mitigation. The significance of adaptation in international and regional climate discussions is underscored by the Paris Agreement, ratified in November 2016, which emphasizes adaptation in addressing climate change, particularly the urgent and immediate needs of countries in the Global South (United Nations, 2015). The Paris Agreement further solidified this commitment by establishing a global goal for adaptation focused on enhancing adaptive capacity, strengthening resilience, and reducing vulnerability to climate change (UNFCCC, 2015, Article 7). This is significant as the Paris Agreement is widely recognized as a pivotal moment for global cooperation in addressing climate change, with 183 parties having ratified the agreement at the time of writing (United Nations, 2018). Most national climate plans submitted before Paris—known as Nationally Determined Contributions (NDCs)—include not only strategies for emission reductions but also detailed adaptation goals, priorities, actions, and financial needs. These objectives are reflected in other international frameworks, such as the 2030 UN Sustainable Development Goals, the Strategic Plan for Biodiversity 2011–2020, and the Sendai Framework for Disaster Risk Reduction (2015–2030).

Adaptation is described in terms of vulnerability reduction, absorptive capacity, and adaptive capacity. These terms represent different ways adaptation efforts can address climate change challenges: reducing vulnerability to specific impacts, enhancing the capacity to cope with a range of threats, and increasing the ability to respond to observed or anticipated climate changes (Ensor & Berger, 2009). Initially, climate change adaptation emphasized conventional, incremental approaches focused on biophysical vulnerability, infrastructure-based responses, and top-down management strategies. Today, adaptation approaches also incorporate social and economic risk drivers, along with other factors not directly related to climate. Additionally, adaptation is recognized as a process aimed at addressing vulnerability by supporting livelihood resilience, preserving ecosystem integrity and services, and building capacity for the most vulnerable populations (Romy, 2017). Contemporary adaptation strategies promote the integration of broader development goals, aiming to better address the complex interdependencies between human societies and their environments. Emerging approaches also emphasize empowering local communities and supporting bottom-up, participatory decision-making and planning within a robust institutional framework.

4.2 Adaptive Capacity

The Intergovernmental Panel on Climate Change (IPCC) defines adaptive capacity as "the property of a system to adjust its characteristics or behavior in order to expand its coping range under existing climate variability or future climate conditions" (Brooks et al., 2005). However, adaptive capacity must also account for political, cultural, and socio-economic factors that can either facilitate or hinder adaptation efforts for individuals and communities.

Adaptive capacity enhances resilience by enabling individuals and communities to modify their behavior and environment to manage and capitalize on changing climatic conditions (Ford et al., 2006). It extends the concept of adaptation beyond merely reducing vulnerability to hazards and disaster preparedness, encompassing an ongoing process in which communities can make informed decisions about their lives and livelihoods in a changing climate. Adaptive capacity is thus understood in terms of the processes that are necessary for communities to effectively respond to emerging environmental changes (Ensor, 2011).

While adaptation can address immediate climate-related problems, enhancing the adaptive capacity of local communities requires adopting a forward-looking approach to adaptation. This involves not only reacting to current challenges but also anticipating and preparing for future changes to ensure long-term resilience.

4.3 Community-Based Adaptation

Community-based adaptation to climate change is a process led by communities themselves, grounded in their priorities, needs, knowledge, and capacities. This approach empowers communities to plan for and cope with the impacts of climate change (Reid et al., 2009a, b). The rise in community-based adaptation stems from the recognition by the international community that traditional top-down initiatives often fail to deliver effective and sustained benefits to vulnerable populations (Dean et al., 2017; McNamara & Buggy, 2017). The process may begin by identifying vulnerable communities in developing countries or by communities seeking assistance on their own (Reid et al., 2009a, b). Community-based adaptation offers a cost-effective and viable approach to addressing climate change by leveraging the extensive knowledge and experience communities have in managing climate variability and change. It builds on human rights-based development approaches, targeting the most vulnerable individuals and ensuring their full participation at all levels of adaptation planning and implementation.

These initiatives aim to address the root causes of vulnerability within communities (Ayers & Forsyth, 2009) while capitalizing on the extensive knowledge and experience that communities have in managing the localized impacts of climate change (Forsyth, 2013; Reid, 2016). Rooted in the expressed needs and perceptions of communities, the focus is on poverty reduction and livelihood improvements while also mitigating vulnerability to climate change and disasters (Reid et al., 2009a, b). Community-based adaptation interventions are designed to assist communities in adapting to climate change and must address impacts that go beyond regular climate variability, whether predicted or experienced (Reid and Schipper, 2014).

Building upon prior work with communities, community-based adaptation incorporates the concepts of risk and uncertainty into its activities, with the goal of enhancing resilience to both current climate variability and future climate change (Reid et al., 2009a, b). Managing uncertainty is a critical component, involving the development of capacity to make decisions that minimize risks to livelihoods and assets from extreme events while also enabling adaptation to ongoing changes. The approach empowers communities to make informed decisions about their livelihoods in a changing climate, addressing underlying social, cultural, and political factors that may impede this process (Berger & Ensor, 2014). Efforts include enhancing livelihood resilience through income diversification and agricultural technologies, disaster risk reduction, strengthening the capacity of government institutions and local civil society, and engaging in advocacy and social mobilization (Girot et al., 2012).

Access to information and knowledge is another crucial element of community-based adaptation (Berger & Ensor, 2014). Initiatives seek to address the potential impacts of climate change on livelihoods and disaster vulnerability by integrating both local and scientific knowledge. This includes incorporating climate information into planning processes, such as long-term predictions from climate models, seasonal forecasts, and trends from nearby weather stations, along with local knowledge of observed trends and past strategies employed by communities to cope with similar shocks or gradual climatic shifts (Reid et al., 2009a, b).

Finally, community-based adaptation must enable communities and individuals to effectively organize for participation in local decision-making processes and to mobilize against factors that impede their ability to adapt to climate change (Berger & Ensor, 2014). The ultimate goal is to mainstream community-based adaptation within policy frameworks and political agendas.

5 Discussion

5.1 Action Research for Community-Based Adaptation

5.1.1 Action Research and Adaptation

Action research is a cyclical process of inquiry designed to describe, interpret, and explain social situations while implementing interventions aimed at improvement and engagement. It is inherently problem-focused, context-specific, and future-oriented. As a collaborative endeavor grounded in clear values, action research involves a partnership between researchers and participants, all of whom are actively engaged in the change process. This participatory approach is both educational and empowering, characterized by a dynamic interplay where problem identification, planning, action, and evaluation are interconnected (Waterman et al., 2001). The dual aims of action research are to address the practical concerns of individuals in challenging situations while simultaneously advancing social science goals (Gilmore et al., 1986). Broadly defined, it is a research methodology where researchers engage with real-world situations to both improve these situations and acquire knowledge (Checkland & Holwell, 1998). Action research approaches emphasize developing and testing theories within practice contexts through intervention experiments.

In the context of climate change adaptation governance, action research faces several challenges and limitations, as with any research method. However, it plays several crucial roles: First, action research facilitates the translation of scientific knowledge for practitioners and policymakers. Second, it contributes to organizing collaborative governance processes and integrating relevant actors. Third, collaborative action research fosters reflection and learning among practitioners, including policymakers and stakeholders. Finally, it aids policymakers in developing concrete strategies and arrangements to address climate change (Van Buuren et al., 2015).

5.1.2 Community-Based Action Research and Adaptation

Action research is a primary method for achieving community-based adaptation, particularly through community-based action research. Community-based action research is considered an equitable research approach operationalized within a social justice framework. It is often viewed as a continuum of research methodologies, ranging from traditional action research to participatory action research (Wilson, 2019).

Community-based action research involves a collaborative approach where all stakeholders participate throughout the research process, from formulating research questions to developing data collection tools, and from analyzing data to disseminating findings. This framework aims to address the practical concerns of community members while fundamentally altering the traditional roles of researchers and participants. It engages those most affected by community issues, often in partnership with individuals possessing research expertise, to conduct research, analyze issues, and develop strategies for resolution. Essentially, community-based action research integrates academic and professional research with research conducted by community members. This ensures that research results are both generated by and directly benefit the people who are most in need and best positioned to utilize them.

Community-based action research operates as a partnership model, involving community members, organizational representatives, researchers, and other stakeholders in all stages of the research process. All partners contribute their expertise and share in decision-making and ownership. The goal of community-based action research is to enhance understanding of specific phenomena and to integrate this knowledge into interventions that drive policy or social change for the benefit of the community (Israel et al., 1998).

For indigenous peoples and local communities, community-based action research can play a crucial role in adapting to climate change at local, national, and international levels.

First, community-based action research serves as a form of empowerment for local people. The right to conduct research extends beyond scholars and scientists to include members of indigenous peoples and local communities. These individuals, particularly local experts, possess valuable insights into climate change adaptation and other issues, even if their research methods differ from traditional scientific approaches.

Second, at the local level, community-based action research can develop more effective climate change adaptation programs. This type of research focuses on indigenous peoples and local communities who have lived in their regions for generations and possess a deep understanding of the challenges and threats posed by climate change. By emphasizing practical, community-based solutions, these programs often prove more effective than externally led initiatives.

Third, community-based action research also serves as a process of capacity building for local people. It does not exclude scientific research but rather incorporates scientific concepts and methods. This integration allows indigenous peoples and local communities to benefit from scientific knowledge and methodologies while developing and implementing climate change adaptation strategies, thereby enhancing their ability to cope with climate change.

Fourth, at the national and international levels, community-based action research contributes to policy advocacy. It provides indigenous peoples and local communities with a platform to express their perspectives, needs, and capabilities in climate change adaptation. Rather than passively enduring the impacts of climate change, these communities actively engage in adaptation measures. The adaptation plans developed through community-based action research offer valuable local insights and experiences for broader climate change strategies. Moreover, by involving policymakers, researchers, and practitioners, community-based action research aims to influence the formulation and implementation of adaptation policies at both national and international levels.

Community-based action research serves as a vital bridge, connecting indigenous peoples and local communities with scientists and policymakers, allowing for mutual understanding and exchange of knowledge on climate change adaptation from diverse perspectives.

One primary function of community-based action research is to enhance knowledge dissemination and the valorization of traditional knowledge. This process not only promotes the recognition and application of indigenous knowledge but also facilitates reflection and learning among policymakers and scientists. By increasing awareness of climate change impacts and adaptation needs at the local level, community-based action research provides valuable insights into alternative strategies and potential tools for addressing climate change at both national and international levels.

Conducting community-based action research in the realm of climate change adaptation serves both practical and scientific objectives. For instance, it can enhance our understanding of decision-makers' behaviors and implicit action frameworks or investigate which governance instruments and approaches might be most effective for implementing adaptation measures.

Community-based action research thus deepens our comprehension of the complex processes involved in climate change adaptation. The close involvement of community members in practice provides scientists and policymakers with insights that might be missed by traditional research methods. The specific practices and data generated through this research can lead to innovative ideas about how climate change adaptation governance functions. These findings can be directly communicated and discussed with scientists and policymakers, thereby aiding in the evaluation of their credibility and applicability for policy development.

5.2 Traditional Knowledge for Community-Based Adaptation

Traditional knowledge refers to the accumulated wisdom and practices that have been passed down through generations, guiding human societies in their interactions with their environment. This body of knowledge is often termed "traditional ecological knowledge" and is defined as "a cumulative body of knowledge, practice, and belief, evolving through adaptive processes and transmitted across generations by cultural means, concerning the relationships between living beings (including humans) and their environment" (Berkes, 2012). Such knowledge systems are perpetuated and renewed by each successive generation, ensuring the well-being of communities globally by providing food security through activities like hunting, fishing, gathering, pastoralism, or small-scale agriculture. Additionally, they contribute to health care, clothing, shelter, and strategies for managing environmental fluctuations and external forces of change (Nakashima & Roué, 2002; Sillitoe, 2007; Sillitoe et al., 2002; Warren et al., 1995).

Various terms are used in the literature to describe this knowledge, including indigenous knowledge, traditional knowledge, traditional ecological knowledge, local knowledge, farmers' knowledge, folk knowledge, and indigenous science. While each term may have specific connotations and reference groups, they often share sufficient overlap in meaning to be used interchangeably (Nakashima & Roué, 2002). In this context, the term "traditional knowledge" will be employed most frequently, as the examples discussed primarily relate to knowledge developed and maintained by indigenous peoples and local communities.

The essence of traditional knowledge lies not in its specific informational content, but in the enduring values of patience, persistence, calmness, respect for elders, and respect for the environment. These values enable indigenous peoples and local communities to remain resourceful and resilient in a changing world, including in the face of climate change (Ford et al., 2007). Consequently, the dynamism of traditional knowledge and its inherent capacity to adapt to climate change must be recognized. Acknowledgment of traditional knowledge has only recently emerged in climate sciences, particularly in understanding climate change adaptation. Traditional knowledge is a dynamic system of understanding, continually questioned and renewed. Through learning-by-doing, experimentation, and knowledge building (Berkes, 2012), knowledge holders can adjust and modify their actions to adapt to climate change.

The value of traditional knowledge has been affirmed in light of the scientific uncertainty surrounding the impacts of climate change, particularly at the local level (Barnett, 2001; Ensor & Berger, 2009). Many communities are already experiencing climate-related impacts and are taking necessary actions at the local level to address these challenges (Rojas Blanco, 2006). As a result, traditional knowledge has been recognized as an integral component in informing future adaptation efforts within the context of climate change (Berkes & Jolly, 2001; Riedlinger & Berkes, 2001). Traditional knowledge can thus provide key insights into the adaptability and resilience of communities within the local context of livelihoods and community life. Indeed, traditional knowledge in natural resource management, familiarity with home territories, and basic skills for safety and survival will become increasingly essential in adapting to climate change.

Recognizing the role of traditional knowledge in climate change adaptation is an important first step. Interdisciplinary collaboration and the development of new methodologies are necessary to achieve a comprehensive and mutually respectful understanding of traditional knowledge related to climate change adaptation. However, basing climate change decision-making and actions on the best available knowledge, both scientific and traditional, presents additional challenges. At times, scientific observations and traditional knowledge may align; in other instances, they may differ but remain complementary; and at still other times, they may be entirely contradictory.

Community-based adaptation seeks to account for the potential impacts of climate change by integrating both traditional and scientific knowledge about climate change and its likely effects (Berger & Ensor, 2014). This approach requires the incorporation of scientific knowledge with traditional knowledge, particularly regarding the trends and changes experienced by communities at the local level, as well as the strategies these communities have historically employed to cope with similar shocks or gradual climatic changes.

Community-based adaptation emphasizes the importance of traditional knowledge and experience as a basis for developing adaptation strategies. It highlights how traditional knowledge can be integrated with scientific knowledge, and the crucial role this integration plays in addressing uncertainty. Traditional knowledge, particularly when combined with scientific knowledge, can help fill information gaps, enabling communities to take action that will facilitate adaptation without unintended consequences.

Knowledge, both scientific and traditional, plays a crucial role in climate change adaptation. Given the complexities involved, there is a pressing need for scientific knowledge, particularly that which can be directly applied to governance practices (Van Buuren et al., 2015). In other words, effective adaptation to climate change requires a collaboration between scientific and traditional knowledge to tackle specific challenges. This collaboration produces knowledge that is both applicable and actionable, which is essential for adapting to climate change. For indigenous peoples and local communities, such knowledge is especially critical in the process of community-based adaptation.

5.3 Ecosystem for Community-Based Adaptation

Ecosystems are intimately linked to climate change. On one hand, the impacts of climate change have affected ecosystems, leading to observable changes in species' abundance and distribution, as well as shifts in ecosystem boundaries, reproductive cycles, and growing seasons. Alterations in the complex interactions among species are modifying the composition of plant and animal communities, with some species already having gone extinct (Booth, 2012; McCarthy, 2012; Reid, 2006; Thomas et al., 2004). The Intergovernmental Panel on Climate Change (IPCC) anticipates further changes to biodiversity and natural systems as climate change progresses (Fischlin et al., 2007). On the other hand, biodiversity and natural resources play a crucial role in helping people adapt to the adverse impacts of climate change. Well-managed, stable, and diverse ecosystems provide significant adaptation benefits (Reid et al., 2009a, b).

For indigenous peoples and local communities, they are on the frontline of a changing climate and are most vulnerable to the impacts of climate change. They live in vulnerable areas and are reliant on natural resources for their subsistence, well-being, and livelihoods. At the same time, they have been using genetic diversity and traditional knowledge about native species to adapt to climate variability for generations. This approach is proving increasingly valuable in the context of climate change adaptation. Sustainable management of these natural resources and the maintenance of genetic, species, and ecosystem diversity will therefore play a key role in helping them adapt to future climate change impacts (Reid, 2011). Well-managed, stable, and diverse ecosystems also provide adaptation benefits (Reid et al., 2009a, b). Consequently, ecosystem-based adaptation has received increasing interest among indigenous peoples and local communities in recent years. There are important differences between community-based adaptation and ecosystem-based adaptation, but both share the goal of increasing the ability of indigenous peoples and local communities to adapt to climate change.

The United Nations Convention on Biological Diversity (CBD) defines ecosystem-based approaches for adaptation as "the use of biodiversity and ecosystem services as part of an overall adaptation strategy to help people adapt to the adverse effects of climate change" (CBD, 2009). This definition was later elaborated by the CBD to include "sustainable management, conservation, and restoration of ecosystems, as part of an overall adaptation strategy that takes into account the multiple social, economic, and cultural co-benefits for local communities" (CBD, 2010). This definition makes it very clear that human adaptation is at the center of ecosystem-based adaptation, which is common to community-based adaptation (Reid, 2014). Community-based adaptation and ecosystem-based adaptation work toward addressing the shortcomings of the mainstream, top-down, hard infrastructure-based approach to adaptation, seeking a balanced and integrated framework to reflect local conditions and community priorities (Girot et al., 2012).

Community-based adaptation focuses on empowering local communities to reduce their own vulnerabilities. It puts community members at the center of adaptation planning, enabling them to identify their own needs, set priorities, and find solutions based on their own knowledge and capacities. A key focus here is to empower communities to adapt. Community-based adaptation needs contextualizing in the wider landscape or ecosystem if it is to be effective (McCarthy, 2012). While ecosystem-based adaptation focuses on harnessing ecosystems to provide goods and services in the face of climate change, it emphasizes protecting and strengthening biodiversity and ecosystem services as a key adaptation strategy. The emphasis is thus on ensuring the sustainable management, conservation, and restoration of ecosystems that people depend on. There is, however, substantial conceptual overlap between community-based adaptation and ecosystem-based adaptation, and in practice, they are often indistinguishable from one another. Good ecosystem-based adaptation approaches complement community-based adaptation as they are people-centered, have a strong community focus, and are participatory and culturally appropriate. Genuine community-based adaptation should also consider ecosystems and natural solutions. In practice, local adaptation activities tend to combine both approaches. The two approaches are increasingly being combined within climate change adaptation policy, planning, and implementation, building on the strengths of both approaches to address the shortcomings of mainstream, top-down strategies.

Under ecosystem-based adaptation, indigenous peoples and local communities can contribute valuable traditional knowledge and practices for adaptation, based on their past experience of coping with climate variability, which can be applied—with full participation and free, prior, and informed consent—to current adaptation with good effect. By incorporating elements of ecosystem-based adaptation, community-based adaptation can learn how to better respect ecological complexity; incorporate ecosystem goods and services in community-led adaptation strategies; adapt management systems to ecosystem or landscape scales; build environmental integrity into monitoring, evaluation, and reporting systems; and ensure that interventions build rather than undermine natural capital. By reducing environmental degradation, it is also possible to minimize some of the indirect negative socio-economic impacts of climate change (Girot et al., 2012). Together, ecosystem- and community-based solutions to climate adaptation and vulnerability equip decision-makers with new choices and options, allowing them to plan for an uncertain future. If their core principles are respected, these approaches, when combined, will promote resilient ecosystems using nature-based solutions to provide benefits to people, especially indigenous peoples and local communities. Community-based adaptation and ecosystem-based adaptation encourage participatory, transparent, accountable, and culturally appropriate solutions to climate adaptation, while actively supporting indigenous peoples and local communities.

6 Process

Our approach to community-based action research on climate change response in the Tibetan community of Deqin County in the Eastern Himalayas of China involves the following steps:

6.1 Community Meetings

Our project team and villagers jointly held community meetings to introduce the concepts and methods of community-based research and advocacy (Fig. 1). I believe that community-based action research on climate change is a process of capacity building for local people. The villagers and I learned through practice. We then helped the villagers analyze the challenges and crises brought about by climate change, listed these crises and challenges, and ranked and scored them, identifying the issues that concerned them most. The community then selected and decided on the action research topics based on these issues. For example, in the "Traditional Semi-Agricultural and Semi-Pastoral Livelihoods of Tibetans in Northwestern Yunnan, China: Climate Change Response Project," we held a community meeting at the beginning of the project. Semi-agricultural and semi-pastoral livelihoods mean that the villagers are both herders and farmers, but with gender differences. Therefore, when organizing discussions at the community meetings, we paid special attention to gender perspectives and organized group discussions. First, we invited herders who frequently graze in the alpine pastures, typically male villagers. Second, we invited farmers who frequently work in the fields, typically female villagers. In the

Fig. 1 Convening community meetings (photoed by Lurong Sinan)

first group discussion, herders identified the challenges and crises posed by climate change to animal husbandry as: premature grass maturation, disrupted pastoral calendars, drought, pests and diseases, increased mosquito populations, decreased nutritional quality of pasture, and heavy snowstorms. In the second group discussion, farmers identified the main impacts of climate change on agriculture as: drought, heavy rainfall, snowstorms, insect pests, disrupted agricultural calendars, decreased crop quality, premature crop maturation, and floods. After the two group discussions, we combined the two groups and held a village meeting. At the meeting, each villager was given five pebbles and asked to rank and rate the crises, challenges, and impacts of climate change listed in their group discussions. This ranking process led to the selection of the top six issues that the villagers considered most serious and urgent, which needed to be addressed: 1. drought, 2. premature grass and crop maturation, 3. pests and diseases, 4. decreased nutritional quality of pasture, 5. floods, and 6. decreased crop quality. These six issues have become the priority issues that we need to help local Tibetan villagers solve in our project "Traditional Tibetan Semi-agricultural and Semi-pastoral Livelihoods in Northwestern Yunnan, China to Cope with Climate Change".

6.2 Establishing a Climate Action Research Team

After determining the action research topic, it's necessary to identify and select individuals to conduct the research. Local experts are undoubtedly the best candidates for community researchers, as they possess accumulated and mastered traditional knowledge specific to the action research topic and often possess extensive experience. Once these local experts are selected as researchers, they will recruit villagers who share a common interest in the topic to form a climate action research team (Fig. 2).

Each action research team typically consists of approximately ten villagers, led by one or two local experts. For example, in the "Conservation Planting of Tibetan Medicinal Plants in the Eastern Himalayas in the Context of Climate Change" project, after our project team identified "Adapting Traditional Knowledge of Tibetan Medicinal Plant Diversity to Climate Change" as the action research theme, we quickly

Fig. 2 Establishing a climate action research team (photoed by Liqing Pingzu)

began investigating and identifying villagers who could lead the project. Through this investigation, we identified three types of local experts among the villagers: First, Tibetan doctors, who had studied traditional Tibetan medicine knowledge, particularly knowledge of medicinal plants, in schools or monasteries, and some continued to practice Tibetan medicine after returning to their villages. Second, village doctors, who had no formal Tibetan medicine education but had learned knowledge of medicine and medicinal plants through family traditions and continued to practice in their villages. Third, herbalists, who had neither formal Tibetan medicine training nor family traditions, had acquired knowledge of medicinal plants through practical experience and accumulation. These three types of local experts possessed distinct and unique traditional knowledge of Tibetan medicinal plants. Consequently, the project team established three climate action research groups based on the specific characteristics of their traditional knowledge of medicinal plants, with each group leader serving as the group leader. Each group then recruited twelve villagers as members based on their traditional knowledge, characteristics, interests, and hobbies.

6.3 Developing a Work Plan

After each climate action research team was established, our project staff assisted them in developing a work schedule and plan for the action research, defining the framework and specific methods. For example, in the "Climate Change Impacts on Pasture Resources in Northwest Yunnan and Community Responses Project", with the assistance of our project team members, the villagers of the Traditional Pasture Action Research Group scheduled their surveys for spring, summer, and autumn based on the pasture growth phase. They also selected survey sites at high, medium, and low altitudes based on the geographic location of pasture growth. They further divided each altitude into shaded and sunny areas based on sunlight exposure and, finally, divided each area into dryland and wetland based on water resource distribution. Once the survey framework was established, the villagers conducted surveys on the impacts of climate change on pasture resources based on season, altitude, sunlight, and water availability.

6.4 Documenting Climate Change

Our project supports each climate action research team by distributing cameras, video cameras, voice recorders, and notebooks to document various climate change phenomena and their impacts on the local natural environment and social life during their fieldwork. In the "Northwest Yunnan Small Watershed Comprehensive Management and Climate Change Response Project", three climate action research teams in the upstream, midstream, and downstream regions used drawing, photography, video recording, audio recording, and notetaking to document the impacts of climate change. The upstream climate action team used drawings and other methods to depict the impacts of climate change on alpine lakes, vegetation, animals, glaciers, and snow in the river's source areas; the midstream climate action team used photography and other methods to document the impacts of climate change on wetlands, forests, springs, alpine meadows, and flora and fauna in the midstream region; and the downstream climate action team used video recording and other methods to document the impacts of climate change and extreme climate phenomena, including landslides, mudslides, floods, droughts, heavy rains, and blizzards. The three teams also used audio recording and notetaking during interviews with elderly villagers in the villages.

6.5 Investigating Traditional Knowledge

After documenting climate change phenomena and their impacts, each climate action research team began using their own research methods to continue organizing and investigating traditional knowledge on related topics, using methods such as drawing, Tibetan writing, oral history, collecting, and traditional plant specimen preparation. For example, in the "Conservation Planting Project for Traditional Tibetan Medicine in Northwest Yunnan under the Context of Climate Change", the climate action research team first mapped representative medicinal plants at different altitudes. They then interviewed elderly villagers, particularly older local experts, and recorded traditional knowledge about these plants in Tibetan, including information about their growing environments and medicinal properties. Following these interviews, the climate action team, based on the traditional knowledge captured in their drawings and written records, conducted field surveys at various altitudes, collecting medicinal plant samples and preparing herbaria using traditional methods. Through drawing based on their own ethnic worldviews, oral history in their own languages, local field surveys, and traditional herbarium preparation methods, the climate action team investigated their own traditional medicinal plant knowledge.

6.6 Investigating Traditional Organizations

In addition to investigating traditional knowledge, we should also be proactive in identifying and leveraging traditional organizations within local Tibetan communities, as these organizations themselves are part of traditional knowledge. These traditional organizations play a vital role in maintaining the relationship between local communities and nature and can also play a crucial role in current efforts to address climate change. For example, in the "Tibetan Women's Climate Change Response Project in the Eastern Himalayas", the Climate Action Research Team first conducted research on two traditional community organizations: the "Sisterhood" and the "Arrow Society". The "Sisterhood" is composed of female villagers, while the "Arrow Society" is composed of male villagers. These two traditional organizations have a long history in the community. Historically, they played an administrative role in managing the local community and continue to play a significant role in village public affairs. However, with most male villagers now working away from home, the role of the "Arrow Society" has gradually declined. In contrast, with the majority of female villagers remaining in the village, the role of the "Sisterhood" has not weakened but has instead grown. Based on this research, the Climate Action Research Team decided to conduct action research on climate change response based on the Tibetan women's organization "Sisterhood". Through capacity building and training, the "Sisterhood" was developed into an indigenous organization for climate change response, thereby further transforming the externally dominated concept and action of "climate change response" into a conscious concept and action within the community.

6.7 Investigating Traditional Mechanisms

When investigating traditional organizations, our project team discovered specific ways in which traditional organizations function within communities—in other words, traditional mechanisms. For example, in the "Tibetan Water Culture in Northwest Yunnan to Response to Climate Change Project", the Climate Action research team, while investigating local Tibetan water culture, discovered a "two-tiered system" for managing water resources within local villages. Villagers divided the village into upper and lower parts. This division wasn't based on geographic location or social status, but rather on the families descended from two important serf owners and their serfs. In other words, all families descended from serf owner A and his serfs were considered "upper village", while all families descended from serf owner B and his serfs were considered "lower village". As the village environment changes, even if families from the "upper village" relocate to the lower part of the village, they are still considered "upper village" families and do not become "lower-village" families simply because of the change in location, and vice versa. Regarding water resource management, each year the village selects three families from each of the "upper-village" and "lower-village" districts. These six families are jointly responsible for allocating water resources for the entire village that year, including drinking water and irrigation water. This "upper-lower division system" for water resource management is particularly effective during dry seasons or droughts. This traditional mechanism allocates water resources between the "upper-village" and "lower-village" districts based on river flows at different times of the day, week, and month, thus avoiding conflicts between villagers caused by water shortages. The Climate Action Research Team subsequently discovered that the "upper-lower division system" plays a significant role not only in the resilience of water resources but also in the resilience of natural resources such as forests, pastures, and farmland to climate change and its associated disasters. Therefore, the Climate Action Research Team decided to leverage this traditional mechanism to address climate change and its associated disasters.

6.8 Innovation of Traditional Knowledge

Based on traditional knowledge, our project team worked with the Climate Action Research Team to discuss and assist community villagers in applying and innovating traditional knowledge to address climate change. For example, in the "Tibetan Women's Climate Change Response Project in the Eastern Himalayas", we assisted the Climate Action Research Team in implementing climate change response actions from a gender perspective based on the traditional knowledge of Tibetan women villagers about forests. The Climate Action Research Team researched and organized women's traditional knowledge about forests, encompassing three broad categories: traditional culture of forest worship, traditional techniques for maintaining forest ecosystems, and traditional wisdom for restoring forest vegetation. In the traditional culture of forest worship, local Tibetan women worship a sacred mountain, the "Miancimu Goddess Mountain", and a sacred forest, the "Bunanzong Goddess Forest". They voluntarily protect the sacred forest on the sacred mountain and strictly prohibit the felling of trees. Building on this traditional culture, we assisted the Climate Action Research Team in establishing the "Miancimu Climate Action Carbon Neutral Forest". In traditional techniques for maintaining forest ecosystems, local Tibetan women have a biosphere reserve for saplings. They use fences and tree maintenance techniques. They surround seedlings with thorny brambles to prevent livestock from gnawing. They also cut back branches at the base of each tree to ensure that nutrients are concentrated in the crown. These techniques ensure the growth of each tree and further maintain the overall forest ecosystem. Following traditional forest restoration practices, local Tibetan women plant trees using local species on barren hills and slopes. They first survey the forested hills for the presence of native tree species. The survey criteria for tree species growth are based on high, medium, and low altitudes, with each altitude divided into shady and sunny slopes, and each slope divided into dry land and wetland. Based on these criteria, the Tibetan women selected eight different tree species for seedling cultivation and then planted trees in the corresponding environments on the treeless hills. This action of restoring forest vegetation embodies the wisdom of local Tibetan women. First, it avoids biological invasion caused by planting foreign tree species. Second, it avoids the monoculture of a single tree species. Third, through the mixing of different tree species, it maximizes the restoration of the future forest layer and biodiversity.

6.9 Innovation of Traditional Organizations

Building on the traditional organizations of the "Sisterhood" and "Arrow Society", our project team and the Climate Action Research Group worked together to assist villagers in establishing modern community management organizations to address climate change. For example, in the "Northwest Yunnan Tibetan Community Integrated Watershed Management Project to Address Climate Change", the Climate Action Research Group conducted research and capacity building with the "Arrow Society", a traditional Tibetan male organization. Focusing on integrated watershed management, a topic of particular interest to male villagers, they established the "Kawagbo Small Watershed Integrated Management Association" within the Arrow Society. In the "Eastern Himalayan Tibetan Women's Climate Change Response Project", the Climate Action Research Group conducted capacity building and training with the "Sisterhood", a traditional Tibetan female organization. Focusing on forest protection and management, a topic of particular interest to female villagers, they established the "Mentsunmo Climate Action SDG Laboratory" within the Sisterhood. The "Kawagebo Small Watershed Comprehensive Management Association" established based on the "Arrow Club" responded to the concerns of male villagers about semi-agriculture and semi-pastoralism. The association led the villagers to carry out action research on how semi-agriculture and semi-pastoralism can cope with climate change; the "Miancimu Climate Action Research Laboratory" established based on the "Sisterhood" responded to the concerns of female villagers about the matsutake mushroom gathering industry. The center led the villagers to carry out action research on maintaining forest ecosystems and responding to climate change.

6.10 Innovation of Traditional Mechanisms

Based on the traditional "upper-lower two-point system" (since the "upper-lower two-point system"), we assisted the Climate Action Research Group in exploring the establishment of community-based mechanisms for responding to climate change and its associated disasters. For example, in the "Tibetan Traditional Knowledge and Climate Disaster Risk Reduction Project in Deqin County, Yunnan Province", the Climate Action Research Group implemented climate disaster risk reduction initiatives that leveraged traditional mechanisms. This traditional mechanism, rooted in the local environment and society, is capable of responding to frequent climate-related disasters, such as droughts, heavy rains, snowstorms, floods, mudslides, and landslides, as well as their secondary hazards and risks. Historically, the "upper-lower two-point system" has served as a de facto disaster response mechanism for local Tibetan villagers. In some major disasters, the "upper-lower two-point system" has proven to be the most effective way to protect lives and property. It plays a crucial role in all three phases of climate disaster management: pre-disaster warning and preparedness, disaster relief and risk reduction, and post-disaster recovery and response. At the same time, the "upper-lower" system has developed both long-term and short-term strategies and actions for reducing climate disaster risks. First, the "upper-lower" system effectively organizes villagers through a dual mechanism of "upper village" and "lower village" to carry out pre-disaster prevention, mutual assistance during disasters, and resource allocation for post-disaster relief, including tree planting, building reservoirs and irrigation systems, and restoring vegetation around the village. This system facilitates long-term management and prevention of climate disaster risks. Second, the "upper-lower" system diversifies livelihoods by organizing "upper-village" and "lower-village" villagers to receive training in modern science and technology tailored to their specific social and natural environments, thereby more effectively reducing climate disaster risks. For example, "upper-village" villagers use renewable energy, install greenhouses, and grow cash crops, while "lower-village" villagers upgrade yak breeds, establish matsutake mushroom sales cooperatives, and build irrigation canals. The innovation and application of the "upper-lower" system has enhanced villagers' resilience to climate disaster risks, minimized the negative impacts and harms of climate disasters on local communities, and become one of the most effective mechanisms for establishing mutual assistance, prevention, and governance mechanisms for climate disasters. In climate disaster risk reduction strategies and actions based on a "top-down system", traditional mechanisms need to be integrated with science and technology to achieve innovation and development, making climate disaster risk reduction more sustainable and effective. By integrating traditional mechanisms with science and technology, local communities can more effectively utilize local natural resources and ecosystems, thereby reducing the threats and challenges posed by climate disaster risks to traditional livelihoods and lives.

6.11 Promoting the Exchange of Traditional and Modern Knowledge

Project staff and the community climate action research team hold regular monthly working meetings to discuss details. Each meeting is attended by at least ten community researchers. They discuss research issues with project staff, and project staff also provide feedback and feedback on the community researchers' methods and findings. This feedback and feedback serves to affirm the work of the climate action research team and community researchers, encourage further research, and assist them in resolving challenges encountered during the research process. For example, in the "Northwest Yunnan Tibetan Traditional Houses Adapt to Climate Change Project", our project staff hold monthly working meetings with traditional Tibetan architects and carpenters from the climate action research team to discuss the risks and challenges of traditional houses facing natural disasters such as heavy rain, drought, and floods. They also assess measures implemented by local architects and carpenters to enhance the resilience of traditional houses to disasters. These measures include constructing triangular bases at the four corners of the houses, increasing the thickness and width of the stone foundations, improving the composition of the earth used for roofs and walls, and refining the wood used in various building components. By recognizing this traditional knowledge, local architects and carpenters are further encouraged to conduct action research to address climate change. The project encourages and supports exchanges and collaborations between local experts and researchers, and provides training to local experts, encouraging them to innovate and further their research based on traditional knowledge and the integration of modern technologies. For example, in the "Northwest Yunnan Tibetan Traditional

Houses Response to Climate Change Project", project staff assisted local architects and carpenters from the Climate Action Research Group in researching new building materials, exploring how traditional Tibetan houses could respond to climate change by integrating traditional building techniques with modern materials. Local architects exchanged ideas with Italian architects, collaborating on the design of bathrooms, toilets, and fecal disposal systems. Architects from the China Academy of Art provided drafting and design training to local architects. Landscape design faculty from Southwest Forestry University provided training to local architects on village landscape layout design.

6.12 Implementing Climate Change Mitigation Actions

Various terrestrial ecosystems and natural resources are important carbon sinks. In Deqin County, terrestrial ecosystems that absorb and store significant amounts of carbon dioxide include forests, meadows, watersheds, and lake wetland ecosystems. Local Tibetan villagers have a close relationship with the surrounding terrestrial ecosystems and natural resources. On the one hand, these terrestrial ecosystems and natural resources provide the material foundation for local villagers' production and livelihoods. On the other hand, through their long-term use of natural resources, local villagers have developed traditional techniques and gradually developed sustainable traditional natural resource utilization and conservation techniques. These traditional natural resource utilization and conservation techniques can objectively enhance the local terrestrial ecosystem's ability to absorb carbon dioxide and increase its carbon sink capacity, thereby contributing to climate change mitigation. Based on this foundation, our project team is assisting the climate action research team in exploring actions to mitigate climate change through ecosystem maintenance (Fig. 3). For example, in the "Tibetan Traditional Forest Knowledge for Climate Change Response Project in Deqin County, Yunnan Province",

Fig. 3 Implementing climate change mitigation actions (photoed by A Lu)

the Climate Action Research Team supported the village's traditional men's organization, the "Arrow Society", to maintain forest ecosystems through a moderate resource utilization system called forest rotation. The "Arrow Society" divided the collective forest, where logging is permitted, into six sections. Each year, the Society rotated through the six sections, one through six, for timber for building and firewood. This cycle repeated itself every six years, and after each cycle, logging resumed from the first section. This prevented concentrated logging in a single area, which could lead to over-logging and consequently damage forest resources and ecosystems. In addition to the forest rotation system, the Climate Action Research Team also assisted villagers in implementing rotational grazing and pasture restoration to maintain alpine meadow ecosystems, as well as integrated watershed management initiatives such as afforestation and vegetation restoration to protect watershed ecosystems. These actions objectively protected the forest, alpine meadow, and watershed ecosystems, as well as their ability to absorb carbon dioxide.

6.13 Implementing Climate Change Adaptation Actions

Traditional livelihoods are dynamic and experience-based, constantly adapting to ongoing environmental and social changes. They are also actively adapting to the impacts of climate change today. For a long time, the Tibetan villagers of Deqin have practiced a semi-agricultural and semi-pastoral lifestyle, developing traditional knowledge. This knowledge enables local villagers to identify and adjust grazing times, planting schedules, crop varieties, livestock breeds, and related technologies, as well as select genetic resources from different crops, to adapt to the impacts of climate change, including changes in soil quality, temperature, slope, orientation, and sunlight. Furthermore, this traditional knowledge related to plant genetic resources can diversify agricultural and animal husbandry livelihoods, helping villagers adapt more effectively to climate change. Our project team is assisting the Climate Action Research Group in exploring actions to adapt to climate change through the adjustment of livelihoods (Fig. 4).

Fig. 4 Implementing climate change adaptation actions (photoed by Kelake)

For example, in the "Adapting Traditional Tibetan Semi-Agricultural and Semi-Pastoral Livelihoods to Climate Change Project in Northwestern Yunnan, China", the Climate Action Research Team supported villagers in selecting genetic resources from local traditional barley varieties and adjusting their farming practices to adapt to climate change. Regarding the selection of genetic resources for traditional barley varieties, the Climate Action Research Team encouraged seed exchange between villages with different climate conditions, altitudes, and regions. Villages at higher altitudes preferred cold-resistant barley varieties, while those in the dry, hot valleys at lower altitudes preferred drought-resistant varieties. Regarding farming adjustments, the Climate Action Research Team assisted villagers in developing mixed and intercropping practices, planting corn, wheat, and beans together on the same plot. This means that even if one crop fails due to climate change, other crops can still be harvested. This agricultural genetic diversity not only ensures food security but also provides a food security insurance policy against climate change. Furthermore, the project team, through the Climate Action Research Team, installed solar water heaters, irrigation systems, greenhouses, and built reservoirs and canals, integrating modern technology with traditional knowledge to improve livelihood security and adapt to climate change.

6.14 Evaluation by All Villagers

Community villagers conduct a final review of the Climate Action Research Team's findings (Fig. 5). At certain stages in the project, the project team convenes a village meeting for the community to conduct a final review and review of the Climate Action Research Team's findings. For example, in the "Integrated Small Watershed Management Project for Tibetan Communities in Northwest Yunnan to Address Climate Change", we convened a village meeting to invite all villagers to review and review the project's findings. These achievements included building flood control embankments along the river using traditional techniques, constructing triangular stone pestles in the center of the river to block flood-

Fig. 5 Evaluation by all villagers (photoed by A Zhu)

waters, stabilizing landslides along the riverbanks, restoring native tree species and vegetation throughout the watershed, maintaining the ecological environment of water sources, protecting springs, and protecting the environment of alpine lakes. All villagers scored and ranked these findings, and the Climate Action Research Team reflected on any findings with lower scores, which will become key areas of focus for the next phase of the project.

6.15 Advocacy and Impact

The promotion and replicability of community-driven action research concepts and methods are key indicators of our project's success. Therefore, our project team has begun applying the same community-driven action research concepts and methods to the Nagani Indigenous Peoples community in Kenya, supporting local communities in implementing climate change mitigation and adaptation initiatives based on traditional knowledge. For example, this project leverages the traditional techniques of local rainmakers and traditional beliefs about sacred forests to mitigate and adapt to climate change. This project is currently ongoing. Our project has also attracted the attention of Native Americans, leading to our project team members being invited to attend the first and second National Tribal and Indigenous Climate Conferences in 2020 and 2022, respectively. At both conferences, our team highlighted community-driven action research conducted by Tibetans in northwestern Yunnan, China, using traditional knowledge and a gender perspective to address climate change, achieving widespread impact. In 2019, our project team's "Tibetan Traditional Knowledge for Climate Change in Eastern Himalayas China" project was selected from 180 projects worldwide and showcased at the Second Paris Peace Forum. In 2020, our project team's "Tibetan Women in Southwest China Response to Climate Change Project" was again selected from 200 projects worldwide and showcased at the Third Paris Peace Forum. At the Second Paris Peace Forum, UN Secretary-General Antonio Guterres highlighted our project, acknowledging its innovative spirit despite being based on traditional knowledge. In 2024, our project team's "Tibetan Women in Southwest China Response to Climate Change Project" was again selected from 170 projects worldwide to receive the Central Asia Economic Cooperation Organization's "Climate Gender Award." These advocacy initiatives demonstrate the value and significance of our project's community-based action research on climate change.

From the perspective of community-based action research methods, if the transition from participant observation to participatory research is a shift in villagers from passive acceptance to active participation, then the shift from participatory research to community-based research is more than a simple role shift; more importantly, it involves a shift in ownership. In other words, in a community-based model, villagers are not only participants but also leaders and decision-makers. Specifically, the shift from village participation to village leadership means that research and advocacy are no longer activities conducted by scholars or non-governmental organizations, but rather the responsibility and business of the villagers themselves.

7 Reflection

It's often assumed that research is the prerogative of academic institutions and research departments. Consequently, scholars and professionals often serve as researchers, leaving local people, the primary holders of information, to become mere subjects of research or mere information providers. However, due to generations of local residence and life experience, these individuals often possess a wealth of experience and local knowledge. Should this experience and knowledge simply be provided as "raw materials" to outsiders, or can it be further processed into "products", empowering locals to become "producers" of knowledge innovation? Community-based research offers a path and model for exploring this question.

First, locality. Unlike previous research models, community-based research recognizes that local people possess a wealth of regional knowledge and information. This knowledge may not be rigorously scientific, but it is generated within the unique natural and cultural environments of their respective regions and reflects local realities. Second, it recognizes the capabilities of local people, assuming they are capable of conducting independent research, rather than merely acting as passive information providers. Third, it emphasizes the role of local people, allowing them to conduct research as active researchers. Fourth, the purpose of research is not to produce academic papers or reports, but to foster local reflection and awareness of change and development, laying the foundation for sustainable, integrated development that prioritizes culture, the environment, and the economy.

Second, comprehensiveness. Community-based research focuses on the perspectives, attitudes, and opinions of local communities, supporting and assisting them in designing action research projects on issues of interest, thereby facilitating a more local perspective on these issues. Community-based research also seeks to engage local people from diverse backgrounds in the research process, drawing on their respective backgrounds and knowledge to reflect local changes and challenges from diverse perspectives and angles. Therefore, through collaborative research, community-based research fosters understanding, dialogue, and exchange, exploring a path for sustainable, integrated development.

Therefore, community-based integrative research focuses on process rather than simply outcomes. This research is an empowering process, empowering local people to conduct research on their ancestral homeland; it is a capacity-building process, helping them understand and master relevant research and analytical methods and approaches; it is a reflective process, encouraging reflection and self-awareness on the issues being studied; and it is a cumulative process, providing theoretical guidance for sustainable development. The interplay between process and outcomes is the perspective and hallmark of community-based integrative research.

In the context of climate change, anthropologists face the impacts of climate change alongside local villagers during their research. Therefore, anthropological field methods are not simply investigations conducted to produce reports or publications, but rather, they should be about mutual learning and influence between anthropologists and community members, a means of communicating, transforming, and developing diverse knowledge systems, and a practical approach to assisting traditional ethnic societies in coping with climate change. Community-based action research methods have two key implications:

First, interdisciplinary, integrated research methods centered on traditional knowledge facilitate the transformation of diverse knowledge systems. Traditional anthropological research methods often only consider the perspective of the investigator, ignoring or failing to address the needs of the communities being surveyed. Anthropologists' fieldwork is essentially a process of seeking knowledge from local communities and conducting research based on this knowledge. In addition to considering issues from the perspective of local people and achieving a mutual transformation between "other-perspective" and "self-perspective", it is crucial to return the knowledge gained through surveys and research to the community. This feedback fosters a mechanism for integrating scientific and traditional knowledge and a knowledge transfer system.

Second, community-based research methods, oriented toward solving practical problems, highlight the dominant position and role of traditional ethnic societies. Currently, anthropological research on climate change should prioritize traditional ethnic societies' own reflections and understandings of climate change, incorporating local perspectives and perspectives into research methods. Community-based action research is based on the culture and values of local traditional ethnic societies and studies the impact of climate change on lifestyle, environment, and culture. Through this research method, anthropologists emphasize the community's independent thinking and practice based on this thinking, helping traditional ethnic societies explore models for responding to climate change that are consistent with their own cultural values.

In summary, anthropological research methods are not only disciplinary but also carry a social responsibility. Faced with the ecological and traditional cultural changes brought about by climate change, anthropologists cannot simply adopt an "objective" or even "indifferent" attitude toward the realities of poverty, loss of traditional culture, and ecological degradation in their field research sites. Admittedly, anthropologists in the traditional sense are merely learners and interpreters of "other" cultures. They cannot impose their own will and choices on "others", nor can they direct the cultural change and social development of "others". However, in today's climate change context, anthropologists should shoulder fundamental social responsibilities in response to the diverse challenges faced by traditional ethnic societies and actively participate in their efforts to address climate change.

Both historically and currently, climate change has a significant and direct relationship with traditional ethnic societies. Climate change has significant impacts on the ecological environment, ethnic culture, lifestyle, and spirituality. Community-based action research on climate change responses can delve deeper into these relationships and impacts. The fundamental principle of this research approach must be to focus on local people and emphasize the interactive relationship between climate change and traditional knowledge. Therefore, community-based action research should be given priority.

First, anthropology recognizes the inextricable relationship between climate change, culture, the environment, and society. Research on ecological and cultural policies in ethnic regions should focus on people, not on objects and events, ignoring the subject of interaction with climate change and simply examining the impact of climate change on the ecosystem. Such research will clearly fail to yield comprehensive and accurate results.

Second, culture is the bond between people and nature, and between people and climate. The relationship between local ethnic groups and climate change that anthropology studies is actually expressed through the lens of social culture. The process by which climate change impacts the ecological environment and traditional culture is precisely the process of interaction and mutual influence between human culture and the natural ecological environment. Third, community-based action research advocates for the borrowing, integration, and transformation of diverse knowledge systems, including social and natural science knowledge, scientific knowledge, and traditional knowledge. Furthermore, community-based action research addresses practical problems in specific contexts, emphasizing the integration of knowledge and practice.

Anthropological research methodology cannot be limited to theoretical discussions or mere theoretical discussions in the study room. Most importantly, it must be rooted in the field, actively engaged in action, and strive for theoretical and practical innovation to explore methods and approaches for knowledge transformation and practical application, thereby promoting the adaptation of traditional ethnic societies to climate change. Therefore, community-based action research is more than a simple disciplinary survey method; it carries the ideals and responsibilities of anthropologists.

8 Conclusion

Due to historical and other reasons, indigenous peoples and local communities often reside in remote geographical areas characterized by fragile ecosystems. They occupy a weakened social status in political and economic dimensions, and their traditional livelihoods are heavily reliant on local natural resources, rendering them particularly vulnerable to the impacts of climate change. However, international, national, and regional adaptation plans frequently overlook their voices, needs, and rights, depriving them of due and sufficient consideration in these relevant adaptation strategies. As a result, these top-down adaptation initiatives are often ineffective for these communities. On the contrary, some adaptation strategies may even exacerbate their already fragile situation, making them more vulnerable.

In response to the impacts of climate change, indigenous peoples and local communities are not passive or inactive. On the contrary, their traditional knowledge, experiences, practices, and even beliefs regarding ecosystems, biodiversity, and climate can serve as potential tools for mitigating and adapting to the threats posed by climate change. However, because traditional knowledge and scientific knowledge belong to different epistemological systems, and much of traditional knowledge is descriptive in nature, it poses challenges for quantitative research and analysis. Consequently, scientific research and models, along with policymaking and adaptation programs based on these studies, often neglect traditional knowledge. Some even dismiss traditional knowledge as "unscientific", deeming it irrelevant to the "scientific" study of climate change and, thus, unworthy of consideration in policymaking and adaptation programs.

To address these challenges, it is imperative to develop adaptation plans and strategies that are led by indigenous peoples and local communities—a concept known as community-based adaptation. Within the current overarching framework and context of adaptation, community-based adaptation enables indigenous peoples and local communities to adapt to the impacts of climate change and enhances their resilience. Thus, community-based adaptation is not merely a framework but also a process for improving the adaptability of indigenous peoples and local communities.

Firstly, community-based action research forms the foundation of community-based adaptation. This approach empowers indigenous peoples and local communities by granting them the agency to study their ecosystems, biodiversity resources, and the impacts of weather and climate. More importantly, it facilitates the exploration of traditional knowledge related to these areas—knowledge that has been accumulated over generations. Community-based action research is not just a research process but also an empowerment process. It strengthens the research capabilities of indigenous peoples and local communities and enhances their awareness of rights, laying the groundwork for a rights-oriented approach to community-based adaptation.

Furthermore, traditional knowledge is central to community-based adaptation. It is the most effective and low-cost tool available to indigenous peoples and local communities for coping with climate change. Whether for mitigation or adaptation, traditional knowledge plays a crucial role. Some scholars and policymakers argue that "tradition" cannot adapt to "modern" contexts and that traditional knowledge is outdated. They contend that, given the unprecedented speed of current climate change, traditional knowledge is no longer sufficient. This perspective is flawed, as it views traditional knowledge as a static and unchanging system. In reality, traditional knowledge has been dynamically evolving through generations of indigenous peoples and local communities. It adapts to ecosystem changes and integrates external knowledge to meet new environmental challenges. Currently, what is needed is a dialogue, exchange, and integration between traditional and scientific knowledge systems to create applicable and actionable knowledge, which is the core of action-oriented community-based adaptation.

Moreover, ecosystems are crucial to the success of community-based adaptation. Ecosystems provide essential natural resources that enable indigenous peoples and local communities to cope with climate change. A healthy environment and sustainable resource management can significantly enhance their resilience and adaptability. Therefore, in community-based adaptation, it is essential to prioritize the local ecosystem and biodiversity to effectively integrate community-based adaptation with ecosystem-based adaptation. Although some environmental practitioners believe that the key to ecosystem-based adaptation lies primarily in ecosystems and the environment, it is evident from the definition of ecosystem-based adaptation that human well-being is at the center, aligning it with the principles of community-based adaptation. The integration of community-based and ecosystem-based adaptation offers more effective strategies for indigenous peoples and local communities, providing them with beneficial solutions and offering new perspectives for policymaking. Thus, ecosystems are essential to the success of community-based adaptation.

Finally, for indigenous peoples and local communities, the ultimate goal of community-based adaptation is not only to implement effective actions to adapt to climate change at the local level in various regions but also to integrate these efforts into national adaptation plans and actively participate in the negotiation processes of international conventions and climate change frameworks. This inclusion allows indigenous peoples and local communities to engage in climate change policymaking at all levels, thereby better protecting their rights and interests.

References

Action Aid International. (2005). Participatory vulnerability analysis: a step-by-step guide for field staff, Action Aid International.

Ayers, J., & Forsyth, T. (2009). Community-based adaptation to climate change. *Environment: Science and Policy for Sustainable Development, 51*(4), 22–31. https://doi.org/10.3200/ENV.51.4.22-31s

Barnett, J. (2001). Adapting to climate change in Pacific Island Countries: The problem of uncertainty. *World Development, 29*(6), 977–993.

Barnett, J., & Campbell, J. (2010). *Climate change and small island states: Power, knowledge and the South Pacific*. Earthscan.

Berger, R., & Ensor, J. (2014). Introduction: Progress in adaptation. In R. Berger, J. Ensor, & S. Huq (Eds.), *Community-based adaptation to climate change: Emerging lessons* (pp. 1–13). Practical Action Publisher.

Berkes, F., & Jolly, D. (2001). Adapting to climate change: Social-ecological resilience in a Canadian Western Arctic Community. *Conservation Ecology, 5*(2), 18–33.

Berkes, F. (2012). *Sacred ecology* (3rd ed.). Routledge.

Booth, T. H. (2012). Biodiversity and climate change adaptation in Australia: Strategy and research developments. *Advances in Climate Change Research, 3*(1), 12–21.

Brooks, N., Adger, W. N., Kelly, P. M. (2005). The determinants of vulnerability and adaptive capacity at the national level and the implications for adaptation. *Global Environmental Change, 15*(2), 151–163.

Byg, A., & Salick, J. (2009). Local perspectives on a global phenomenon—climate change in Eastern Tibetan Villages. *Global Environmental Change, 19*, 156–166.

Campbell, J. (2009). Islandness. *Shima: The International Journal of Research into Island Cultures, 3*, 85–97.

Convention on Biological Diversity. (2009). Connecting biodiversity and climate change mitigation and adaptation: Report of the second ad hoc technical expert group on biodiversity and climate change. CBD Technical Series No. 41. Secretariat of the Convention on Biological Diversity, Montreal, Canada.

CCSP. (2008a). Preliminary review of adaptation options for climate-sensitive ecosystems and resources. a report by the U.S. climate change science program and the subcommittee on global change research. In S. H. Julius & J.M. West (Eds.), J. S. Baron, L. A. Joyce, P. Kareiva, B. D. Keller, M. A. Palmer, C. H. Peterson, & J. M. Scott. U.S. Environmental Protection Agency, Washington, D.C., USA, p. 873.

Checkland, P., & Holwell, S. (1998). Action research: Its nature and validity. *Systemic Practice and Action Research, 11*(1), 9–21.

Dean, A., Green, D., & Nunn, P. D. (2017). Too much sail for a small craft? Donor requirements, scale, and capacity discourses in Kiribati. In E. Stratford (Ed.), *Island geographies: Essays and conversations* (pp. 67–88). Routledge.

Ensor, J. (2011). *Uncertain futures: Adapting development to a changing climate*. Practical Action Publishing.

Ensor, J., & Berger, R. (2009). *Understanding climate change adaption: Lessons from community-based approaches*. Practical Action Publishing.

Fischlin, A., Midgley, G. F., Price, J. T., Leemans, R., Gopal, B, Turley, C., Rounsevell, M. D. A., Dube, O. P., Tarazona, J. & Velichko, A. A. (2007). Ecosystems, their properties, goods, and services, in M. L. Parry, O. F. Canziani, J. P. Palutikof, P. J. van der Linden, & C. E. Hanson, (Eds.), *Climate Change 2007: Impacts, Adaptation and Vulnerability. Contribution of Working Group H to the Fourth Assessment Report of the Intergovernmental Panel on Climate Change*, Cambridge: Cambridge University Press.

Ford, J. D., Smit, B., & Wandel, J. (2006). Vulnerability to climate change in the Arctic: A case studyfrom Arctic Bay Canada. *Global Environmental Change, 16*(2), 145–160.

Forsyth, T. (2013). Community-based adaptation: A review of past and future challenges. *Wiley Interdisciplinary Reviews: Climate Change, 4*(5), 439–446. https://doi.org/10.1002/wcc.231

Glick, P., Staudt, A., & Stein, B. (2009). A new era for conservation: review of climate change adaptation literature. National Wildlife Federation, Washington, D.C., USA.

Gilmore, T., Krantz, J., & Ramirez, R. (1986). Action based modes of inquiry and the host researcher relationship. *Consultation, 5*(3), 160–176.

Girot, P., Ehrhart, C., & Oglethorpe, J. (2012). Integrating community and ecosystem-based approaches in climate change adaptation responses. http://careclimatechange.org/files/adaptation/ELAN_Integrated Approach_150412.pdf

Huq, S., Roberts, E., & Fenton, A. (2013). Commentary: Loss and damage. *Nature Climate Change, 3*, 947–949.

IPCC. (2007). Summary for policymakers. In M. L. Parry, O. F. Canziani, J. P. Palutikof, P. J. van der Linden, & C. E. Hanson (Eds.), *Climate change 2007: impacts, adaptation and vulnerability. Contribution of working group II to the fourth assessment report of the intergovernmental panel on climate change*. Cambridge: Cambridge University Press, pp. 7–22.

IPCC. (2012). Managing the risks of extreme events and disasters to advance climate change adaptation (SREX). Special Report of the Intergovernmental Panel on Climate Change (IPCC). Geneva: IPCC Secretariat.

IPCC. (2014). Climate change 2014: Impacts, adaptation, and vulnerability. Contribution of working group II to the fifth assessment report of the intergovernmental panel on climate change. In C. B. Field, V. R. Barros, D. J. Dokken, K. J. Mach, M. D. Mastrandrea, T. E. Bilir, M. Chatterjee, K. L. Ebi, Y. O. Estrada, R. C. Genova, B. Girma, E. S. Kissel, A. N. Levy, S. MacCracken, P. R. Mastrandrea, & L. L. White. Cambridge, United Kingdom and New York, NY, USA: Cambridge University Press, vol. 1785, pp. 1–1131.

Israel, B. A., Schulz, A. J., Parker, E. A., & Becker, A. B. (1998). Review of community-based research assessing partnership approaches to improve public health. *Annual Review of Public Health., 19*, 173–202. https://doi.org/10.1146/annurev.publhealth.19.1.173

McCarthy, P. D. (2012). Climate change adaptation for people and nature: a case study from the U.S. southwest. *Advances in Climate Change Research, 3*(1), 22–37.

Macchi, M., Oviedo, G., Gotheil, S., Cross, K., Boedhihartono, A., Wolfangel, C., Howell, M. (2008). Indigenous and traditional peoples and climate change. IUCN Issues.

McNamara, K. E., & Buggy, L. (2017). Community-based climate change adaptation: A review of academic literature. *Local Environment, 22*(4), 443–460. https://doi.org/10.1080/13549839.2016.1216954

Nakashima, D., & Roué, M. (2002). Indigenous knowledge, peoples and sustainable practice. In T. Munn (ed.), *Encyclopedia of global environmental change*. Chichester: Wiley, pp. 314–324.

Ott, H. E., Sterk, W., & Watanabe, R. (2008). The Bali roadmap: New horizons for global climate policy. *Climate Policy, 8*(1), 91–95.

Petheram, L., et al. (2010). Strange changes: Indigenous perspectives of climate change and adaptation in NE Arnhem Land (Australia). *Global Environmental Change, 20*, 681–692.

Pielke, R. A., Jr., et al. (2007). Climate change 2007: Lifting the taboo on adaptation. *Nature, 445*, 597–598.

Reid, H. (2006). Climatic change and biodiversity in Europe. *Conservation and Society, 4*(1), 84–101.

Reid, H., Alam, M., Berger, R., Cannon, T., Huq, S., & Milligan, A. (2009a). Community-based adaptation to climate change: An overview. *Participatory Learning and Action, 60*, 11–33.

Reid, H., Phillips, J., & Heath, M. (2009b). *Natural resilience: Healthy ecosystems as climate shock insurance (IIED Briefing)*. IIED.

Reid, H. (2011). Improving the evidence for ecosystem-based adaption, IIED opinion: Lessons from adaptation in practice series, London: International Institute for Environment and Development (IIED).

Reid, H., & Huq, S. (2014). Community based adaptation to climate change: scaling it up. In E. Lisa, & F. Schipper. Routledge. ProQuest Ebook Central, https://ebookcentral-proquest-com.ezproxy.library.uq.edu.au/lib/uql/detail.action?docID=1600474

Reid, H. (2014). A natural focus for community-based adaptation. In J. Ensor, R. Berger, & S. Huq (Eds.), *Community-based adaptation to climate change: Emerging lessons* (pp. 35–54). Practical Action Publishing.

Reid, H. (2016). Ecosystem- and community-based adaptation: Learning from community-based natural resource management. *Climate and Development, 8*(1), 4–9. https://doi.org/10.1080/17565529.2015.1034233

Riebsame, W. E. (1990). Anthropogenic climate change and a new paradigm of natural resource planning. *Professional Geographer, 42*, 1–12.

Riedlinger, D., & Berkes, F. (2001). Contributions of traditional knowledge to understanding climate change in the Canadian Arctic. *Polar Record, 37*(203), 315–328.

Rojas Blanco, A. V. (2006). Local initiatives and adaptation to climate change. *Disasters, 30*(1), 140–147.

Romy Chevallier. (2017). Integrated community-and ecosystem-based approaches to climate change adaptation. SAIIA Policy Insights No 49.

Salick, J., & Byg, A. (Eds.). (2007). *Indigenous peoples and climate change*. Tyndall Centre.

Sillitoe, P. (Ed.). (2007). *Local science vs. global science: Approaches to indigenous knowledge in international development*. Berghahn Books.

Sillitoe, P., Bicker, A., & Pottier, J. (Eds.). (2002). *Participating in development: Approaches to indigenous knowledge*. Routledge.

Thomas, C. D., Cameron, A., Green, R. E., Bakkenes, M., Beaumont, L. J., Collingham, Y. C., Erasmus, B. F. N., Ferreira de. Siqueira, M., Grainger, A., Hannah, L., Hughes, L., Huntley, B., van Jaarsveld, A. S., Midgley, G. F., Miles, L., Ortega-Huerta, M. A., Peterson, A. T., Phillips, O. L. & Williams, S. E. (2004). Extinction risk from climate change. *Nature, 427*, 145–148.

United Nations Framework Convention on Climate Change (UNFCCC). (2015). Adoption of the Paris Agreement (FCCC/CP/2015/L.9/Rev.1), https://unfccc.int/resource/docs/2015/cop21/eng/l09r01.pdf

United Nations. (2015). Paris Agreement 25, http://unfccc.int/files/essential_background/convention/application/pdf/english_paris_agreement.pdf

United Nations. (2018). Paris Agreement-Status of Ratification, https://unfccc.int/process/the-paris-agreement/status-ofratification

Van Buuren, A., Eshuis, J., & van Vliet, M. (2015). Action research for climate change adaptation: Developing and applying knowledge for governance. Routledge: Abingdon, Oxon, UK; New York, NY, USA.

Warren, D. M., Slikerveer, L. J., & Brokensha, D. (Eds.). (1995). *The cultural dimension of development: indigenous knowledge systems*. Intermediate Technology Publication.

Waterman, H., Tillen, D., Dickson, R., & de Koning, K. (2001). Action research: A systematic review and assessment for guidance. *Health Technology Assessment, 5*(23), 1–166.

Wilson, E. (2019). Community-based participatory action research. In P. Liamputtong (Ed.), *Handbook of research methods in health social sciences.* Springer, Singapore. https://doi.org/10.1007/978-981-10-5251-4_87

Xu, J., & Rana, G. M. (2005). Living in the mountains. In T. Jeggle, (Ed.), *Know risk. U. N. interagency secretariat of the international strategy for disaster reduction* (pp. 196–199). Geneva.

Traditional Knowledge in Water Resource Management for the Climate Change Adaptation

3

1 Introduction

The Himalayas are Asia's water towers and are the source of the continent's largest rivers. These rivers and their tributaries support approximately 1.4 billion people (Immerzeel et al., 2010). Variations in river flow distribution and groundwater recharge over time and space are influenced by changes in temperature, evaporation, and, crucially, precipitation (Chiew, 2007). Some impacts of climate change on hydrological processes have already been observed (Rosenzweig et al., 2007). Rivers in the Himalayas are likely to experience consequences due to climate change in the region (Immerzeel et al., 2010).

Climate change impacts are already occurring in the Himalayas (Beniston, 2003; Cruz et al., 2007). Limited studies on temperature and precipitation in a few localized areas show that warming in the Himalayas is three times greater than the global average. Thus, the Himalayas are among the region's most vulnerable to climate change (J. Xu et al., 2009) and are undergoing rapid environmental change (Bawa et al., 2010). Upstream snow and ice reserves in the Himalayas, which are crucial for sustaining seasonal water availability, are likely to be substantially affected by climate change (Immerzeel et al., 2010). A shift in winter precipitation from snow to rain, as temperatures rise, leads to changes in the timing of streamflow peaks. The spring snowmelt peak is either brought forward or eliminated entirely, and winter flows increase. As glaciers retreat due to warming, river flows initially increase but decline once the glaciers disappear (Z. W. Kundzewicz et al., 2008). Ongoing climate change over the coming decades will likely have additional negative impacts across these mountains, including significant cascading effects on river flows, groundwater recharge, natural hazards, and biodiversity; ecosystem composition, structure, and function; and human livelihoods (Nijssen et al., 2001; Parmesan, 2006; Bates et al., 2008; Ma et al., 2009). Additionally, lakes are forming as glaciers retreat from prominent Little Ice Age moraines in the Himalayas. These lakes thus have a high potential for glacial lake outburst floods (B.C. Bates et al., 2008).

Traditional ecological knowledge held by local communities about climate change can benefit climate science and policy (J. Cruikshank, 1981; Reidlinger & Berkes, 2001; Moller et al., 2004). Such knowledge can contribute to improved understanding of the impacts of climate change and provide insights for the development of equitable and effective climate change adaptation strategies (Leonard et al., 2013). Recent studies on traditional knowledge have examined perceptions of climate change in Tibet, the Eastern Himalayas, and the Western Himalayas (Vedwan & Rhoades, 2001; Salick & Ross, 2009; Byg & Salick, 2009; Sharma et al., 2009; Tse-ring et al., 2010). However, these studies do not examine the impacts on water resources.

Y. Lun, *Climate Change and Traditional Knowledge*, Sustainable Development Goals Series,
https://doi.org/10.1007/978-3-032-12015-1_3

In this article, we report, for the first time, on traditional ecological knowledge about climate change and its consequences for water in the Himalayas. We explore the critical role that traditional ecological knowledge and the culture of water play in community-based water resource management in a Tibetan village in northwestern Yunnan, China, and how this knowledge and culture may influence future decision-making regarding water resources and adaptation to climate change. This paper begins with a brief overview of recent research on traditional ecological knowledge and the culture of water, community-based water resource management, and climate change. We then discuss the historical, geographical, and social context of this research, as well as the methodology employed. Finally, we present our research findings, which are organized into four themes: beliefs, customary law, ecological knowledge, and management techniques.

2 Methodology

The research methodology consisted of a combination of ethnographic research methods and community-based action research applied.

2.1 Ethnographic Research and Participatory Rural Appraisal (PRA)

This research draws on ethnographic data collected in Tibetan villages in Deqin between June 2010 and September 2015. A mixed-methods approach was employed, including semi-structured interviews with individuals and groups, participant observation, oral history recordings, literature reviews, and participatory rural appraisal (PRA).

The research considers traditional knowledge as a lens for observing the influence of climate change on water resources. Therefore, the research questions were not explicitly designed to focus on climate change, but rather on traditional knowledge and its impact on water resources.

Since some traditional knowledge is strongly linked with water resources at the community level, the classification of traditional knowledge about water should be framed within community-based water resource management. Based on the classification system for traditional knowledge categories (Xue & Guo, 2009), and extensive work in investigating, organizing, and documenting traditional knowledge in Tibetan villages in Deqin over the past years, this research divides traditional knowledge about water into four categories:

(1) Knowledge for the use of water resources.
(2) Customary law governing the use of water resources.
(3) Technical innovations for water resource use and traditional farming and lifestyle practices.
(4) Knowledge of traditional cultures related to the conservation and sustainable use of water.

We used PRA to investigate the questions based on the above classification. Considering the location of each village, including altitude, climate conditions, and most importantly water resources, the villagers' answers could be related to differences in watershed location and water resource conditions of each village. Thus, a total of nine villages were grouped into three types, as shown in the flowchart below (Fig. 1).

In each of the nine villages, 30 interviews were conducted, and a questionnaire survey was also carried out. The interviews were with villagers over 40 years old, with a focus on household heads. This age group was selected because traditional knowledge is closely related to one's length of residence in the area, accumulated natural environmental experience, and social and cultural background. Villagers over 40 years of age can provide significant and valuable comparisons between past and present climate and environmental changes, and can also offer insights into the variability of traditional knowledge based on long-term environmental experience and social-cultural background. Additionally, gender equality was considered in selecting respondents to ensure a balanced gender perspective. Interviews were conducted in Tibetan and Chinese with the assistance of interpreters. The interviews focused on

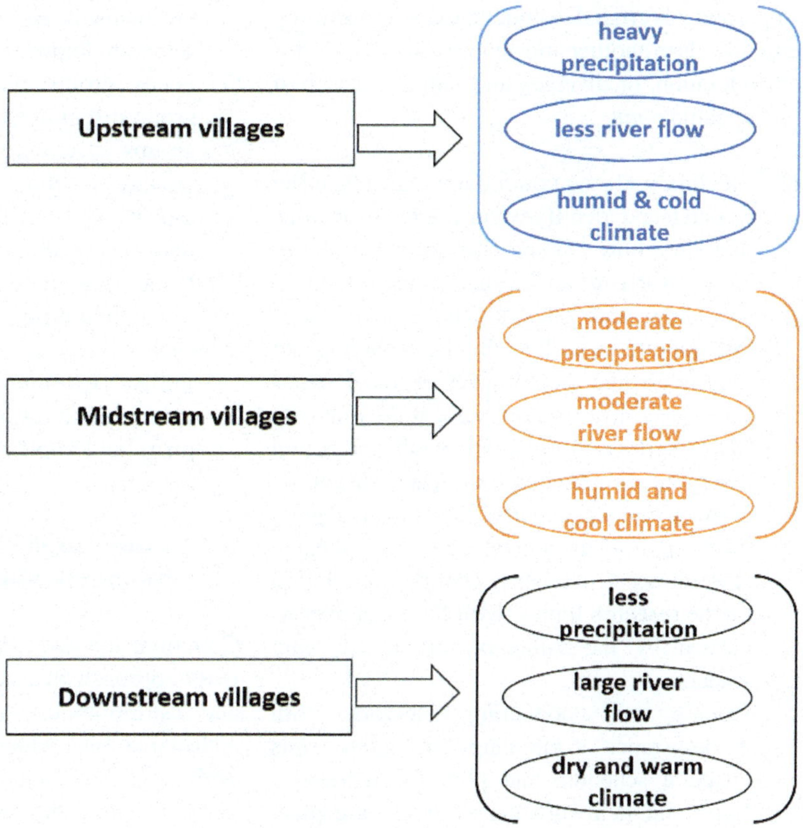

Fig. 1 Simple flowchart illustrating differences in river discharge and associated environmental conditions among upstream, midstream, and downstream villages. The information on precipitation, river flow, and climate conditions is based on surveys of local villagers and reflects their perceptions rather than quantitative measurements

local villagers' traditional knowledge about water, including its association with climate change phenomena (e.g., rain, snow, river flow, and glaciers) and disasters (e.g., floods, avalanches, landslides, and droughts). Villagers were also asked about the reasons for these changes, their impact on water resources and daily lives, and the role of traditional knowledge in adapting to these changes.

2.2 Participatory Action Research

Action science attempts to bridge the gap between social research and social practice by building theories that explain social phenomena, inform practice, and adhere to the fundamental criteria of science (Friedman, 2001). Action research is driven by reflection on interventions aimed at changing specific situations (Argyris et al., 1985) and is particularly suited for initiating and guiding change processes. It is a productive method for fostering more reflective practices within and between organizations through direct collaboration between practitioners and researchers (Duijn et al., 2010). The participatory action research methodology was used in this study because it actively involved villagers in exploring new insights and feasible arrangements for alternative strategies in community-based water resource management.

The participatory action research methodology was designed as an iterative interaction process involving three stakeholders: villagers, local experts, and researchers. Researchers facilitated this iterative interaction among the stakeholders. The participatory action research methodology empowers local villagers and experts to co-design community-based water resource management and supports villagers in formulating local adaptation strategies to climate change. The participatory action research methodology includes eight steps, and it is crucial to understand how each step either leads to another, introduces a new aspect for consideration, or diverges and requires action at that specific point.

(1) Meeting with the community: Introduce the research and identify water-related problems, including how climate change or variability affects water resources such as precipitation, river flow, and glaciers. This also includes indirect effects such as floods, droughts, and landslides, and their impact on local livelihoods and infrastructure. Assess the intensity of these effects through villagers' scoring and determine the research focus accordingly.
(2) Identifying local experts: Select local experts with traditional knowledge and village researchers. Establish research teams based on the research topics, with the local experts elected by the villagers serving as team leaders.
(3) Planning and conducting fieldwork: With the assistance of facilitators, research teams create a schedule and plan for fieldwork. Carry out fieldwork according to the plan, collecting data through photographs, video recordings, notes, and other methods. This data will support each research topic.
(4) Post-fieldwork discussion: After the initial fieldwork stage, convene a meeting to discuss the details. Facilitators organize the meeting with research team leaders, members, and 10 to 15 village representatives to review the fieldwork results, consolidate data, and address any investigation shortcomings.
(5) Drafting and planning: Facilitators prepare a first draft for review by local experts and village researchers, incorporating their comments and revisions. Collaboratively develop a plan for the second stage of fieldwork.
(6) Second stage fieldwork: Collect additional data through in-depth interviews with key stakeholders. Involve researchers from academia and the local government's water conservancy department in the research teams. Work together with local experts and village researchers to create scenarios and mechanisms for integrating traditional and scientific knowledge.
(7) Focus group discussions: Organize focus group discussions with different community groups differentiated by age, gender, and village structure. Document socio-cultural aspects and verify the final results of the fieldwork with the community.
(8) Establishing a water committee: Based on the research findings, set up a water committee in each village, involving various stakeholders. Initiate community-based water resource management at the village level, develop plans, and implement activities.

2.3 Community-Based Water Resource Management

Community-based water resource management is an approach that enables individuals, groups, and institutions to participate in identifying and addressing local water-related issues (Ali, 2011). Led by local communities, community-based water resource management empowers individuals to cope with climatic variability. In this system, local priorities, knowledge, needs, and capacities are crucial for developing adaptation plans (Prabhakar et al., 2010).

Agro-pastoralism is a weather-dependent livelihood, making the Tibetan community in Deqin particularly vulnerable to climate change. Climate change-related water disasters, such as droughts, floods, and landslides, are occurring more frequently and severely affecting the lives of local people. Although villagers, especially local experts with extensive traditional knowledge, possess valuable experience in coping with unexpected climatic events, the accelerating impacts of climate change may exceed their current knowledge and capabilities. For instance, the uncertainty of rainfall affects crop selection, sowing time, irrigation schedules, and harvesting times in agriculture. Droughts also impact the pastoral environment, influencing grass growth timing, grass maturity, grass nutrition, and the disappearance of wetlands in pastoral areas. In this context, community-based water resource management offers an opportunity for integrating traditional knowledge with scientific technology and provides a platform for knowledge exchange among different stakeholders.

As a consequence, it is expected that villagers will choose different approaches to varying extents based on their village's specific conditions, such as altitude, climate, location, and water resources. Moreover, each village may require a tailored approach to water resource management depending on its unique social and environmental circumstances.

To implement more effective water resource management, we have developed three community-based water resource management models based on survey data and the results of the participatory action research in different villages:

(1) Water Environment Conservation Model. This model is primarily based on traditional cultural knowledge related to the conservation and sustainable use of water. Villages located in the upstream areas of a watershed adopt this model to protect the environment of the water source.
(2) Water Resource Allocation Model. This model relies on knowledge about the use of water resources and customary laws governing water use. Villages situated in the midstream of a watershed typically use this model to equitably distribute water resources among different parts of the village and among households.
(3) Water Disaster Risk Reduction Model. This model is based on knowledge of water resource use, technical innovations, and traditional farming and lifestyle practices. Villages located in the downstream areas of a watershed employ this model to reduce the risk of water-related disasters such as droughts, floods, and landslides.

3 Results and Discussion

Based on participatory action research and community-based water resource management, we selected thirty villagers aged forty and above from each of the nine villages in the upstream, midstream, and downstream areas for investigation. The sample comprised half males and half females, and the primary focus was on "The Perspective of Water in Climate Change", encompassing two aspects: "The Perspective of Water Change" and "The Perspective of Water Disasters".

3.1 The Perspective of Water in Climate Changes

3.1.1 The Perspective of Water Change

All the villagers interviewed perceived changes related to water, including rain, snow, river flow, and glacier.

From Fig. 2, we observe the following perceptions among villagers regarding climate-related changes:

- **Upstream Villages:** Villagers are most aware of the reduction in glacier size and its impact on river flow due to glacier melt. However, changes in snow and rainfall are less noticeable to them.
- **Midstream Villages:** Villagers have a clear perception of changes in river flow, snowfall, and rainfall. Many have also observed a decrease in glacier size.
- **Downstream Villages:** Villagers prominently perceive changes in river flow and rainfall, but are less concerned about the reduction in glacier size and changes in snowfall (Fig. 3).

3.1.2 The Perspective of Water Disasters

Also all the villagers interviewed perceived hazards and disasters related to water, including flood, avalanche, landslides, and drought (Fig. 4).

By comparing the villagers' perceptions of disasters, we observe the following:

- **Upstream Villages:** Villagers are particularly sensitive to avalanches but show little awareness of other types of disasters.

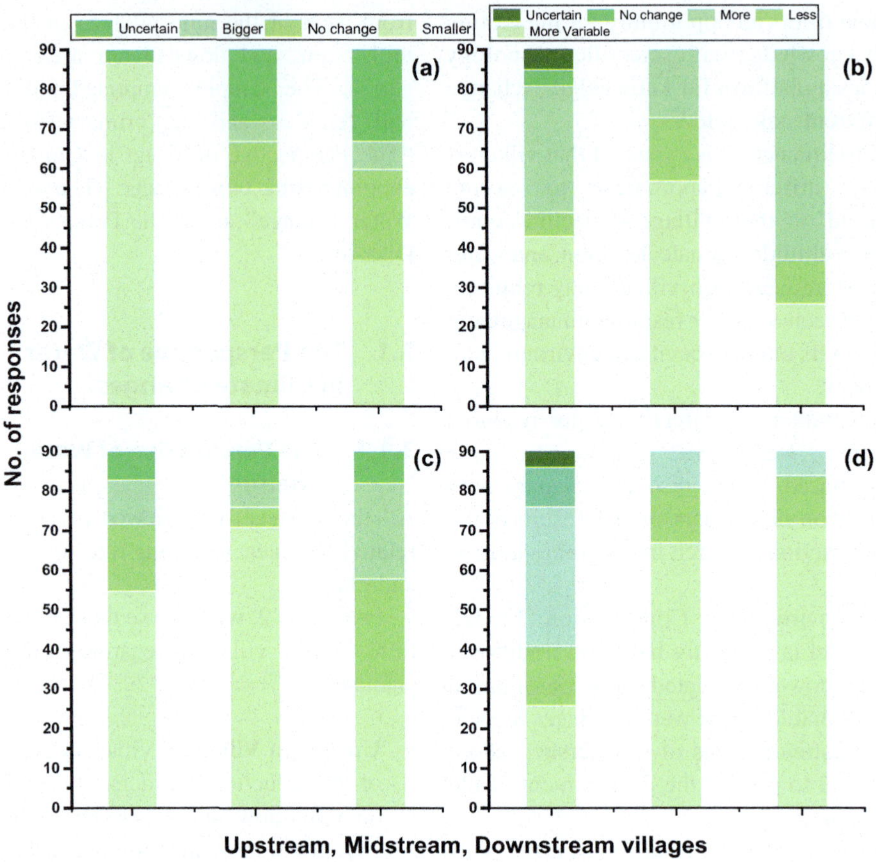

Fig. 2 Changes in climate- and water-related features as perceived by villagers in upstream, midstream, and downstream areas. Panels (a–d) represent glacier extent, river flow, snowfall, and rainfall, respectively. Columns correspond to upstream, midstream, and downstream villages. The data are derived from a survey of 30 villagers aged 40 years and above in each of the nine villages

- **Midstream Villages:** Villagers have a notable awareness of droughts, floods, and landslides and also have some perception of avalanches.
- **Downstream Villages:** Villagers are most sensitive to droughts, floods, and landslides, but have limited awareness of avalanches (Fig. 5).

3.2 The Traditional Knowledge of Water and Climate Change

The analysis of the PAR results according to the four dimensions of traditional knowledge of water.

3.2.1 Knowledge and Practice for Use of Water Resources

Through field surveys of local villagers, herdsmen, monks, Tibetan doctors, veterinarians, and other groups, we have learned that for the local Tibetans, water is categorized into four distinct types: sacred water, domestic water, agricultural water, and medical water. Each type of water has different sources, including lakes, rivers, rain, and springs.

- **Sacred Water:** This water originates from alpine lakes and waterfalls formed by glacier and snowmelt at high altitudes. Villagers consider these lakes and waterfalls to be trea-

Fig. 3 Mountain streams formed by snowmelt from snow-capped mountains. Photograph by Yin Lun

sures of the sacred mountain, and thus, the water from them is deemed sacred. It is used in Buddhist rituals and ceremonies honoring the sacred mountain.

- **Domestic Water:** Sourced from rivers, domestic water is stored in large cylinders placed in a special house, which symbolizes family wealth. Families use this stored water for daily needs.
- **Agricultural Water:** This water comes from precipitation and rivers and is used to irrigate farmland and provide water to livestock. Villagers believe that rainwater is superior to river water for crop growth.

As one senior villager stated: "*In normal rainfall years, we rely mainly on rainwater for crop irrigation, using river water to supplement as needed. In drought years, we depend on river water for irrigation, often up to 20 times a year. Crops irrigated with rainwater are stronger and more delicious.*"

- **Medical Water:** This water is sourced from springs and is used in traditional Tibetan medicine for treating diseases, alongside plant, animal, and mineral medicines. Different springs are believed to treat different diseases.

An elderly Tibetan practitioner explains: "*We determine the medicinal properties of the water based on the surrounding rocks, soil, and plants. The taste, color, and smell of the spring also indicate its medicinal value. High-altitude spring water is considered more effective than low-altitude water. We often mix medical water with sacred water to enhance its therapeutic effects.*"

The timing of water collection, as well as associated hazards and disasters, is linked to climatic conditions and phenological events. For example, during the dry season, villagers predict droughts by observing cloud shapes, wind sounds, and cedar tree leaves. In the wet season, they predict floods by monitoring river flow velocity and sound.

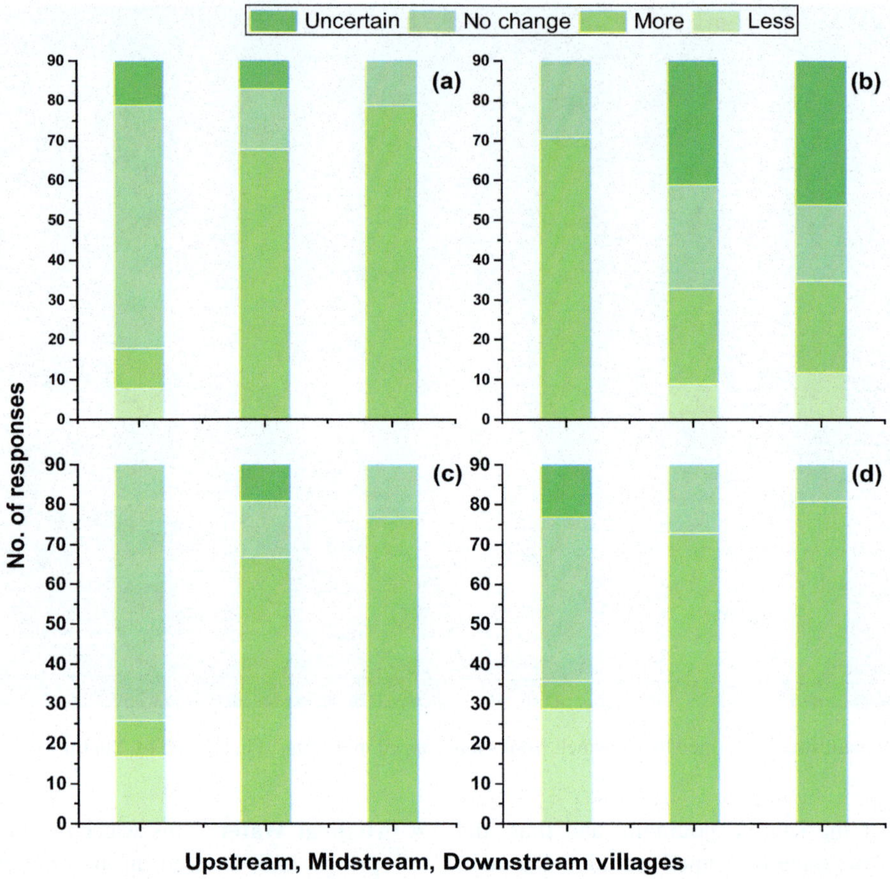

Fig. 4 Perceived changes in water-related disasters reported by villagers in upstream, midstream, and downstream areas. Panels (a–d) represent floods, avalanches, landslides, and droughts, respectively. Columns correspond to upstream, midstream, and downstream villages. The data are derived from a survey of 30 villagers aged 40 years and above in each of the nine villages

Tibetan villagers' knowledge and practices related to water resources involve identifying and classifying various components of the watershed environment. This empirical knowledge includes observations and information about water sources, which are crucial for ensuring drinking water, managing livelihoods, and reducing risks associated with hazards and disasters. This traditional knowledge, accumulated over time and reinforced by the observations and experiences of local experts, offers valuable insights into climate change at a local level. It reflects adaptations to changing climate and environmental conditions, highlighting the depth of understanding inherent in Tibetan traditional water knowledge (Fig. 6).

3.2.2 Customary Law for Use of Water Resources

In addition to knowledge for use of water resources, the traditional knowledge of water also contains customary law for use of water resources as well as management of watershed, which includes methods to prepare the anticipated hazards and reduce the risk. The villagers have developed a variety of customary law and strategies to use the water resources and to cope with the water risk which affected by the climate change, such as water source sanctifying, watershed conserving, water resource sharing, and water disaster resisting.

Fig. 5 Lancang River (Mekong River) flowing through Tibetan villages. Photograph by Yin Lun

Water source sanctifying is related to the belief of Buddhism and especially the sacred mountain worship of local villagers. The Tibetan villagers think, just like other natural resources, water belongs to the god of sacred mountain, so whether it is the water melted by glaciers and snow, or the water flowing out of the mountain, after the formation of rivers, lakes, waterfalls, and springs, it is a gift from the gods of sacred mountain. So the villagers thought that the water source area was sacred and sanctified the place; they hang the Buddhist Scripture flags and build the mani heap (prayer stones) in the water source, at the same time prohibit any damage and pollution to water sources (Fig. 7).

Watershed conserving is a customary law related to the forest conservancy. On the one hand, the villagers forbid the felling of the forest vegetation on both sides of the river. On the other hand, the villagers also plant trees on the riparian slopes prone to landslides and debris flows. There are special villagers in the village who are responsible for the care of the forest. If anyone is found to have felled trees, those who cut trees will be fined. The purpose of this customary law is to protect the watershed environment and prevent the risk of landslides and debris flows during the rainy season.

The customary law of water resource sharing is usually used in drought. When the dry season comes, without rainfall, the amount of water in the river is also decreasing. At this time, for one village, a collective meeting of villagers will be held inside the village to discuss how to allocate water resources. The local Tibetan villages in Deqin are divided into two parts according to their geographical environment. The customary law arranges farmers who live in the upper part of the village to irrigate farmland in the morning and farmers in the lower part of the village in the afternoon. For several villages, village meetings are also required to arrange the use of water resources, so that the upstream and downstream villages can take turns and make fair use of water resources. The customary law of water resource sharing avoids the conflict of water use during

Fig. 6 Tibetan villagers crossing the river using a zip line. Photograph by Yin Lun

drought and makes the most efficient use of water resources.

The customary law of water disaster resisting is mainly for villages that located in the middle and downstream of river. During the rainy season, especially when heavy rainfall occurs, the risks of flood and debris flow are increased, and the villagers will build and reinforce the river dykes and dams. At the same time, they are divided into two groups according to the customary law. They will check the situation of the river in the daytime and night and help the villagers living in the higher-risk areas to move away.

3.2.3 Traditional Cultures Related to the Water

The final aspect of Tibetan traditional water knowledge discussed in this article is the integration of traditional cultures and value systems, including worldviews and cosmologies. For local Tibetan villagers, their traditional water knowledge is deeply embedded in their belief system, particularly the concept of the sacred mountain. Water is perceived as sentient, and it serves as a medium of communication between the deities of the sacred mountain and the people.

Dingzheng Duji, a local Tibetan Buddhist monk, explains:

Fig. 7 Tibetan women cleaning the sacred river. Photograph by Yin Lun

> The sacred lakes, waterfalls, and rivers possess feelings. The gods associated with these sacred waters are very sensitive, and people must show them great respect. For example, one should not damage the surrounding environment of a sacred lake, pollute its waters, or even speak loudly and frolic near the lake. Such behavior is considered disrespectful, and the lakes might respond with rainstorms, floods, or debris flows.

However, during the dry season, especially when droughts occur, this respect is transformed into a ritual for rain. The monk continues:

> In times of drought, people visit monasteries to seek blessings from Buddha and the sacred mountain. The monks provide them with three deity stones. People then collect water from the Lancang (Mekong) River, along with a stone and a small piece of wood from the riverbank. They take these items to the sacred lake in the high mountains, where they throw everything into the lake, disrupting its calm. This act is believed to provoke the lake's deity, leading to rainfall. Once sufficient rain has fallen, people return to the sacred lake to chant scriptures and perform rites, praying for the deity's forgiveness. If the deity is appeased, the rain will cease.

Due to their sacred mountain beliefs, villagers are convinced that their actions can influence the environment, including climate and water conditions. For example, inappropriate behaviors such as hunting or logging in the sacred mountains are believed to anger the deities, leading to punishments in the form of avalanches, floods, droughts, rainstorms, and other water-related disasters or adverse weather conditions. Conversely, positive actions such as planting trees, releasing flowers, and undertaking pilgrimages are thought to earn favorable responses from the deities, resulting in appropriate precipitation and beneficial climate conditions (Fig. 8).

Such knowledge, customary laws, and cultural practices related to water within the local Tibetan community help protect water environments—such as lakes, rivers, and springs—

Fig. 8 Tibetan Tibetan villagers praying for rain by a sacred lake. Photograph by Yin Lun

mitigate water disaster risks, and meet water needs for livelihoods. Additionally, these practices form the foundation for community-based water resource management (CBWM) at the village level (Table 1).

3.2.4 The Management of Water to Adapt to Climate Change

Local Tibetan villagers are acutely aware of changes in water resources and climate conditions. They observe these changes and respond to them using their traditional knowledge of water, which imbues these changes with cultural and spiritual significance. As a result, villagers and other stakeholders collaborate to develop and implement a community-based water resource management plan that is both acceptable and effective for adapting to climate change.

Each adaptation strategy is grounded in a set of values that determine what merits the effort of adaptation (Adger et al., 2009a, b; Dumaru, 2010; Ford, 2009; Ford et al., 2010). For Tibetan villagers, these values are closely tied to their traditional water knowledge. This knowledge helps villagers select appropriate adaptation plans tailored to the specific conditions of their villages, thereby defining and shaping the goals of the adaptation strategy.

In this context, villagers have developed and implemented various community-based water resource management plans, each with distinct trends and orientations. For villages located upstream in the watershed, the villagers have adopted a water environment conservation model. This model focuses on conserving and enhancing the ecosystem of rivers and lakes. The conservation efforts are deeply rooted in traditional culture; villagers invite monks from local monasteries to perform Buddhist ceremonies, which both sanctify the water sources and render the surrounding forests sacred. As part of this sanctification, monks are also invited to construct white pagodas, which are considered sacred objects in Buddhism. The erection of these pagodas symbolizes the protection of water, forest, and land by the sacred mountain and Buddha. Actions that harm the environment or pollute water sources are strictly prohibited, as they are believed to incur

Table 1 Classification of water. (The classification information regarding water in the table is derived from field surveys conducted among local villagers, herdsmen, monks from monasteries, Tibetan doctors, veterinarians, and other groups)

Type	Conditions	Use	Management
Sacred water	Water from sacred lake, falls, and river	Water for ritual of Buddhism and sacred mountain; Water for ceremony to pray for good weather; Water for ritual of avoiding disaster	Forbidding polluting the water resources; Forbidding speaking loudly by the water
Domestic water	Water from river	Drinking, making food, bathing, and washing	Built the canal to divert the water to households in the upper reaches of the river to avoid water contamination; Storage of water in the household
Livelihood water	Water from precipitation and river	Farmland irrigation, livestock drinking water	Building the canal to divert the water to the farmland; Building the reservoir to store water resources, to irrigate farmland in the dry season; Building the river embankment to prevent floods from destroying farmland
Medical water	Water from spring	Treatment of gynecopathy, gastropathy, and other diseases	Protecting the environment around the spring

divine punishment. Additionally, villagers plant trees within the watershed to appease the god of the sacred mountain, hoping to secure favorable water and climate conditions.

For villages located in the midstream of the watershed, the water resource allocation model is employed. The primary goal of this model is to distribute water resources equitably among farmland belonging to different households during the dry season. This allocation is guided by traditional customary laws.

Initially, villagers have adapted their customary laws to ensure a fair distribution of potable water resources. They constructed a water diversion channel at the upper part of the village, which now delivers water directly to every household, including those in the lower part of the village. This change has replaced the previous practice of separate water usage in the upper and lower parts of the village, allowing households in the lower part to access cleaner water and facilitating a more equitable distribution of resources.

Additionally, villagers have reinforced customary laws to manage livelihood water during the dry season. They maintain a traditional practice of separate water use with a specific schedule: on Mondays, Wednesdays, and Fridays, households in the upper part of the village irrigate their farmland in the morning, while those in the lower part do so in the afternoon; on Tuesdays, Thursdays, and Saturdays, the schedule is reversed. Villagers believe that the temperature of river water varies between morning and afternoon, which impacts crop growth and harvest quality. Cold water is thought to enhance crop growth, while warm water improves harvest quality. Therefore, this schedule is considered a fair approach to allocating water resources between households in the upper and lower parts of the village.

For the downstream villages, the water disaster risk reduction model is employed. The primary goal of this model is to mitigate the risk of water-related hazards and disasters, including floods, droughts, landslides, and debris flows. This approach is grounded in traditional knowledge and practices related to water use.

During periods of drought, in addition to equitable distribution of water resources, villagers construct reservoirs to store water. These reservoirs are built and managed collectively to ensure that every household has access to water. The distribution of water from these reservoirs follows the same rules as river water distribution.

In the rainy season, to address risks such as floods, mudslides, and landslides, villagers conduct regular inspections of rivers, reinforce embankments, and construct diversion channels. Residents living in high-risk areas move their property and livestock to safer locations in advance of the rainy season to protect them.

This integration of traditional water knowledge into community-based water resource management allows Tibetan villagers to draw on their historical practices and adapt them to modern challenges. The community-based water resource management plans developed from this knowledge reflect the villagers' aspirations to enhance their capacity to manage water resources effectively in the face of climate change.

4 Conclusion

This research examines three aspects of traditional water knowledge in the context of climate change: local perceptions of climate change impacts on water, traditional knowledge about water and climate change, and water management strategies to adapt to climate change.

Climate change manifests differently across various locations. Even within a small geographic area, variations in altitude and environment can lead to diverse impacts on local water resources. To comprehend these localized phenomena, the observations and experiences of local inhabitants are crucial. For Tibetan villagers, only they can accurately describe the local manifestations of climate change and its impacts on water resources, as well as observe, feel, and assess these changes. Moreover, villagers in different parts of a watershed perceive and experience water changes and water-related disasters differently. These perceptions and observations, in turn, shape their traditional water knowledge and influence their strategies for managing water resources to adapt to climate change.

Local Tibetan villagers possess extensive knowledge of water. Through participatory action research, these villagers have studied their own water knowledge, and this article highlights the value and significance of water within their society: water is not merely a resource issue but also a social and cultural phenomenon. Additionally, the impact of climate change on water resources extends beyond environmental concerns to encompass spiritual and belief systems. Villagers have observed that climate change has affected their traditional water knowledge; some knowledge has become inaccurate or obsolete. However, in many cases, villagers have diligently utilized traditional water knowledge to address the effects and risks of climate change, thereby laying the foundation for community-based water resource management.

The local perception and the traditional knowledge can improve understanding of the influence of climate change to water and provide ideas to develop an equitable and effective water resource management plan to adapt the climate change. For local Tibetan villagers, compared with the externally driven adaptation options, they are more willing to take the community-based water resource management plan, because such a plan enables them to apply their experience and traditional knowledge according to the geographical location and environment of their village, so that their strategies and actions to adapt to climate change are more effective.

References

Adger, W. N., Barnett, J., Ellemor, H. (2009a). Unique and valued places at risk. In: chneider, S., Rosencranz, A., Mastrandrea, M. (Eds.), Climate Change Science Policy. Island Press, Washington, DC, 131–138.

Adger, W. N., Dessai, S., Marisa, G., Hulme, M., Lorenzoni, I., Nelson, D. R., et al. (2009b). Are there social limits to adaptation to climate change? *Climatic Change, 93*, 335–354. https://doi.org/10.1007/s10584-008-9520-z

Ali, M. H. (2011). *Practices of Irrigation and On-Farm Water Management*. Springer. https://doi.org/10.1007/978-1-4419-7637-6

Argyris, C., Putnam, R., & McLain Smith, D. (1985). *Action research: Concepts, methods, and skills for research and intervention.* Jossey-Bass Publishers.

Bates, B. C., Kundzewicz, Z. W., Wu, S., & Palutikof, J. P. (Eds.). (2008). *Climate Change and Water* (p. 210). Technical Paper of the Intergovernmental Panel on Climate Change.

Bawa, K. S., Koh, L. P., Lee, T. M., Liu, J., Ramakrishnan, P., et al. (2010). China, India, and the Environment. *Science, 327*, 1457–1459. https://doi.org/10.1126/science.1185164

Beniston, M. (2003). Climatic change in mountain regions: A review of possible impacts. *Climatic Change, 59*, 5–31. https://doi.org/10.1023/A:1024458411589

Byg, A., & Salick, J. (2009). Local perspectives on a global phenomenon-climate change in eastern Tibetan villages. *Glob. Environ. Change, 19*, 156–166. https://doi.org/10.1016/j.gloenvcha

Chiew, F. H. S. (2007). Estimation of rainfall elasticity of streamflow in Australia. *Hydrological Sciences Journal, 51*(4), 613–625. https://doi.org/10.1623/hysj.51.4.613

Cruikshank, J. (1981). Legend and landscape: Convergence of oral and scientific traditions in the Yukon Territory. *Arctic Anthropol, 18*, 67–93.

Cruz, R., et al. (2007). Asia. In M. Parry, et al. (Eds.), *Climate change 2007: Impacts, adaptation and vulnerability* (pp. 469–506). Contribution of Working Group II to the Fourth assessment report of the Intergovernmental Panel on Climate Change. Cambridge University Press.

Duijn, M., Rijnveld, M., & van Hulst, M. J. (2010). Meeting in the middle: Joining reflection and action in complex public sector projects. *Public Money and Management, 30*(4), 227–233. https://doi.org/10.1080/09540962.2010.492183

Dumaru, P. (2010). Community-based adaptation: Enhancing community adaptive capacity in Druadrua Island, Fiji. *WIRE Climate Change, 1*, 751–763. https://doi.org/10.1002/wcc.65

Ford, J. D. (2009). Dangerous climate change and the importance of adaptation for the Arctic's Inuit population. *Environmental Research Letters, 4*, 1–9. https://doi.org/10.1088/1748-9326/4/2/024006

Ford, J. D., Pearce, T., Duerden, F., Furgal, C., & Smit, B. (2010). Climate change policy responses for Canada's Inuit population: The importance of and opportunities for adaptation. *Global Environmental Change, 20*, 177–191. https://doi.org/10.1016/j.gloenvcha

Friedman, V. (2001). Action science: Creating communities of inquiry in communities of practice. In P. Reason & H. Bradbury (Eds.), *The handbook of action research* (pp. 159–170). Sage.

Immerzeel, W. W., Van Beek, L. P. H., & Bierkens, M. F. P. (2010). Climate change will affect the Asian water towers. *Science, 328*, 1382–1385. https://doi.org/10.1126/science.1183188

Kundzewicz, Z. W., Mata, L. J., Arnell, N. W., DÖLL, P., Jimenez, B., Miller, K., Oki, T., Şen, Z., & Shiklomanov, I. (2008). The implications of projected climate change for freshwater resources and their management, 53, 1, 3–10.

Leonard, S., et al., (2013). The role of culture and traditional knowledge in climate change adaptation: Insights from East Kimberley, Australia. Global Environ. Change. https://doi.org/10.1016/j.gloenvcha.2013.02.012

Ma, X., Kang, Y., & Khan, S. (2009). Climate change impacts on crop yield, crop water productivity and food security–a review. *Progress in Natural Science, 19*(12), 1665–1674. https://doi.org/10.1016/j.pnsc.2009.08.001

Moller, H., Berkes, F., O'Brian, L. P., & Kislalioglu, K. (2004). Combining science and traditional ecological knowledge: Monitoring populations for co-management. *Ecology and Society, 9*, 2. https://doi.org/10.5751/ES-00675-090302

Nijssen, B., O'Donnell, G. M., Hamlet, A. F., & Lettenmaier, D. P. (2001). Hydrologic sensitivity of global rivers to climate change. *Climate Change, 50*, 143–175. https://doi.org/10.1023/A:1010616428763

Prabhakar S. V. R. K., Binaya Rai Shivakoti, Bijon Kumer Mitra. (2014). Climate change adaptation in water, In Sangam Shrestha, Mukand S. Babel and Vishnu Prasad Pandey (eds) Climate change and water resources. NW: Taylor & Francis Group, LLC, 209–238.

Reidlinger, D., & Berkes, F. (2001). Contributions of traditional knowledge to understanding climate change in the Canadian Arctic. *Polar Record, 37*, 315–329. https://doi.org/10.1017/S0032247400017058

Rosenzweig, C., et al. (2007). Assessment of observed changes and responses in natural and managed systems. In: Climate Change 2007: Impacts, Adaptation and Vulnerability. Contribution of Working Group II to the Fourth Assessment Report of the Intergovernmental Panel on Climate Change (ed. by M. L. Parry, *et al.*). UK: Cambridge University Press, 79–131.

Salick, J., Anderson, D., Woo, J., Sherman, R., Cili, N., Ana, Dorje, S. (2004). Bridging scales and epistemologies: linking local knowledge and global science in multiscale assessments. *Millenium Ecosystem Assessment*, Alexandria, Egypt.

Salick, J., & Ross, N. (2009). Traditional peoples and climate change. *Global Environmental Change, 19*, 137–139. https://doi.org/10.1016/j.gloenvcha

Salick, J., Yang, Y., & Amend, A. (2005). Tibetan land use and change near Khawa Karpo, Eastern Himalayas. *Economic Botany, 59*, 312–325. https://doi.org/10.1663/0013-0001(2005)059[0312:TLUACN]2.0.CO;2

Sharma, E., Chettri, N., Tse-ring, K., Shrestha, A. B., Jing, F., Mool, P. &Eriksson, M. (2009). Climate change impacts and vulnerability in the Eastern Himalayas. *Kathmandu, Nepal: ICIMOD.*

Tse-ring, K., Sharma, E., Chettri, N. & Shrestha, A. (eds). (2010). Climate change vulnerability of mountain ecosystems in the Eastern Himalayas; climate change impact an vulnerability in the Eastern Himalayas—synthesis report. Kathmandu, Nepal: ICIMOD.

Vedwan, N., & Rhoades, R. E. (2001). Climate change in the Western Himalayas of India: A study of local perception and response. *Clim. Res, 19*, 109–117. https://doi.org/10.3354/cr019109

Xu, J., Grumbine, R., Shrestha, A., Eriksson, M., Yang, X., et al. (2009). The melting Himalayas: Cascading effects of climate change on water, biodiversity, and livelihoods. *Conservation Biology, 23*, 520–530.

Xue, D. Y., & Guo, L. (2009). On the concept and protection of traditional knowledge. *Biodiversity Science, 17*(2), 135–142. (in Chinese).

Traditional Knowledge and Medicinal Plant Species Shifts Under Climate Change Adaptation

4

1 Introduction

In the face of anthropogenic climate change, species must either adapt, migrate, or face extinction. Evidence shows that climate change is affecting vegetation patterns, including phenology and distribution (Cleland et al., 2007). Plant species are shifting their ranges in both altitude and latitude in response to regional climate changes. Climate warming has now reached a point where significant ecological impacts are readily observable, particularly in alpine and arctic environments worldwide.

Research highlights the endemic nature of species in regions or ecosystems that are especially vulnerable to climate change, such as the Arctic and alpine regions, which may be at the greatest risk (Cavaliere, 2008). After the polar regions, alpine environments are among those most affected by global climate change (Kullman, 2004). Notably, alpine plant diversity is higher than the global average (Körner, 1999). Some cold-adapted plant species in alpine regions have begun migrating further up mountain summits, a phenomenon directly correlated with rising temperatures (Held et al., 2005). In certain cases, these plants continue migrating upward until there are no higher elevations left to inhabit, at which point they may face extinction. Moreover, this upward migration of plant species intensifies competition for space and resources, adding further stress to alpine plant populations (Das et al., 2016). Differential migration rates induced by climate change may lead to the formation of novel plant assemblages, potentially altering ecosystem functioning (Root et al., 2003).

Climate change is having noticeable effects on the life cycles and distributions of global vegetation, including wild medicinal plants. Some medicinal plants are endemic to geographic regions or ecosystems that are particularly vulnerable to climate change, putting them at increased risk (Neilson et al., 2005). Climate change is impacting medicinal plants worldwide and could ultimately result in the loss of some key species (Cavaliere, 2009). Shifts in seasonal timing and geographical ranges driven by climate change may endanger wild medicinal plant populations (Das et al., 2016). Consequently, many medicinal plant populations are gradually declining as a result of climate change.

A global study found that useful Tibetan plants, particularly medicinal species, comprised 62% of all plant species identified in alpine Himalayan sites (Bhardwaj et al., 2007). This high percentage highlights the critical importance of the Himalayas for Tibetan medicine and underscores the threats posed by potential plant losses due to climate change. For instance, the snow lotus (Saussurea laniceps), a medicinal species native to the Tibetan mountains, faces significant threats from climate change (Das et al., 2016). However, there remains a significant lack of data on the effects of climate change on Himalayan biodiversity, including

Y. Lun, *Climate Change and Traditional Knowledge*, Sustainable Development Goals Series,
https://doi.org/10.1007/978-3-032-12015-1_4

medicinal plant species. Even climate envelope modeling, which has been widely applied in other parts of the world, has been less utilized in the Himalayas due to insufficient data on both climate and plant distributions (Heikkinen et al., 2006).

The Eastern Himalayas in northwestern Yunnan, China, is a biodiversity hotspot that also harbors a rich abundance of medicinal plants. Local Tibetans refer to the mountains in this region as "Mainri", meaning "Medicine Mountains." The mountains, with their tremendous plant diversity derived from steep gradients in both elevation and precipitation, have long been a vital source of Tibetan medicinal plants (Salick et al., 2009). The greatest plant diversity and richness is found in alpine environments between 4,200 and 4,500 m, with Tibetan medicinal plants being most abundant in alpine meadows (Salick et al., 2004). Research on the threats climate change poses to Himalayan alpine vegetation and biodiversity, including medicinal plant species, has been conducted in this region. These studies utilize complex data, formulas, and ethnobotanical methods to demonstrate how alpine plants respond to environmental and climate variables such as elevation, precipitation, and biogeography (Salick et al., 2004, 2009).

Drawing from the area's abundant medicinal plant resources, local Tibetans have developed a deep body of traditional medicinal knowledge. This knowledge encompasses theoretical medical concepts (including drug theory, prescription theory, disease and treatment theory), medicinal biological resources (such as a vast array of medicinal herbs, animals, and microbes), traditional herbal processing techniques, medicinal herb cultivation and breeding practices, traditional prescriptions, treatment methods, healthcare techniques, and unique signs and symbols specific to Tibetan medicine (Yin et al., 2018). Climate change not only affects biodiversity but also influences the traditional knowledge tied to biodiversity. Simultaneously, traditional knowledge serves as a critical tool for indigenous peoples to observe and measure the effects of climate change on biodiversity. Indigenous communities frequently use this knowledge to implement climate change adaptation strategies (Yin et al., 2018).

In this study, we hypothesize whether indigenous indicators can effectively measure the impact of climate change on biodiversity. How do local populations perceive the effects of climate change on biodiversity, particularly on medicinal plant species? Can traditional knowledge be used as an indicator to assess the impact of climate change on medicinal plant resources? Additionally, how do local Tibetans apply their traditional knowledge to adapt to climate change?

2 Methods

2.1 Study Zones

Based on the topographic and elevation characteristics of Deqin, three research zones were selected for this study: the alpine area at elevations between 3,500 and 4,700 m, the mountainous area at elevations between 2,500 and 3,500 m, and the valley area at elevations below 2,500 m. In each zone, we identified two research sites based on the relationship between the mountain and sunlight: one on the shady slope and the other on the sunny slope. Each research site was further divided into two research areas—dry and wet—according to the local environmental conditions.

2.2 Community-Based Action Research

Both research and action research methods were employed to achieve the objectives of this study. The research methods followed participatory rural appraisal (PRA) approaches, including key informant interviews, group discussions, and historical mapping of various climate-related events. The action research methods involved the development of community-based climate change adaptation plans. These methods are designed to empower local villagers to create and implement strategies for adapting to climate change. An outline of this process is presented below:

I. Investigation on the Tibetan Medicinal Plants

Among the villagers, there are experts in local Tibetan medicine who possess extensive knowledge in this field. Some have studied Tibetan medicine in schools, some in monasteries, and others have learned it from folk traditions. They are all referred to as Tibetan doctors. These Tibetan doctors have profound experience and understanding of the distribution of medicinal plants, and they have established local Tibetan medicine associations. Therefore, the investigation of medicinal plants in this study is undertaken by these local Tibetan doctors and their Tibetan medicine associations. In this study, the Tibetan doctors investigated medicinal plants in different altitude zones, survey locations, and slopes (see Appendix A) (Figs. 1, 2 and 3).

II. Investigation on the Traditional Knowledge of Medicinal Plants

Since Tibetan doctors with knowledge of Tibetan medicine, especially knowledge about medicinal plants, this part of the study utilized ethnoecological methods. Alongside the investigation of medicinal plants, traditional knowledge regarding these plants was recorded. This included information on the uses of medicinal plants, their growth environments, growth cycles, phenology, and adaptations to climate change.

III. Selection of Indigenous Indicators to Measure the Impact of Climate Change on Species Move

Indigenous indicators, including indicator plants and the associated traditional knowl-

Fig. 1 Local experts investigate Tibetan medicinal plant resources (photoed by Qun Lu)

Fig. 2 Local experts are drawing a distribution map of medicinal plant resources (photoed by Danzhen Leci)

edge, will be used in this study. Representative plants will be selected at different sites as indicators to quantify the number of collection points and to gather data on plants at the same and varying altitudes. This approach will help assess the impact of climate change on species migration (Fig. 4).

IV. Selection of Indigenous Indicators to Measure the Impact of Climate Change on Shift of Traditional Knowledge

Tibetan doctors document traditional knowledge about medicinal plants, including their growth environments, cycles, phenology, and habitats. Based on this knowledge, certain medicinal plants have become sacred due to ritualistic beliefs. This study will use these rites as indicators to analyze how traditional knowledge shifts in response to climate change.

Fig. 4 Medicinal plants collected by local experts (photoed by Danzhen Leci)

V. Climate Change Adaptation

Climate change adaptation based on traditional knowledge includes information on plant varieties, gathering methods, and land management.

Fig. 3 Distribution map of medicinal plant resources (photoed by Danzhen Leci)

Climate change affects not only Tibetan medicinal plants but also the associated knowledge and culture. As a result, local biocultural conservation practices are being implemented to adapt to climate change.

3 Results

3.1 Traditional Classification of Tibetan Medicinal Plants Species

Observation and understanding of medicinal plants is a prerequisite for local Tibetan doctors to understand and utilize this resource, and also an important part of Tibetan medicine knowledge. In recent years, with the impact of climate change and extreme drought disasters, many of the original rich resources of local medicinal plants have become increasingly scarce. Climate change has not only led to changes in medicinal plants, but also has an impact on traditional methods of collecting. Local Tibetan doctors' knowledge of the classification of meidicinal plants and the change of collection methods provide a new perspective for the study and analysis of the impact of climate change on the shift of medicinal plants.

Local Tibetan doctors classify medicinal plants into different categories according to their growing regions and environments (Table 1).

Observation and understanding of medicinal plants are prerequisites for local Tibetan doctors to effectively utilize these resources and are integral to Tibetan medical knowledge. In recent years, the impacts of climate change and extreme droughts have led to a significant decline in the once-abundant local medicinal plant resources. Climate change has not only altered the characteristics of medicinal plants but also affected traditional collection methods. The knowledge of local

Table 1 Traditional classification of medicinal plants. (The classification knowledge of medicinal plants presented in the table is derived from field surveys conducted with local Tibetan doctors and the Tibetan Medical Association.)

Classification basis	Category	Represent medicinal plants (Tibetan name)
1. According to the place of growth	Alpine meadow, forest, and driftstone beach	oubeisai bao, qiagaosuba, re duo na bao
	Forests and hillsides	song cha, ba duo la, bi zhui
	River valley	pa le ga, jiawa, tan rao he
2. According to altitude	High altitude	honglin, can ma er, yi mu nani
	Middle altitude	xing di, wu mu zi, ta rang
	Low altitude	ban re mu, re re, Saiya
3. According to the growing environment	Dry land	mei duo lanna, hong da xia,gang gaqiong man ba
	Wetland	xiangzhe se bao, can ma er
4. According to the slope of the mountain	Shady slope	oubeisai bao, da mu qi wa
	Sunny slope	hong da xia, gang gaqiong man ba

Tibetan doctors regarding the classification of medicinal plants and the adaptation of collection methods offers a valuable perspective for studying and analyzing the effects of climate change on the distribution of medicinal plants.

Local Tibetan doctors classify medicinal plants into different categories based on their growing regions and environments (Fig. 5).

3.2 Shift of Tibetan Medicinal Plants Species

3.2.1 Shift of the Collection Points and Collecting Plants Within Different Altitudes

Over the past six years, community-based research involving local Tibetan doctors has led to a statistical analysis of the number of long-term fixed medicinal plant collection points and indicator plant species in the Baima Snow Mountain area. This analysis examines changes in these collection points and species, reflecting the impact of climate change on the distribution of medicinal plants.

Table 2 shows that medicinal plant diversity is greatest in the middle elevation area, followed by the high elevation area, with the lowest diversity in the low elevation area. At the same elevation, medicinal plants are more abundant on sunny slopes compared to shady slopes and in wet areas compared to dry areas. Overall, the number of medicinal plant collection points and indicator species has decreased across all elevation zones, although this trend is not uniform year by year:

1. The number of medicinal plant collection points and indicator species declined slowly until 2008, after which there was a sharp decrease. Following 2008, numbers began to rise slightly before declining again from 2010 onward.
2. Within this general trend, changes in the middle- and high-altitude regions are more pronounced than those in low-altitude regions.
3. At the same elevation, declines are more significant on shady slopes compared to sunny slopes.
4. Similarly, within the same altitude and slope conditions, declines are more marked in wet areas compared to dry areas.

The shifts in collection points and medicinal plants at different altitudes correspond with local climate changes and extreme weather events. Specifically, the drought of 2011 and the snow disaster of 2013 significantly impacted the habitats of medicinal plants, particularly in the middle- and high-altitude areas. From 2014 to 2016, worsening extreme weather further degraded the

Fig. 5 Medicinal plant specimens made by local experts (photoed by Danzhen Leci)

growing environments of medicinal plants on shady slopes and in humid areas. This analysis demonstrates that local climate changes, especially extreme weather events, have a profound effect on medicinal plants. Conversely, changes in the number of medicinal plant collection points and indicator species also reflect the impact of climate change on the distribution of medicinal plants.

3.2.2 Shift of the Collection Points and Collecting Plants Among Different Altitudes

In this part of the study, Tibetan doctors selected three types of medicinal plants from high, middle, and low altitudes as indicators. Data variables were measured by the number of collection points and the quantity of collected plants. Through six years of observation and data collection, changes in collection points and plant quantities at the same altitudes were analyzed, revealing the impact of climate change on the distribution of medicinal plants.

Based on Table 3 and the selected three indicator plants, it is evident that plants primarily distributed in high-altitude areas have fewer collection points and lower quantities of collected plants, although they are also found to some extent in middle-altitude areas. In contrast, plants from middle-altitude areas show the highest number of collection points and quantities of collected plants, and these plants are also distributed across high and low altitudes. For plants in low-altitude areas, the number of collection points and collected plants ranks second among the three indicators, and they are also found in middle-altitude areas.

Table 2 Shift of the number of collection points and collecting plants within different altitudes. (The information is sourced from community-based action research conducted by local Tibetan doctors.)

Location of collection points			Collection season	Indicator plants (Tibetan name)	Year											
					2011		2012		2013		2014		2015		2016	
					N1	N2	N1	N2	N1	N2	N1	N2	N1	N2	N1	N2
High altitude	Sunny slope	Dry land	Summer	hong da xia	58	132	58	144	45	116	49	123	48	127	45	118
		Wetland		da mu qi wa	63	147	64	152	58	117	52	115	47	107	44	104
	Shady slope	Dry land		oubeisai bao	48	113	48	117	42	107	44	102	44	103	46	99
		Wetland		xiangzhe se bao	55	157	52	149	50	143	47	117	42	104	38	101
Middle altitude	Sunny slope	Dry land	Spring	ba li ga	66	163	64	167	56	136	59	157	62	146	62	153
		Wetland		lujia mu bao	74	177	78	185	70	173	70	154	68	141	65	133
	Shady slope	Dry land		suogawa	68	127	72	136	65	133	58	123	54	138	51	132
		Wetland		ba li ga	77	148	73	141	62	113	65	104	61	101	56	95
Low altitude	Sunny slope	Dry land	Autumn	zhawa	52	97	52	93	50	87	48	91	45	83	45	81
		Wetland		sang di	48	117	46	107	42	81	38	73	35	77	34	78
	Shady slope	Dry land		re re	45	104	45	113	42	92	41	84	41	88	38	92
		Wetland		sa ma dong	42	83	42	86	36	73	34	67	31	63	31	64

N_1 = number of medicinal plants collection points N_2 = number of indicator collecting plants

Table 3 Shift of the number of collection points and collecting plants among different altitudes. (The information is sourced from community-based action research conducted by local Tibetan doctors.)

Indicator plants (Tibetan name)	Collection season	Location of collection points	Year											
			2011		2012		2013		2014		2015		2016	
			N_1	N_2	N_1	N_2	N_1	N_2	N_1	N_2	N_1	N_2	N_1	N_2
ban jianga bao	Summer	High altitude	47	107	43	93	42	87	32	73	33	77	30	63
		Middle altitude	31	63	33	70	34	73	27	56	24	51	21	48
		Low altitude	0	0	0	0	0	0	0	0	0	0	0	0
xing di	Spring	High altitude	19	33	21	37	23	41	20	34	26	49	29	51
		Middle altitude	63	134	55	127	50	117	56	131	65	139	67	145
		Low altitude	23	49	26	57	22	51	20	43	18	35	15	31
ye gexinga bao	Autumn	High altitude	0	0	0	0	0	0	0	0	0	0	0	0
		Middle altitude	22	51	26	62	28	67	23	61	34	79	39	84
		Low altitude	69	155	64	157	62	153	57	144	65	161	67	169

N_1 = number of medicinal plants collection points N_2 = number of indicator collecting plants

The changing trends in the number of collection points and collected plants reflect the impact of climate change on the distribution of medicinal plants:

1. For plants primarily distributed in high and middle altitudes, the number of collection points shows a downward trend.
2. For plants mainly found in middle altitudes but also in high and low altitudes, the changes in collection points and collected plants are more complex. The number of collection points and collected plants decreases gradually in low-altitude areas but increases in middle- and high-altitude areas, particularly in high-altitude regions.
3. For plants primarily distributed in low and middle altitudes, the number of collection points and collected plants increased rapidly in middle-altitude areas, while showing a decreasing and unstable trend in low-altitude areas.

These observations highlight that the resilience of medicinal plants to climate change varies by altitude.

Medicinal plants primarily distributed in high-altitude areas exhibit the least resilience to climate change due to their restricted habitat space. As temperatures rise, glaciers and snow gradually melt, causing tree lines to move upward and alpine meadows to be overtaken by forests. While this shift in habitat temporarily offers new living conditions for high-altitude medicinal plants, these environments are unstable and limited. The new habitats are susceptible to extreme climatic events, such as snowstorms and rainstorms, and require the presence of pioneer species for colonization. Not all medicinal plants can adapt quickly to these changes. Additionally, once the tree line reaches areas with drift stones and glaciers, forest vegetation can no longer migrate upward.

Medicinal plants primarily distributed in middle-altitude areas demonstrate the greatest resilience to climate change, as they have the broadest habitat range and are found at both high and low altitudes. Although rising temperatures and vegetation migration to higher altitudes cause these plants to shift their distribution—decreasing in low-altitude areas while increasing in high- and middle-altitude regions—they generally maintain sufficient living space and favorable conditions.

Medicinal plants mainly found in low-altitude areas also show considerable resilience to climate change because they can migrate to the more extensive habitats available in middle-altitude areas. Despite recent extreme weather events, such as increased temperatures and drought, significantly impacting their traditional low-altitude habitats, these plants are expanding their distribution in middle-altitude areas.

3.3 Shift of Traditional Knowledge of Tibetan Medicinal Plants Species

In local traditional beliefs, Tibetan villagers consider plants and animals, including Tibetan medicinal plants, to be associated with sacred mountains such as Baima Snow Mountain. Consequently, Tibetan doctors perform ceremonies to seek the mountain's consent before collecting medicinal plants. The timing of plant collection is intricately linked to their growth cycles, with many medicinal plants being collected during their germination and growth phases. Therefore, the timing of these ceremonies can serve as an indicator to observe the effects of climate change on plant growth cycles by tracking changes in the ceremonial timing over recent years.

From Table 4, it is evident that over six years of observation and recording, the timing of medicinal plant gathering ceremonies has advanced to varying degrees across high-, middle-, and low-altitude areas. This shift in ceremonial timing objectively indicates that the growth periods of medicinal plants, such as germination and maturation, are occurring earlier, reflecting the impact of climate change on their growth cycles. Moreover, the extent of this shift varies by altitude: the advance is most pronounced in high-altitude areas, where the ceremonial timing has moved forward by 17 days over six years, followed by

Table 4 Shift of the ritual time of collecting medicinal plants. (The information is sourced from community-based action research conducted by local Tibetan doctors.)

		High altitude		Middle altitude		Low altitude	
		D	A	D	A	D	A
Year	2011	September 3rd	0	June 17th	0	April 19th	0
	2012	September 1st	2	June 14th	3	April 17th	2
	2013	August 28th	4	June 12th	2	April 16th	1
	2014	August 25th	3	June 11th	1	April 13th	3
	2015	August 21st	4	June 8th	3	April 9th	4
	2016	August 17th	4	June 4th	4	April 8th	1
E		17		13		11	

D = ritual date A = days ahead of the previous year E = days ahead of the six years later

middle-altitude areas with a 13-day advance, and low-altitude areas with an 11-day advance. These differences highlight that climate change affects the growth cycles of medicinal plants differently at various altitudes.

3.4 Traditional Knowledge and Climate Change Adaptation

Due to the local Tibetan villagers' belief in sacred mountains, many areas of Baima Snow Mountain are regarded as sacred. These sacred areas, located at high, middle, and low altitudes, hold significant cultural and spiritual value and include sites such as sacred lakes, springs, waterfalls, and forests. In these areas, hunting, logging, and collection are prohibited or strictly regulated due to religious and customary laws, which helps to protect biodiversity, including Tibetan medicinal plants. Consequently, sacred areas effectively function as protected zones or "gene banks" for preserving species resources.

Building on this traditional belief, local Tibetan doctors have implemented protective planting of Tibetan medicinal plants in sacred areas to mitigate the impacts of climate change. They have conducted interviews with local elders and monks to document traditional knowledge and, based on this information, collected and nurtured medicinal plant seeds and seedlings. They have established semi-wild protective cultivation of 40 Tibetan medicinal plant species in sacred areas across different altitudes (see Appendix B).

Given the limited number and size of sacred areas, Tibetan doctors have enlisted monks to conduct ceremonies in non-sacred, plant-abundant areas to establish their sacred status. White pagodas have been erected in these areas to prohibit tree cutting and collection, thereby designating them as sacred. Once these areas are recognized as sacred, Tibetan doctors commence the cultivation of medicinal plants.

This protective planting initiative in sacred areas helps to mitigate the impacts of climate change and represents a localized strategy for adaptation. Rooted in traditional knowledge and beliefs, this measure enhances the resilience of local communities in facing climate change.

4 Discussion

The traditional knowledge and culture of the Tibetan people in the eastern Himalayas offer a valuable perspective for studying and analyzing the impact of climate change on species migration. Additionally, this traditional knowledge and culture play a crucial role in developing conservation strategies for biodiversity in response to climate change.

4.1 Traditional Knowledge as an Indicator to Measure the Species Move Caused by Climate Change

The question of whether traditional knowledge can be quantitatively measured and how it can be used to assess the impacts of climate change on biodiversity is a topic of discussion in climatic anthropology. Traditional knowledge is often described qualitatively, making it challenging for local experts to provide convincing evidence of climate change impacts in a format that is acceptable to scientists and policymakers. This is because traditional knowledge is difficult to quantify, leading to its dismissal as "unscientific."

This paper seeks to advance the quantitative analysis of traditional knowledge, using Tibetan medicinal plants as a case study. Firstly, it examines how climate change affects the distribution of Tibetan medicinal plants by analyzing changes in the number of collection points and collecting plants. Secondly, it uses shifts in the timing of rituals to reflect the impact of climate change on the traditional knowledge of Tibetan medicinal plants, including insights into their growth cycles and phenology.

4.2 Traditional Culture as Biocultural Conservation Practices to Adapt to Climate Change

Through the quantitative research of traditional knowledge, this paper attempts to make it as an index to measure the shift of Tibetan medicinal plants under the influence of climate change. We can understand climate change from the local point of view and provide a new perspective for the research of traditional knowledge, species move and climate change. Of course, this research is still in its infancy and needs to be improved in theory and methodology, but it can still draw a conclusion on the impact of climate change on species shift to a certain extent.

Meanwhile, it is undeniable that the traditional culture of indigenous and local communities can enhance the resilience of local people in the face of climate change. Therefore, in the research of climate change, the interdisciplinary cooperation between natural science and social science is needed. Similarly, in the process of formulating climate change mitigation and adaptation strategies, the policies of international organizations and governments need to be combined with local traditional culture and knowledge.

5 Conclusion

This paper employs quantitative research methods to assess traditional knowledge as an index for measuring the shift of Tibetan medicinal plants in response to climate change. By incorporating local perspectives, the study offers a novel approach to understanding climate change and provides new insights into the relationship between traditional knowledge, species migration, and climate impacts. Although this research is still developing and requires further theoretical and methodological refinement, it provides valuable preliminary conclusions about the effects of climate change on species distribution.

Furthermore, the traditional culture of indigenous and local communities plays a crucial role in enhancing resilience to climate change. Therefore, effective climate change research and policy development must integrate interdisciplinary approaches, combining natural and social sciences. In formulating climate change mitigation and adaptation strategies, it is essential to align international and governmental policies with local traditional knowledge and cultural practices.

Appendix A

Medicinal plants species. (The information is sourced from community-based action research conducted by local Tibetan doctors.)

No	Tibetan name	Species	Habitat	Location
001	tan rao he	*Ricinus communis*	Dry-hot valley	Low altitude
002	cha zhuo	*Lepidium apetalum*	Hillside or terrain	Low, Middle, Highaltitude
003	Gougulu	*Ailanthus vilmoriniana*	Dry-hot valley	Low altitude
004	Buxieze	*Dendrobium hookerianum*	Forest	Low altitude
005	Gawa	*Verbena officinalis*	Forest	Low altitude
006	sai xi	*Verbascumthapsus*	Grassland slopes and wasteland	Low altitude
007	Zhawa	*Anaphalium affine*	Sunny slope and valley	Low altitude
008	Suogawa	*Capsella bursa-pastoris*	Hillside or terrain	Low altitude
009	cha gawa	*Thlaspiarvense*	Hillside or terrain	Low altitude
010	qi rou	*Elsholtziadensa*	Hillside, wetland, forest, or terrain	Low altitude
011	Kanba	*Artemisia sieversiana*	Hillside or terrain	Low altitude
012	na bao	*Bidens pilosa* var.*radiata*	Forest	Low altitude
013	sa ma dong	*Pteris cretica*	Forest	Low altitude
014	Aomaosai	*Dysosmatsayuensis*	Hillside, wetland, forest, or terrain	Low altitude
015	xintuo li	*Salvia mekongensis*	Hillside or terrain	Low altitude
016	cha bei	*Lepisorussoulianus*	Hillside or terrain	Low altitude
017	qi cai	*Xanthium sibiricum*	Hillside or terrain	Low altitude
018	Jiawa	*Angelica sinensis*	Riverside	Middle, High altitude
019	Saiwa	*Rosa omeiensis*	Hillside or terrain	Low altitude
020	wen bu	*Myricariagermanica*	Hillside or terrain	Low altitude
021	Jiaqiong	*Hypericum bellum*	Forest	Low altitude
022	sang di	*Swertiamussotii*	Hillside, wetland, forest, or terrain	Low altitude
023	Zhaqiong	*Rohden japonica*	Forest	Low altitude
024	xi ge chi	*Lonicera japonica*	Hillside, forest, or terrain	Low altitude
025	jie ji na bao	*Gentianacrassicaulis*	Hillside, wetland, forest, meadow, or terrain	Low altitude
026	me ya	*Taxus chinensis*	Forest	Low altitude
028	qi dang ga	*Embelalaeta*	Forest	Low altitude
029	sai ji mei duo	*Herpetospermumpenduculosum*	Hillside, wetland, forest, or terrain	Low altitude
030	cha jia ha wu	*CorallodiscusKingianus*	Hillside	Low altitude
031	baerba da	*Hypecoumleptocarpum*	Forest, meadow	Low altitude

(continued)

(continued)

No	Tibetan name	Species	Habitat	Location
032	Saiya	*Chaenomelesthibetica*	Hillside, wetland, forest, or terrain	Low altitude
033	du re gezha	*Paris polyphylla*	Hillside, wetland, forest, or terrain	Low altitude
034	Ganzhagari	*Rubusirritans*	Hillside	Low altitude
035	wang dong chi	*Platycodon grandiflorum*	Hillside, wetland, forest, meadow, or terrain	Low altitude
036	cai tun	*Ephedra Saxatilis*	Hillside or terrain	Low altitude
037	na bao chi	*Salvia brachyloma*	Hillside or terrain	Low altitude
038	Keermang	*Taraxacummongolicum*	Hillside, wetland, forest, or terrain	Low altitude
039	xintuo li ga bao	*Chelonopsisalbiflora*	Hillside, wetland, forest, or terrain	Low altitude
040	re re	*Drynariasinica*	Hillside, wetland, forest, or dry-hot Valley	Low altitude
041	ban re mu	*Polygonum sphaerostachyum*	Hillside, meadow	Low altitude
042	pa le ga	*Aristolochiamacrocarpa*	Hillside, wetland, forest, or dry-hot valley	Low altitude
043	xu da	*Acorus calamus* L	Wetland	Low altitude
044	ye gexinga bao	*Senecio solidagineus*	Hillside, forest	Low altitude
045	Luqiong	*Aster poliothamnus*	Meadow	Low altitude
046	xi saiga bao	*Astragalus yunnanensis*	Meadow	Middle altitude
047	Geni	*Polygonatumverticillatum*	Hillside, wetland, forest	Middle altitude
048	duo mu niu	*Cynanchumforrestii*	Hillside, wetland, forest	Middle altitude
049	bai ma sai bao	*Paeonia delavayi*	Hillside	Low, Middle altitude
050	re ni	*Polygonatumcirrhifolium*	Hillside, wetland, forest	Middle altitude
051	re le qiong	*Adiantum pedatum*	Forest	Middle altitude
052	mu qia la	*Quercus pannosa*	Forest	Middle altitude
053	Juewa	*Sophoramoocroftiana*	Hillside, wetland, forest	Middle altitude
054	nixin cha ma meiba	*Asparagus filicinus*	hillside, wetland, forest	Middle altitude
055	jie jig a bao	*Gentianacrassicaulis*	Hillside, wetland, forest	Middle altitude
056	ga di hong bao	*Androsacestenophylla*	Hillside	Middle altitude
057	di da	*Swertiacincta*	Meadow	Middle altitude
058	ji ziga bao	*Salvia roborowskii*	Meadow	Middle altitude
059	sanziga bao	*Galiumaparine*	Meadow	Middle altitude
060	Suogawa	*Anemone rivularis*	Hillside, wetland, forest, meadow, or terrain	Middle altitude
061	wo zeng	*Thalictrum atriplex*	hillside, forest	Middle altitude
062	Chegui	*Rosa sertata*	Hillside, wetland, forest	Middle, High altitude
063	lujia mu bao	*Pedicularissiphonantha*	Hillside, wetland, forest	Middle altitude

(continued)

(continued)

No	Tibetan name	Species	Habitat	Location
064	ba li ga	*Aristolochiamoupinensis*	Hillside, wetland, forest	Middle altitude
065	xin cha	*Cinnamomum cassia*	Wetland, forest	Middle altitude
066	Bazhe	*Mirabilis himalaica*	Hillside, wetland, forest	Middle altitude
067	bawuga bao	*Phytolaccaacinosa*	Hillside, forest	Middle altitude
068	Shazhu	*Urtica triangularis* ssp. pinnatifida	Hillside, forest	Middle altitude
069	lan dan ze	*Hyoscyamusniger*	Hillside, wetland, forest	Middle altitude
070	ga la ba qi tuo	*Stephania delavayi*	Hillside, wetland, forest	Middle altitude
071	da remei duo	*Lilium lophophorum*	Forest, meadow	Middle altitude
072	song di	*Saxifragasignata*	Hillside	Middle altitude
073	wo zi li	*Portulacaolearacea*	Wetland, forest	Middle altitude
074	gelu gong	*Carthamustinctorius*	Wetland, forest	Middle altitude
075	sendeng ma bao	*Pterocarpus indicus*	Forest	Middle altitude
076	ma le	*Inularacemosa*	Hillside, wetland, forest, meadow, or terrain	Middle altitude
077	senxinna ma	*Daphne tangutica*	Forest	Middle altitude
078	tan congna bao	*Anisodustanguticus*	Shrub, meadow	Middle altitude
079	da wa	*Arisaemaaridum*	Shrub, meadow	Middle altitude
080	bi zhui	*Rubiacordifolia*	Shrub, meadow, forest	Middle altitude
081	Gecuo	*Melandriumviscidulum*	Meadow	Middle altitude
082	suoba cha le jian	*Sabina wallichiana*	Shrub, meadow, forest	Middle altitude
083	zhi re a	*Cuscutachinensis*	Shrub, meadow, forest	Middle altitude
084	Hongbaxiaga	*Corydalis tongolensis*	Shrub, meadow	Middle altitude
085	Suanmo	*Rumexacetosa*	Shrub, meadow	Middle altitude
086	xi ka chi	*Aleuritopterisargentea*	Shrub, meadow	Middle altitude
087	xia gong ba	*Delphinium caeruleum*	Shrub, meadow, screes	Middle altitude
088	li zhi	*Tinosporasinensis*	Hillside, forest	Middle altitude
089	ba duo la	*Belamcandachinensis*	Meadow, forest	Middle altitude
090	song cha	*Cirsium bolocephalum*	Shrub, meadow, forest	Middle altitude
091	wu mu zi	*Sinopodophyllumhexandrum*	Shrub, meadow, forest	Middle altitude
092	ta rang	*Plantagodepressa*	Shrub, meadow, forest, wetland	Middle altitude
093	qi buga re	*Hippuris vulgaris*	Wetland	Middle altitude
094	jie cha	*Ranunculus brotherusii* var.*tanguticus*	Meadow, wetland	Middle altitude
095	ba duo la	*Bletillastriata*	Meadow, forest, wetland	Middle altitude
096	gong tuba	*Erysiumhieracifolium*	Hillside, meadow	Middle altitude
097	xing di	*Phamnellagilgitica*	Shrub, forest	Middle altitude
098	ba le ga	*Aristolochiagriffithii*	Shrub, meadow, forest, wetland	Middle altitude
099	zhi yang ge	*Dracocephalumtanguticum*	Meadow, forest	Middle altitude

(continued)

(continued)

No	Tibetan name	Species	Habitat	Location
100	cha gu	*Allium forrestii*	Meadow, screes	High altitude
101	da li	*Rododendronintricatum*	Shrub, meadow	High altitude
102	ban na	*Aconitum pendulum*	Shrub, meadow, screes	High altitude
103	suoluoga bao	*Pegaeophytonscapiflorum*	Shrub, meadow, screes	High altitude
104	le dui du ji	*Codonopsis nervosa*	Meadow, forest	High altitude
105	ban jianga bao	*Gentianaalgida*	Meadow, screes	High altitude
106	cha ma	*Caragana brevifolia*	Shrub, meadow, screes	High altitude
107	mei duo lanna	*Pedicularisintegrifolia*	Forest, meadow, screes	High altitude
108	gang gaqiong man ba	*Gentianaszechenyi*	Meadow, screes	High altitude
109	gourigou mu	*Carthamnustinctorius*	Shrub, meadow, screes	High altitude
110	yi mu nani	*Paraquilegiamicrophylla*	Shrub, meadow, screes	High altitude
111	can ma er	*Rhodiolaalgida*	Wetland	High altitude
112	gong baga ji	*Saussureajaponica*	Meadow, screes	High altitude
113	lurisai bao	*Pedicularis longiflora* var. *tubiformis*	Wetland, forest, meadow, screes	High altitude
114	da mu qi wa	*Caltha scaposa*	Wetland, forest	High altitude
115	hong da xia	*Oxytropistragacanthoides*	Screes	High altitude
116	suo gong ba	*Soroserisgillii*	Shrub, meadow, screes	High altitude
117	a zhongga bao	*Arenarialancangensis*	Meadow, screes	High altitude
118	ban ga	*Aconitum tanguticum*	Wetland, meadow, screes	High altitude
119	ban can burou	*Eriophytonwallichii*	Screes	High altitude
120	zhi da sazeng	*Fragirianilgerrensis*	Forest, meadow	High altitude
121	ga tun	*Bergenia purpurascens*	Forest, meadow, screes	High altitude
122	ba bao saibao	*Veratrillabaillonii*	Meadow	High altitude
123	min jianna bao	*Cremanthodiumplantagineum*	Shrub, meadow	High altitude
124	xiangzhe se bao	*Primula sikkimensis*	Wetland, meadow	High altitude
125	si li ma bao	*Rhodiolacrenulata*	Shrub, meadow, screes	High altitude
126	bo ka chi	*Saussureaquercifolia*	Meadow	High altitude
127	Quzha	*Rheum nobile*	Screes	High altitude
128	Honglin	*Picrorhizascrophulariiflora*	Meadow, screes	High altitude
129	re duo na bao	*Saussureaobvallata*	Meadow, screes	High altitude
130	xiagui bi	*Delphinium thibeticum*	Meadow, forest	High altitude
131	a bi ka	*Fritillaria cirrhosa*	Shrub, meadow, forest	High altitude
132	Dioxin	*Fraxinus suaveolans*	Forest	High altitude
133	ji wa	*Berberis dictyophylla*	Shrub, meadow, forest	High altitude
134	baya cha wa	*Roscoeatibetica*	Shrub, meadow, forest	High altitude
135	da zhi	*Schisandra rubiflora*	Meadow, forest	High altitude
136	ya ji ma	*Chrysospleniumforrestii*	Shrub, meadow, screes, forest	High altitude
137	Qiagaosuba	*Saussureamedusa*	Screes	High altitude
138	oubeisai bao	*Meconopsispunicea*	Meadow	High altitude

Appendix B

Semi-wild protective cultivation of 40 Tibetan medicinal plants in sacred areas. (The information is sourced from community-based action research conducted by local Tibetan doctors.)

No	Species	Location of sacred areas
1	***Seneciosp.***	High altitude
2	*Holarrhenaantidysenterica*	High altitude
3	*Clematis rehderiana*	High altitude
4	*Drynariasinica*	High altitude
5	*Taraxacumtibetanum*	High altitude
6	*Lepidium apetalum*	High altitude
7	*Solanum nigrum*	High altitude
8	*Rumexnepalensis*	High altitude
9	*Rosa* sp.	High altitude
10	*Hippophaerhamnoides*	High altitude
11	*Myricariagermarica*	High altitude
12	*Vitis*sp.	Middle altitude
13	*Rubus* sp.	Middle altitude
14	*Berberis jamesiana*	Middle altitude
15	*Aster flaccidus*	Middle altitude
16	*Plantago* sp.	Middle altitude
17	*Euphorbia nematocypha*	Middle altitude
18	*Fragaria sp.*	Middle altitude
19	*Drynariapropinqua*	Middle altitude
20	*Asparagus* sp.	Middle altitude
21	*Gnaphalium affine*	Middle altitude
22	*Mirabilis jalapa*	Middle altitude
23	*Rheum palmatum*	Middle altitude
24	*Rhodiolacrenulata*	Middle altitude
25	*Bergenia purpurascens*	Middle altitude
26	*Gentianastraminea*	Low altitude
27	*Rhododendron sp.*	Low altitude
28	*Rhododendron sp.*	Low altitude
29	*Ephedra* sp.	Low altitude
30	*Sinolimprichtiaalpina*	Low altitude
31	*Gymnadeniaconopsea*	Low altitude
32	*Polygonum polystachyum*	Low altitude
33	*Rheum pumilum*	Low altitude
34	*Iris lactea*var.*chinensis*	Low altitude
35	*Fritillaria* sp.	Low altitude
36	*Chrysospleniumnudicaule*	Low altitude
37	*Gentianaurnula*	Low altitude
38	*Cardamine macrophylla*	Low altitude
39	*Lonicera maackii*	Low altitude
40	*Primula sikkimensis*	Low altitude

References

Bhardwaj, J., Singh, S., & Singh, D. (2007). Hailstorm induced crop losses in India: some cast studies. In *Abstract for presentation at 4th European Conference on Severe Storms in Trieste*, Italy, 10–14, September 2007

Cavaliere, C. (2008). Drought reduces 2007 saw palmetto harvest. *HerbalGram, 77*, 56–57.

Cavaliere, C. (2009). The effects of climate change on medicinal and aromatic plants. *HerbalGram, 81*, 44–57.

Cleland, E. E., Chuine, I., Menzel, A., Mooney, H. A. & Schwartz, M. D. (2007). Shifting plant phenology in response to global change. *Trends in Ecology and Evolution, 72*(7), 357–64. https://doi.org/10.1016/j.tree.2007.04.003PMid:17478009

Manish, D., Jain, V., & Malhotra, S. K. (2016). Impact of climate change on medicinal and aromatic plants. *Indian Journal of Agricultural Sciences, 86*(11), 1375–1382.

Heikkinen, R. K., Luoto, M., Araújo, M. B., Virkkala, R., Thuiller, W., & Sykes, M. T. (2006). Methods and uncertainties in bioclimatic envelope modeling under climate change. *Progress in Physical Geography, 30*, 1–27. https://doi.org/10.1177/0309133306071957

Held, I. M., Delworth, T. L., Lu, J., Findell, K. L., & Knutson, T. R. (2005). Simulation of Sahel drought in the 20th and 21st centuries. *PNAS, 105*(50), 17891–17896. https://doi.org/10.1073/pnas.0509057102

Körner, C. (1999). Alpine plant life: Fundional Piant ecology of high mountain ecosystems. Springer, Berlin. https://doi.org/10.1007/978-3-642-98018-3

Kullman, L. (2004). The changing face of the alpine world. *IGBP Global Change Newsletter, 57*, 12–14.

Lun, Y., Misiani, Z., Yanyan, Z. (2018). The traditional knowledge associated to biodiversity in an age of climate change. *Earth Sciences, 7*(5), 209–215. https://doi.org/10.11648/j.earth.20180705.12

Yin, L., & Zachary, M. (2018). Traditional knowledge and community-based research to support climate change adaptation in Eastern Himalayas. In C. Change (Ed.), *Suiven John Paul Tume and ViecentItai Tanyanyiwa* (pp. 159–170). Vernon Press.

Neilson, R. P., Pitelka, L. F., & Solomon, A. M. (2005). Forecasting regional to global plant migration in response to climate change. *Bio Science, 55*(9), 749–759. https://doi.org/10.1641/0006-3568(2005)055[0749:FRTGPM]2.0.CO;2

Root, T. L., Price, J. T., Hall, K. R., Schneider, S. H., Rosenzweig, C., & Pounds, A. (2003). Fingerprints of global warming on wild animals and plants. *Nature, 421*, 57–60. https://doi.org/10.1038/nature01333

Salick, J., Anderson, D., Woo, J., Sherman, R., Norbu, C., Na, A., & Dorje, S. (2004). Tibetan Ethnobotany and Gradient Analyses, Menri (Medicine Mountains), Eastern Himalayas. Millenium Ecosystem Assessment, Alexandria, Egypt.

Salick, J., Fang, Z. D., & Byg, A. (2009). Tibetan ethnobotany and climate change in the Eastern Himalayas. *Global Environmental Change, 19*, 147–155. https://doi.org/10.1016/j.gloenvcha.2009.01.008

5 Traditional Knowledge in Agro-Pastoralism Livelihood for the Climate Change Adaptation

1 Introduction

Climate change refers to variations in climate characterized by rising temperatures and associated phenomena, including increased frequency and intensity of extreme weather events such as heatwaves, droughts, and floods, as well as heightened climatic uncertainty (IPCC, 2011). The impacts of climate change are evident globally (Oppenheimer et al., 2014). Evidence indicates that natural climatic variability, exacerbated by climate change, will negatively affect millions of livelihoods worldwide (IPCC, 2007).

Climate change adversely impacts crop farming through heightened water stress, altered runoff patterns, shifts in seasonality, and temperature variations (FAOSTAT, 2015). It threatens agricultural productivity by increasing temperatures, changing precipitation patterns, and intensifying occurrences of extreme weather events (Nelson, 2009). Additionally, climate change may introduce new pests and diseases, limit irrigation water availability, and worsen soil erosion (Adams et al., 1998). The emergence of new conditions favorable to weeds, insects, and pathogens could further reduce agricultural productivity. Competition for space, water, and nutrients between weeds and crops has already been linked to significant crop losses globally—approximately 34 percent (Oerke, 2006)—and may be exacerbated by ongoing climate change.

For livestock husbandry, climate has become increasingly unpredictable, with erratic rainfall leading to further reductions in space and pastoral mobility (Miller, 1995; Nori & Davies, 2007). The nature of pastoralists' livelihoods, often described as "wealth on the hoof," renders them highly vulnerable to both natural and anthropogenic impacts, including climate change, with poorer households (Jones & Thornton, 2009; Mulder et al., 2010) and women (Watson, 2010) being disproportionately affected. Declining pasture availability and reduced crop yields due to climate change present serious challenges to pastoralists (Cruz et al., 2007). The productivity of the livestock sector is declining as a result of heat stress, poor nutrition (Muntifering et al., 2006), and water shortages, all of which can be attributed to climate change. Livestock play a crucial role in enhancing the resilience of vulnerable, impoverished populations by diversifying risks and increasing assets, which is essential given their exposure to climatic, market, and disease shocks (Freeman et al., 2008; Krisna et al., 2004). Climate change can affect the livelihoods, food security, and health of these vulnerable groups through its impacts on livestock and livestock systems, including changes in water and feed availability, biodiversity, and

Y. Lun, *Climate Change and Traditional Knowledge*, Sustainable Development Goals Series,
https://doi.org/10.1007/978-3-032-12015-1_5

animal health (Luseno et al., 2003; McPeak, 2006; Thornton et al., 2007, 2009).

Although climate change is a global phenomenon, its impacts and adaptation capacities, preferences, and strategies vary locally (Sewando et al., 2016). Adaptation refers to "initiatives and measures to reduce the vulnerability of natural and human systems against actual or expected climate change effects" (IPCC, 2007) and "adjustments in natural or human systems in response to actual or expected climatic stimuli or their effects, which moderate harm or exploit beneficial opportunities" (IPCC, 2001). At the rural level, adaptation to climate change requires farmers to first recognize that the climate has changed. They then need to identify and implement potentially useful adaptations (Silvestri et al., 2012). Adaptation is increasingly viewed as the management of climate risk (Carter et al., 2015), which involves building adaptive capacity to reduce vulnerability. The IPCC defines vulnerability to climate change as a function of a system's exposure to and sensitivity to climate change hazards, as well as its adaptive capacity to mitigate potential damages or capitalize on associated opportunities (Watson et al., 2001). Factors influencing vulnerability at the local level can vary significantly depending on cultural, political, economic, and ecological characteristics, as well as the institutional environment in which adaptation takes place. Over the past decade, several studies have shown that climate change adaptation encompasses a variety of approaches, particularly through community capacity building (Archer et al., 2015). One advantage of community-based approaches is their ability to engage a broad range of stakeholders and recognize the roles of local political, cultural, economic, institutional, and technological forces in shaping people's ability to anticipate and respond to climate impacts (van Aalst et al., 2008).

Traditional knowledge has been recognized as a valuable asset for climate change adaptation (Swart et al., 2003). Incorporating traditional knowledge into climate change policies can lead to the development of effective, cost-efficient, participatory, and sustainable mitigation and adaptation strategies (Hunn, 1993; Robinson & Herbert, 2001). Significant progress could be achieved in climate-proofing the agricultural sector by implementing local adaptation strategies and/or strengthening existing ones. Given that climate change adaptation is influenced by factors such as perception, culture, and socio-economic status, context-specific local adaptation strategies are likely to address agricultural issues more effectively (Lambi2001; Ngwa & Balgah, 2016).

Current adaptation options in the agricultural sector include crop diversification, livestock farming, mixed cropping systems, the use of different crop varieties, adjustments to planting and harvesting dates, and the integration of drought-resistant varieties with high-yield, water-sensitive crops (Oyekale et al., 2009). For livestock producers, key adaptation strategies involve integrating crop and livestock production, destocking, diversifying livestock feeds, changing animal breeds, and relocating animals to alternative sites. Desired adaptations also include introducing new breeds and increasing herd size. Indigenous adaptation techniques, often practiced over generations by specific communities, should be identified, strengthened, and promoted to mitigate the impacts of climate change on agriculture-dependent populations in Sub-Saharan Africa (UNISDR, 2015; UNFCCC, 2016).

Agro-pastoralists' livelihood strategies are based on crop farming (Sewando et al., 2016) and livestock husbandry (Goldman & Riosmena, 2013). The complementarity of crop and livestock systems for agricultural sustainability, particularly in developing countries, has long been recognized (Scoones & Wolmer, 2000; Wolmer, 1997). Local communities practicing either crop production or animal husbandry have increasingly adopted integrated approaches to secure livelihoods, adapt production systems to contemporary lifestyles, and cope with uncertainties and challenges arising from demographic growth, globalization, and climate change (Wolmer, 1997). Livestock increasingly provide food, manure, and draft power for crop farming, while crop-livestock systems offer income, insurance, and a buffer against climate-related shocks (Wane et al., 2009; Seo, 2010; Balgah and Buchenrieder, 2011).

Fig. 1 Traditional agricultural and pastoral livelihoods (photoed by Yin Lun)

Agro-pastoralism is a traditional livelihood for the Tibetan people in the eastern Himalayas of China, and it offers local insights into climate variability and change based on traditional practices (Yin & Zachary, 2018). This paper focuses on the impacts of climate change and the local adaptation strategies of the Tibetan people in Yunnan Province, China. It aims to understand local perceptions of climate change and how community-based adaptation strategies are formulated based on traditional agro-pastoral knowledge. The paper demonstrates how Tibetans accumulate baseline information about weather and climate to guide their agro-pastoral practices, highlights the major impacts of climate change on agro-pastoralism, and examines the vital role of traditional knowledge in local adaptation strategies. This knowledge forms a foundation for innovating agro-pastoral practices to better adapt to the impacts of climate change and variability.

2 Background

For many ethnic groups living in the mountainous regions of the Eastern Himalayas, agro-pastoralism is crucial not only as a livelihood but also for biodiversity resource management and cultural preservation. For the Tibetan people in Deqin, agro-pastoralism plays a significant role in their way of life (Fig. 1).

In the agro-pastoral system, agriculture and herding are interdependent. Both practices are essential for the Tibetan people, providing basic and necessary food. Livestock herding supplies manure needed for crops, while crop straw serves as important fodder for the livestock. Additionally, after harvest, farmland can be used as grazing land during the winter season (Fig. 2).

Fig. 2 Fertilization of farmland (photoed by Yin Lun)

The Tibetan people primarily raise yaks, cattle yaks, yellow cattle, and goats. They manage their herding practices according to the seasons and physical locations, classifying grazing areas into three types: summer grazing land (alpine pasture), spring–autumn interim grazing land, and winter grazing land. In Tibetan, the summer grazing land

Fig. 3 Yaks on alpine pastures (photoed by Yin Lun)

is called "Ru La", meaning grassland covered by snow, located at an altitude of around 4,000 m and referring to alpine meadows. The spring–autumn interim grazing land is referred to as "Ru Mei", meaning grassland at a middle altitude, around 3,000 m, and includes meadows and sloping land. The winter grazing land, known as "Ru Bo", means grassland near villagers' houses, at an altitude of about 2,000 m, and refers to sloping land close to the village (Fig. 3).

Villagers typically herd yaks in the alpine pastures and goats in the winter grazing land. Cattle yaks and yellow cattle are rotated among the three grazing areas. Each year, from March to April, cattle yaks and yellow cattle are moved from winter grazing land to spring interim grazing land. From May to July, they graze in the summer alpine pastures. In August, they move to autumn interim grazing land, and from September to February, they return to winter grazing land.

In Deqin, the main crops grown by local people include highland barley, wheat, corn, potatoes, buckwheat, and creeping wood. They classify their farmland into two types based on its irrigation capability: "Qing Wan" and "Ri Xing" (Fig. 4).

> "Qing Wan" refers to irrigated farmland that can draw water directly from the river. This type of land is flat and allows for two planting cycles per year. Villagers practice intensive cultivation on this land. In contrast, "Ri Xing", also known as "Ni La", refers to non-irrigated sloping land. This type of farmland is located at higher altitudes and is steep, allowing only one planting cycle per year. "Ri Xing" is associated with slash-and-burn cultivation and retains traditional practices such as relying on climate and crop rotation. Typically, villagers plant large areas but harvest smaller yields from "Ri Xing".

Each year in May (April in the lunar calendar), villagers plant corn on "Qing Wan", and in October (November in the lunar calendar), they plant highland barley and wheat. On "Ri Xing", they cultivate potatoes, some buckwheat, and creeping wood. Additionally, they grow vegetables, beans, and chili peppers near their homes (Yin & Zachary, 2018).

Fig. 4 Harvest in farmland (photoed by Yin Lun)

3 Methodology

Considering the importance of traditional agro-pastoralism to local livelihoods and culture, as well as its unique value for local biodiversity, we have conducted a literature review and primary data collection to assess the impact of climate change on local livelihoods and responses. Based on this, we have developed community-based adaptation strategies to climate change that draw on traditional knowledge.

3.1 Literature Review

The extensive literature review included published research papers and relevant reports on local perceptions, traditional knowledge, and community-based adaptation strategies within agro-pastoralist livelihoods globally, with a particular focus on the Himalayan region.

The review also covered approaches to data collection in ecology and natural resource management, including the local ecological knowledge of resource users (White et al., 2005). Four main types of local indicators derived from this knowledge were used to explain climate change: local observations of climate change, including changes in temperature, precipitation, and wind, and their impacts on physical, biological, and socio-economic systems (Reyes-García et al., 2015).

Both quantitative and qualitative participatory rural appraisal (PRA) methods were employed in this research. These methods included key informant interviews, focus group discussions, household interviews, historical mapping of climate-related events over recent years, wealth ranking of socio-economic groups based on local criteria, and direct field observations through transect walks (Kangalawe et al., 2005; Lema & Majule, 2009; Yanda et al., 2005).

3.2 Data Collection

The study employed a mixed-methods approach to collect primary data, utilizing multiple techniques including structured questionnaires, semi-structured surveys, in-depth interviews, key informant interviews, historical mapping, and focus group discussions (FGDs).

Primary data collection was conducted across six villages with varying altitudes. The altitudes of the study villages ranged approximately from 2,200 to 3,800 m above sea level. Summer grazing occurs in alpine meadows at altitudes up to 4,500 m, while winter grazing takes place in areas as low as 2,500 m, located in temperate and subtropical forests. These villages were selected because the residents practice agro-pastoralism and depend heavily on livestock for their livelihoods. The study also aimed to compare the impacts of climate change across different altitudes on agro-pastoralism.

A structured questionnaire was used to gather data on local perceptions of climate change through face-to-face interviews. Three hundred questionnaires were distributed across the six villages, with fifty questionnaires per village to obtain basic information. Based on these data, a semi-structured survey was conducted with 180 households engaged in agro-pastoralism, with 30 households surveyed in each village. Both the structured questionnaires and semi-structured interviews, which included open-ended questions, were administered in the local language at the villagers' residences to maximize the response rate.

Through structured questionnaires and semi-structured household interviews, we identified local experts with extensive experience and traditional knowledge of agro-pastoralism among villagers. A total of 75 local experts were identified across the six villages. Based on their expertise and the characteristics of agro-pastoralism livelihoods, these experts were categorized into two groups: pastoral experts and agricultural experts. We found that the distribution of local experts and their knowledge correlates with altitude; specifically, high-altitude villages tend to have more experts in pastoralism, while low- and medium-altitude villages have more experts in agriculture. These local experts were interviewed using an in-depth interview guide.

In addition to local experts, some villagers were selected as key informants based on their age and gender to understand diverse perspectives on the impact of climate change. Historical mapping was also conducted to document climate-related events over the past 20 years. This mapping captured villagers' perceptions of important climate change indicators, including changes in rainfall, temperature, and extreme weather events, and how these changes have affected agro-pastoralism, particularly crop and livestock production.

Furthermore, focus group discussions were held in each village, totaling six groups. Each group comprised villagers engaged in both agricultural and pastoral activities, with participants aged between 30 and 60, including 25 older villagers and 30 women, recognizing the increasingly significant role of women in agro-pastoralism. These discussions provided both quantitative and qualitative insights into varying perceptions and responses to climate change.

3.3 Community-Based Action Research

Based on the data collection and field survey above, we try to analyze the adaptation strategies to cope with climate change of the villagers based on their agro-pastoralism livelihood. In the process of long-term livelihood practice, local villagers have been sensing and adapting to the surrounding environmental risks, and they know and understand the change of climate conditions and its threat to their livelihood. The local villagers have observed long-term increased temperatures, declining and pattern change in precipitation, and increased in drought frequencies changes in rain patterns as results of climate change. The vulnerability and perception to climate change are influenced by factors such as soil fertility, lack of finance, access to water for irrigation and access to climate information, etc., especially villagers with rich traditional knowledge are likely to have high awareness of climate risks and better adaptive capacity.

Local villagers' perceptions of changes in climatic conditions and traditional knowledge shape their response and adaptation practices to risks associated with climate change. So after the determination of villagers' perceptions on climate change, the impact factors of climate change and traditional knowledge of agro-pastoralism livelihood, we conducted the community-based action research with villagers, especially the local experts, to develop the adaptation strategies to climate change in rural level. This gave a better insight on the situation and guides the development of the local adaptation framework.

4 Results and Discussion

Historically, Tibetans were nomadic. However, upon migrating from the Qinghai-Tibet Plateau to Deqin in the Eastern Himalayas, they encountered new geographical and climatic conditions and were influenced by the surrounding farming ethnic groups. As a result, the local Tibetans gradually transitioned to a settled lifestyle and began developing agriculture, while retaining their tradition of animal husbandry. This evolution led to the formation of a unique agro-pastoral system that combines both agriculture and animal husbandry.

The animal husbandry practiced within this agro-pastoral system differs from traditional nomadism and from livestock breeding in settled agriculture. It has specific requirements for grazing areas, seasonal management, and labor organization, while also integrating traditional knowledge linked to climate and weather conditions.

In recent years, local villagers have observed increasingly evident signs of climate change and its impacts on their traditional livelihoods. In response, they have begun adopting adaptation strategies rooted in their traditional knowledge to mitigate the effects of climate change.

4.1 Local Perception on the Climate Change

The field survey results indicate that most villagers have observed the impact of climate change on their agro-pastoral livelihoods, with many effects being negative. These negative impacts include rising temperatures, unstable precipitation, erratic weather patterns, and changes in wind direction and intensity. These climate phenomena have led to changes in phenology, such as earlier growth of pasture grasses due to increased temperatures and earlier crop maturity due to an earlier onset of the rainy season. Additionally, villagers have noted extreme climate events, including prolonged droughts, heavy rainfall, and blizzards, which have resulted in natural disasters such as droughts, floods, debris flows, and avalanches.

4.1.1 Local Perception on the Climate and Phenology Change

The most noticeable effect of climate change observed by villagers is the rise in temperature, which has caused significant changes in phenology. This temperature increase is the most profound change mentioned by the villagers. Zhaba, a villager in his 50s, shared his observations:

> The temperature is much higher now than when I was a child and even compared to ten years ago. The most noticeable change is that I don't need to wear as many clothes in summer, and winter isn't as cold as it used to be. Additionally, there are more mosquitoes and flies, which I attribute to the hotter weather. The rise in temperature has also caused the grass on the pasture to grow earlier, so I have to move to the alpine summer pasture sooner than before. Similarly, some grasses mature earlier, requiring me to adjust the herd's movements to adapt to these changes.

Danzhen, a female villager, recounted her perceptions of climate change:

> Rainfall has become increasingly unpredictable. When it should rain during the normal rainy season, it often stays dry for weeks or even months. When it does rain, it's usually very heavy. This instability greatly impacts crop growth. If it doesn't rain when needed for irrigation, or if heavy rains flood the farmland, it affects both the harvest and the quality of the crops.

Other villagers also agreed that they had never witnessed such dramatic climate changes in the past. They noted that their traditional weather prediction knowledge, including proverbs related to phenology, now seems to be challenged by these unpredictable weather patterns.

4.1.2 Local Perception on the Extreme Weather and Disaster

Over the past decade, continuous high temperatures, low rainfall, and resulting droughts have been the primary extreme weather events and natural disasters faced by local villagers, with a severe negative impact on agro-pastoralism. Peichu, a herdsman with extensive grazing experience, shared his observations on these extreme weather events and natural disasters:

> Now, it's much hotter than before, and there is less snow in winter, particularly with prolonged dry spells. I believe these weather changes are accelerating the melting of glaciers and snow, leading to more natural disasters such as avalanches and floods in the alpine pastures. Last year, my herd and I encountered three avalanches on the way to the alpine pastures—one large and two small—events that used to be rare in this area. There are rivers in the alpine pastures formed by melting glaciers and snow. Over the past ten years, I've noticed that the water levels in these rivers are very unstable. Sometimes, sudden floods caused by increased water volume destroy parts of the pastures and pose threats to the herds.

The increase in temperature and ongoing drought have also led to a rise in diseases and insect pests, particularly affecting animal husbandry. Many livestock are now afflicted with diseases that were either previously unknown or rare. Gezong, a herdsman and local veterinarian with over 30 years of experience, explained:

> I have been treating livestock for more than 30 years. Recently, I have become much busier because there are more diseases affecting livestock, including some that I haven't encountered before. When faced with these unfamiliar diseases, I attempt traditional treatments, but sometimes I have no effective remedies. I believe these issues are due to factors such as rising temperatures, persistent drought, and prolonged periods without rain. In the past, winter was very cold with thick snow, which helped kill many pests and pathogens. Now, with the heat, diseases and pests from outside are more likely to enter.

Many herdsmen share this perspective, agreeing that their herds face increased threats due to higher temperatures and more frequent natural disasters. They also report a rise in livestock diseases and insect pests.

4.2 Impacts of Climate Change to Agro-Pastoralism

The climate change phenomena have seriously affected the agro-pastoralism livelihoods, both in agriculture and animal husbandry.

4.2.1 Impacts of Climate Change to the Varieties of Agro-Pastoralism

(1) Agricultural crop varieties

Our investigation identified ten main crop species, with highland barley being notably diverse in subspecies. Highland barley, known for its cold resistance, is represented by five subspecies across the six villages. Data analysis reveals that global warming is affecting crop species, leading to observable changes. Although pre-2005 data is unavailable, a comparison between 2005 and 2017 shows a shift in the types of crops planted.

Firstly, the number of households planting cold-resistant highland barley decreased significantly from 2005 to 2015. Conversely, the number of households growing other, less cold-resistant crops increased by 2017.

Secondly, we categorized the six villages into three altitude ranges: high, medium, and low. The analysis shows that the number of households growing cold-resistant highland barley was significantly lower in the low- and medium-altitude villages compared to the high-altitude villages. In the lower- and medium-altitude areas, some villagers have stopped planting certain highland barley subspecies. In contrast, the number of households growing non-cold-resistant crops in these lower- and medium-altitude villages has

Table 1 Comparison of crop plantation. (The information is sourced from field surveys conducted among villagers)

Crop Species	Numbers of households from high-altitude villages (n = 60 households)		Numbers of households from medium-altitude villages (n = 60 households)		Numbers of households from low-altitude villages (n = 60 households)	
	2005	2017	2005	2017	2005	2017
Black highland barley	44	34	36	28	34	26
Green highland barley	46	36	30	16	28	14
Red highland barley	44	30	37	25	34	22
White highland barley	36	32	19	3	16	0
Yellow highland barley	30	22	16	4	12	0
Wheat	50	50	50	50	50	50
Corn	50	50	50	50	50	50
Potato	28	38	29	41	24	36
Buckwheat	26	18	21	9	16	4
Creeping wood	34	26	30	18	24	12
Vegetable	10	38	27	50	20	50

Data is from investigation in 2005 and 2017

increased compared to those in high-altitude villages (Table 1).

(2) Pastoral livestock breeds

Our study examines the effects of climate change on traditional livestock breeds over the past two decades. The semi-nomadic lifestyle of local villagers has undergone significant changes, with a noticeable decline in the scale and number of cattle-based livestock breeding.

Among the five main livestock breeds raised by local villagers, yak and cattle yak are known for their cold resistance. Our investigation across six villages reveals that global warming is also affecting livestock species. A comparison of livestock numbers between 2005 and 2017 shows distinct trends.

Firstly, the number of yak and cattle yak per household decreased in 2017 compared to 2005, while the number of yellow cattle increased. Conversely, the numbers of pigs and chickens remained unchanged.

Secondly, an altitude comparison indicates that the number of yak and cattle yak per household decreased in the four low- and medium-altitude villages compared to the two high-altitude villages. In contrast, the numbers of yellow cattle, pigs, and chickens increased in the lower- and medium-altitude villages. Notably, some villagers in the low-altitude areas have ceased raising yak altogether (Table 2).

However, the changes in the above data do not show that there is a direct relationship between the annual decrease in the number of cattle raised and climate change. If we do not know the traditional knowledge about livestock breeds, we cannot see how climate change affects the local herdsmen to choose cattle varieties.

4.2.2 Impacts of Climate Change to the Resources and Phenological Knowledge of Agro-Pastoralism

(1) Impacts of climate change on management of agricultural land resource and phenological knowledge of corps

Deqin, situated in the eastern Himalayas, is characterized by its fragmented farmland, consisting mostly of small, flat, or sloping plots scattered along valley sides and hillsides. A significant manifestation of climate change's impact on agriculture in this region is the alteration of crop phenology. Our research focuses on how farmers'

Table 2 Comparison of livestock raised. (The information is sourced from field surveys conducted among villagers)

Livestock Species	Numbers of livestock raised in high-altitude villages		Numbers of livestock raised in medium-altitude villages		Numbers of livestock raised in low-altitude villages	
	2005	2017	2005	2017	2005	2017
Yak	642	327	557	262	474	153
Cattle Yak	597	261	523	302	457	284
Yellow cattle	523	302	457	284	384	257

Data is from investigation in 2005 and 2017

phenological knowledge and farming technologies have adapted in response to these changes.

Farmers' observations of crop phenology—key components of traditional agricultural knowledge—are crucial for selecting appropriate crop varieties and optimizing land use. As climate change affects these phenological patterns, farmers have adjusted their traditional knowledge and farming practices to better align with the evolving conditions. Thus, the impact of climate change on agriculture can be assessed through the lens of local farmers' management practices, land resource classification, and phenological understanding of crops.

Local farmers categorize their land based on altitude and irrigation mode into three types: dry land, alternate land, and irrigated land (Table 3).

1. Dry Land:

Dry land is found on sloped terrain or within alpine forests at altitudes exceeding 2500 m. This type of land does not have direct access to river water and relies on irrigation through constructed canals. It predominantly depends on rainfall and is characterized by broad planting areas with relatively low yields. Crops grown here include highland barley, corn, potatoes, nettles, and other cold- and drought-resistant species. Local farmers note that the high altitude and forested environment impart unique taste and nutritional qualities to the grain.

2. Alternate Land:

Alternate land is situated between 2300 and 2800 m above sea level, on hillsides away from rivers, and relies entirely on rainfall for irrigation. This land type is traditionally associated with shifting agricultural practices, characterized by rotational cultivation. Farmers grow crops such as wheat, buckwheat, and vines, which are harvested annually. After each growing season, the land is left fallow for a year before being replanted.

3. Irrigated Land:

Location and Characteristics: Irrigated land is situated on terraces along both sides of river valleys, with altitudes ranging between 2000 and 2500 m above sea level. The terrain is relatively flat, allowing for intensive cultivation and

Table 3 Traditional classification of land. (The information is sourced from field surveys conducted among villagers)

Type	Location	Local name	Meaning	Altitude and topography	Climatic conditions	Crops
Dryland	Hillside	Ri Xing	Sloping land without irrigation	2800–3500m above sea level	Drought	Potato, buckwheat, and sorghum
Alternate land	Hillside	Re ren	Rotation land without irrigation	2300–2800m above sea level	Drought	Wheat, buckwheat, turnip
Irrigated land	Valley	Qing Wan	Farmland that can directly draw water from river	2000–2500m above sea level	Moist	Corn, highland barley, and wheat

high yields. This land type benefits from direct irrigation with river water, either through direct access or via diversion canals. Farmers utilize irrigated land to grow two crops per year, including wheat, corn, highland barley, various beans, and vegetables.

Over 20 years ago, dry land and alternate land were crucial to local agriculture. According to elder villagers, ample rainfall at that time allowed crops in these areas to grow well, nearly matching the output of irrigated land.

In recent years, reduced precipitation and frequent droughts have led to the abandonment of dry land and alternate land. The challenges of managing drought and the high cost of irrigation have made these lands less viable. Crops like sorghum and buckwheat, once grown in these areas, have been largely discontinued. Potatoes have been relocated to courtyards near homes. Consequently, farmers have concentrated their efforts on irrigated land, which continues to be productive due to its ability to access water for irrigation.

Corn, highland barley, and wheat are cultivated in rotation on irrigated land. Spring plowing begins from May to June each year to plant corn, a process referred to as "Gai" in the local Tibetan language. Barley and wheat are then planted from September to October, a period known locally as "Nan." Between August and September, the irrigated land also serves as winter pasture for animal husbandry, where herders graze yaks and yellow cattle. This practice exemplifies the complementary relationship between agriculture and animal husbandry in agro-pastoralism.

Prior to spring plowing each year, local farmers visit the temple to request a ceremonial date from the monks. This ceremony, known as "Gawandan" in the Tibetan language, translates to "predicting the days of farming." Although the ceremony holds symbolic significance, villagers traditionally set the plowing date based on their experience and traditional knowledge. The ceremony is considered an auspicious ritual, aligning the start of plowing with the will of the sacred mountain and the Buddha, thereby seeking protection and blessings for the crops.

Table 4 Date and change of the spring plowing ceremony in high-altitude villages. (The information is sourced from field surveys conducted among villagers)

		High-altitude villages	
		Spring plowing ceremony	
		D	A
Year	2012	9th June	0
	2013	4th June	5
	2014	7th June	-3
	2015	3rd June	4
	2016	26th May	7
	2017	23rd May	3
Days ahead after 6 years in total		17	

Note D represents the date of the ceremony; A represents the number of days ahead of the previous year

Farmers use phenological indicators and traditional knowledge to determine the precise timing of spring plowing and to conduct the "Gawandan" ceremony. By comparing the dates of the ceremony over the years, we can observe changes in phenology and traditional knowledge in the context of climate change and assess the impact of these changes on agricultural practices. Our analysis indicates that the date of the ceremony has been advancing progressively. From 2012 to 2017, the ceremony's date moved forward by approximately half a month (Tables 4 and 5).

Compared to lower-altitude villages, the spring plowing ceremony in high-altitude villages occurs later. However, in all villages—whether at high, medium, or low altitudes—the start of spring plowing has progressively advanced each year. The increase in temperature due to climate change has caused soil and river water temperatures, which are crucial for irrigation, to rise. This, in turn, has led to the earlier ripening of barley and wheat. Consequently, climate change has altered key phenological phenomena and subsequently shifted the traditional knowledge that farmers have relied upon. As a result, farmers have been compelled to adjust their plowing schedules annually, with the start of plowing moving progressively earlier. These shifts can be clearly observed through the annual comparison of the spring plowing ceremony dates.

Table 5 Date and change of the spring plowing ceremony in medium- and low-altitude villages. (The information is sourced from field surveys conducted among villagers)

		Medium-and low-altitude villages	
		Spring plowing ceremony	
		D	A
Year	2012	5th June	0
	2013	2nd June	3
	2014	4th June	-2
	2015	29th May	5
	2016	24th May	5
	2017	20th May	4
Days ahead after 6 years in total		16	

Note D represents the date of the ceremony; A represents the number of days ahead of the previous year

(2) Impacts of climate change on management of pastoral meadow resource and phenological knowledge of herbage

Located in the Eastern Himalayas, Deqin County is home to extensive alpine meadows. Scientific research has shown that climate change significantly affects the phenology of herbage in this region. Our study analyzes the impact of climate change on the traditional phenological knowledge and grazing practices of local herdsmen. The ability to observe and understand shifts in herbage phenology is crucial for herdsmen as they manage pasture resources to sustain their livestock. Climate change has not only altered the phenology of herbage but has also influenced traditional grazing knowledge and techniques. The local herdsmen's classification of pastures and their grazing practices offer a new perspective for examining the effects of climate change on traditional livelihoods. Based on seasonal temperatures and altitude, herdsmen categorize pastures into three main types: alpine pastures for summer, transition pastures for spring and autumn, and valley pastures for winter (Table 6).

To fully utilize the advantages of the local pasture's vertical distribution and forage resources, herdsmen have developed a traditional grazing system based on moving between pastures at different altitudes to adapt to seasonal changes, temperature variations, and forage availability. The spring migration begins in mid-May each year when the weather warms, and the pastures in the transitional areas begin to turn green. The herdsmen drive their cattle from the valley pastures to graze in the transitional pastures for about a month. Once the weather further warms and the alpine pastures begin to green, they move the cattle to the alpine pastures, where they remain until August. In late August, as temperatures drop and alpine pastures wither, the herdsmen return with their cattle to the transitional pastures for another month before eventually moving back to the valley pastures in October, where they stay until the following May.

On the first day of the spring migration, the herdsmen perform a sacrificial ceremony known as Song La Deng, meaning "the ceremony of asking for permission from the sacred mountain for herding." Upon arrival at the alpine pasture, they hold another ritual called Song Zhe La, or "the ceremony of asking the sacred mountain to bless the grazing." Before the winter migration, they

Table 6 Traditional classification of pastures. (The information is sourced from field surveys conducted among villagers)

Type	Season	Local name	Meaning	Altitude	Climatic conditions	Grazing livestock
Alpine pastures	Summer	Rula	Grassland with snow	4000m to 5000m above sea level	Cold	Yak, Cattle Yak
Transition pastures	Spring and autumn	Rumei	Grassland in the middle	3000m to 4000m above sea level	Warm	Yak, Cattle Yak, Yellow Cattle
Valley pastures	Winter	Rubo	Grassland near home	2000m to 3000m above sea level	Dry hot	Cattle Yak, Yellow Cattle

perform the Song La Deng ceremony again, this time as a gesture of gratitude to the sacred mountain for allowing them to graze during the year.

Tracking the migration ceremonies in six villages over the past six years reveals a progressive advancement of the ceremony dates. Using 2012 as a baseline, by 2017, the ceremonies had advanced by approximately half a month (Tables 7 and 8).

From the data in the table, it is evident that the timing of the three key ceremonies—spring, grazing, and winter—has progressively advanced, indicating that the seasonal transitions are also occurring earlier. In low- and mid-altitude villages, the transfer to alpine pastures occurs earlier than in higher-altitude villages due to the greater distance between the pastures.

The rising temperatures have prompted earlier transitions, altering the traditional organization of animal husbandry activities and practices among local herdsmen. This shift has led to difficulties in accurately determining the optimal times for cattle mating, calving, and milk production. Consequently, the timing for producing dairy products such as butter and milk curds has changed, with shorter storage periods that increase the risk of spoilage.

The purpose of the spring transition is to ensure that cattle can graze on fresh alpine pasture grass in a timely manner. The earlier transitions are

Table 7 Date and change of the transition ceremony in high-altitude villages. (The information is sourced from field surveys conducted among villagers)

		High-altitude villages					
		Spring transition ceremony		Grazing ceremony		Winter transition ceremony	
		D	A	D	A	D	A
Year	2012	27th May	0	24th June	0	28th September	0
	2013	24th May	3	20th June	4	26th September	3
	2014	20th May	4	18th June	2	24th September	2
	2015	19th May	1	15th June	3	20th September	4
	2016	16th May	3	14th June	2	18th September	2
	2017	13th May	3	11th June	3	14th September	4
Days ahead after 6 years in total		14		14		15	

Note D represents the date of the ceremony; A represents the number of days ahead of the previous year

Table 8 Date and change of the transition ceremony in medium–and low-altitude villages. (The information is sourced from field surveys conducted among villagers)

		Medium-and low-altitude villages					
		Spring transition ceremony		Grazing ceremony		Winter transition ceremony	
		D	A	D	A	D	A
Year	2012	25th May	0	21st June	0	26th September	0
	2013	23rd May	2	17th June	4	24th September	2
	2014	19th May	4	16th June	1	21st September	3
	2015	17th May	2	13th June	3	17th September	4
	2016	14th May	3	9th June	2	17th September	0
	2017	12th May	2	8th June	1	14th September	3
Days ahead after 6 years in total		13		13		12	

Note D represents the date of the ceremony; A represents the number of days ahead of the previous year a result of the earlier onset of the grass growth period, which is closely linked to climate change. Similarly, the winter transition occurs due to the withering of forage in the high-altitude pastures and overgrazing. The advance of the forage withering period, driven by climate change, has caused the winter transition to move earlier as well. Thus, the shift in the timing of these ceremonies serves as evidence of phenological changes in herbage from the perspective of traditional knowledge, while also confirming the impact of climate change on phenological understanding and grazing practices.

4.2.3 Impacts of Climate Change on Diseases and Insect Pests of Agro-Pastoralism

(1) Impacts of climate change on diseases and insect pests of corps

The most common crop diseases in Deqin are wheat rust and ear seedless disease, known locally as "Sani" and "Gena" in Tibetan. In recent years, the frequency of wheat rust outbreaks has increased significantly, affecting larger areas of agricultural land. Nearly all villages, regardless of altitude, have been impacted, with low-altitude villages suffering the most severe losses—some experiencing near-total crop failure. Farmers believe these trends are closely linked to climate change.

Firstly, villagers attribute the rise in crop diseases and pests to the shifting phenological patterns caused by climate change. They believe that crops sown earlier in the season have a diminished ability to resist diseases and pests, leaving them more susceptible to infection. Damaji, a 60-year-old farmer with extensive experience, noted:

> More and more crop diseases and pests have emerged over the past decade, and this is closely related to climate change. The traditional planting times have shifted, and crops sown earlier are now much more vulnerable to disease.

Secondly, rising temperatures have led to the introduction of exotic crop varieties, which are less resistant to local diseases and pests, making them more prone to infection. Furthermore, the warming climate has facilitated the spread of non-native diseases and pests, some of which were previously unseen in the region. Anama, a 55-year-old farmer, explained:

> The temperature rise has encouraged farmers to grow exotic crops that couldn't survive the colder climate before, but these new crops bring with them diseases and pests. In response to natural disasters like drought and wind, we've introduced new crop varieties, but these are more vulnerable to diseases and pests.

Additionally, warmer winters and reduced snowfall have diminished the natural control over pests in farmland. Lamu, a 50-year-old female farmer, shared her observations on the rise in crop pests:

> The number of pests in crops is noticeably higher than when I was younger. I believe the main cause is the warming weather. Winters used to be much colder, with thick snow that killed pests hiding in the soil. Now, the winters are warmer, and snowfall is rare. In our valley village, it snows maybe once a year, and it melts quickly, which doesn't kill the pests in the soil at all.

Climate change may also disrupt the balance between crop pests and their natural predators. Lachu, a 65-year-old female farmer, explained:

> There's a native bird called Sara, which is golden and larger than pigeons. These birds specialize in eating pests on crops, especially during the semi-mature stage of wheat and highland barley. When I was young, these birds were common in our village's farmlands. Now, they are much rarer. These birds prefer cool, humid environments, but with the weather getting hotter and drier, they've moved to higher altitudes.

Furthermore, extreme weather events and disasters have exacerbated the occurrence of crop diseases and pests. For instance, heavy rains increase the susceptibility of corn, wheat, and barley to rust disease, while prolonged droughts contribute to more frequent outbreaks of crop pests.

Beyond the rise in plant diseases and pests, extreme weather conditions also negatively impact the planting and growth of crops. For example, increasingly severe windstorms have significantly affected the cultivation of traditional tall-stalk corn and highland barley, causing large areas to collapse and resulting in drastic yield reductions. Similarly, snowstorms during the grain-filling period have caused widespread lodging of highland barley and wheat. Extreme droughts, meanwhile, have led to the large-scale death of crops due to water shortages, further reducing yields.

(2) Impacts of climate change on diseases and insect pests of livestock

Livestock, including cattle, are susceptible to various diseases and pests throughout their lives. Without timely and effective treatment, these conditions can worsen, potentially leading to severe complications or death. The emergence, transmission, and outbreaks of many such diseases are closely linked to climatic conditions.

According to investigations conducted by the Deqin Veterinary Station and local veterinary experts, several common cattle diseases and pests have been identified, including toxic diarrhea, parainfluenza, emphysema, gangrene, tapeworm infection, myiasis, acariasis, and omasum obstruction. Among these, three are directly connected to climate factors (Table 9).

Three cattle diseases and pests are particularly influenced by weather conditions. For example, parasitic diseases like myiasis are typically controlled when winter temperatures are cold enough to kill off parasite eggs. However, with milder winters, fewer eggs are being eliminated. Concurrently, the increasingly hot weather, along with sudden and prolonged droughts, has led to a rise in diseases such as emphysema and omasum obstruction, which are related to respiratory and digestive system complications. These diseases have become more prevalent in recent years. The table below reflects the trend in the incidence of these three cattle diseases from 2007 to 2017 (Table 10).

The data in the above table indicate that despite an overall decline in the cattle population from 2007 to 2017, the incidence of three climate-related diseases and pests—emphysema, myiasis, and omasum obstruction—continues to rise, underscoring the significant impact of climate change on livestock farming.

4.3 Traditional Knowledge of Agro-Pastoralism and Climate Change Adaptation

4.3.1 Traditional Knowledge of Corps Varieties and Climate Change Adaptation

There is a close relationship between agriculture and climate, as farmers select appropriate crop varieties based on varying climatic condi-

Table 9 Diseases and insect pests are directly related to climate factors. (The information is sourced from field surveys conducted among villagers)

Name	Category	Symptom	Climate-related etiology
Emphysema and gangrene	Infectious diseases	In case of acute air swelling in the muscle, the skin of the swelling part is dark red or black, the cow is mentally depressed, unwilling to exercise, and does not eat or drink, resulting in dyspnea, lying on the ground, temperature drop; if not treated in time, it will die in 1–2 days	It often occurs during continuous high temperature, hot weather, and drying
Myiasis	Parasitosis	Emaciation, hypoplasia, and decrease of milk quantity	In winter, the larvae can be frozen to death. In summer, adults reproduce, mate, and lay eggs in large numbers
Omasum obstruction	Diseases of digestive system	Dry nostrils, fever, dyspepsia, bad appetite, sleepy every day	The disease is caused by drought, hot weather, and insufficient drinking water

Table 10 Changes in the number of cattle diseases and insect pests in six villages. (The information is sourced from field surveys conducted among villagers)

Types of diseases	Year and case of illness										
	2007	2008	2009	2010	2011	2012	2013	2014	2015	2016	2017
Emphysema and gangrene	31	42	47	44	53	67	71	74	67	81	93
Myiasis	113	106	125	117	123	137	149	153	144	157	164
Omasum obstruction	34	47	68	53	75	83	97	109	105	101	117

tions. Through this selection process, less suitable varieties are gradually phased out, while those well-adapted to the climate are cultivated. Over time, this process has led to the accumulation of extensive experience in seed selection, resulting in traditional knowledge related to crop varieties. This traditional knowledge now serves as a foundation for developing effective adaptation strategies in the face of climate change.

Highland barley, a traditional crop in Deqin with a long cultivation history, exemplifies this accumulated knowledge. Compared to wheat and corn, Tibetan farmers possess more extensive traditional knowledge about highland barley. The essence of this traditional knowledge lies in the classification of highland barley. Unlike scientific classifications, local farmers categorize highland barley in various ways: by color (white, black, and yellow highland barley), by shape (long-stalked and short-stalked highland barley), by planting location (mountain and valley highland barley), and by climate adaptability and resistance to disasters (drought-resistant and cold-resistant highland barley).

Among these classification methods, the one based on climate adaptability and disaster resistance is the most detailed. In this classification, highland barley is divided into 17 varieties across six categories: drought resistance, barren resistance, cold resistance, wind resistance, moisture resistance, and high-temperature resistance (Table 11).

From the table above, it can be seen that:

- Cuwa, Jiezhigeluna, He Li, and Jiongnai are drought-resistant varieties that thrive in warm climates and are cultivated in medium- and low-altitude farmlands.
- Ge Nie, Ge Jiang, and Ge Si are barren-resistant varieties that adapt to both cold

Table 11 Relationship between traditional classification of highland barley and climate condition. (The information is sourced from field surveys conducted among villagers)

Types of highland barley			Characteristic	Climate condition and altitude	
No	Local Tibetan name	Latin name		Temperature	Altitude
1	Cuwa, Jiezhigeluna, He li, Jiongnai	Hordeum vulgare	Drought resistance	Warm	Medium and low altitude
2	Ge nie, Ge jiang, Ge si		Barren resistance	Cold, warm	High, medium, and low altitude
3	Jiu ge, Helugeme, Ge na		Cold resistance	Cold	High altitude
4	Ma lu, Rige, Ge le		Wind resistance	Cold, warm	High and medium altitude
5	Gecai, Ni wo		Moisture resistance	Cold, warm	High, medium, and low altitude
6	Naina, Shi na		High-temperature resistance	Warm	Low altitude

and warm climates and are planted in high-, medium-, and low-altitude farmlands.

- Jiu Ge, Helugeme, and Ge Na are cold-resistant varieties suited to cold climates and are grown in high-altitude farmlands.
- Ma Lu, Rige, and Ge Le are wind-resistant varieties that adapt to both cold and warm climates and are cultivated in high-, medium-, and low-altitude farmlands.
- Gecai and Ni Wo are moisture-resistant varieties that are suitable for both cold and warm climates and are grown in high-, medium-, and low-altitude farmlands.
- Naina and Shi Na are high-temperature-resistant varieties that thrive in warm climates and are planted in low-altitude farmlands.

The impact of climate change on traditional knowledge of crop varieties can be observed through the planting trends of 17 highland barley varieties in six research villages over the past six years:

1. High-altitude Villages: There is an increasing preference among farmers for drought-resistant highland barley varieties, particularly short-stalked highland barley. This shift is due to the worsening wind disasters in high-altitude areas, which have caused significant lodging of traditional highland barley, thereby severely affecting yields.
2. Low-and Medium altitude Villages: In these regions, villagers are increasingly choosing highland barley varieties that are resistant to high temperatures and barren conditions. This change is prompted by rising temperatures and soil warming, which have led to decreased land fertility.
3. Villages Across All Altitudes: In both high- and low- to medium-altitude villages, there is a notable increase in the adoption of drought-resistant highland barley varieties. This trend is attributed to the frequent droughts over the past decade, which have posed major risks and challenges to agriculture.

Simultaneously, traditional highland barley varieties with other characteristics are being gradually phased out, and in some villages, they are no longer cultivated.

In response to the crop diseases and pests exacerbated by climate change, local farmers continue to rely on traditional knowledge. For example, to combat wheat rust, farmers plant cedar stumps around affected fields and apply lime and stove ash to prevent further infection. For seedless disease, which causes highland barley, wheat, and corn ears to turn black and remain seedless, farmers dry the seeds before sowing to prevent the disease.

The control of crop pests is also intertwined with traditional beliefs. Local farmers attribute pest damage to the unfulfilled desires of the deceased, believing that these spirits turn into pests. To address this, they hold a sacrificial ceremony called "Duduo" led by monks. During the ceremony, farmers collect over 100 types of grass and various soils, mix them with butter or walnut oil, and burn the mixture into scorched earth. After chanting sutras and receiving blessings from the monks, the scorched earth is scattered on infested fields to repel the pests.

4.3.2 Traditional Knowledge of Animal Husbandry Varieties and Climate Change Adaptation

Climate change significantly influences changes in herdsmen's breeding knowledge. In this context, the local herdsmen's classification system for different cattle varieties offers a new perspective for analyzing the impact of climate change on traditional livelihood technologies and methods, such as grazing and breeding.

Cattle are the primary large livestock in local animal husbandry, including yak, cattle yak, and yellow cattle. However, according to the local herdsmen's classification system, there are seven distinct types of cattle: Ra (male yak), Nega (female yak), Yong (male yellow cattle), Ba (female yellow cattle), Zong (male cattle yak), Gele (female cattle yak), and Nezu (male cattle yak). Additionally, Ai, Dele, and Jimo are also included in their classification (see Fig. 5).

According to the different climate and temperature of living environment, the local herdsmen

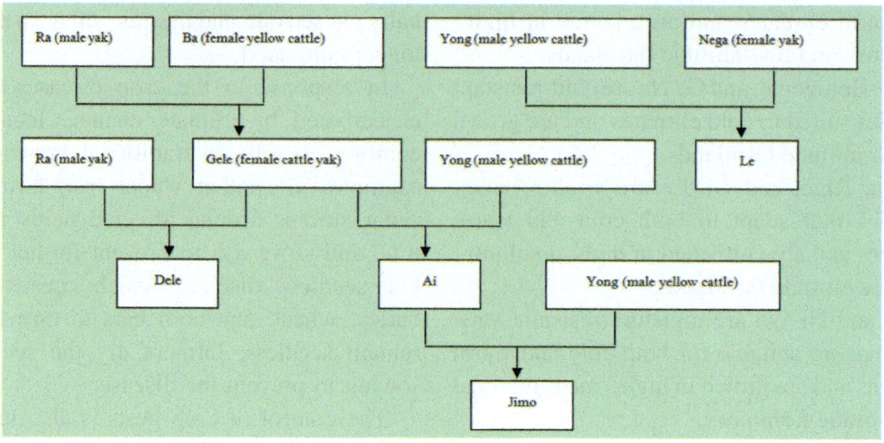

Fig. 5 Local herdsmen's cognition and classification system of cattle. (The information is sourced from field surveys conducted among villagers)

Table 12 Relationship between traditional classification of cattle and climate condition. (The information is sourced from field surveys conducted among villagers)

Types of cattle			Characteristic	Climate condition and altitude	
No	Local Tibetan name	English name		Temperature	Altitude
1	Ra, nega	Yak	Cold resistance	Cold	High altitude
2	Yong, ba	Yellow cattle	Drought tolerance	Warm	Low altitude
3	Zong, gele	Cattle yak	Cold resistance, Drought tolerance	Cold, warm	High, medium, and low altitude
4	Nezu, le		Cold resistance	Cold	High altitude
5	Ai		Cold resistance, Drought tolerance	Cold, warm	High, medium, and low altitude
6	Dele		Cold resistance	Cold	High altitude
7	Jimo		Drought tolerance	Warm	Low altitude

grazed these seven kinds of cattle at different altitudes (Table 12).

From the table above, it can be observed that Ra, Nega, Nezu, Le, and Ai are classified as cold-resistant varieties that thrive in cold climates and graze in high-altitude alpine pastures year-round. In contrast, Yong, Ba, and Jimo are also cold-resistant but prefer warm climates, and they graze in low-altitude valley pastures throughout the year. Zong, Gele, and Ai are categorized as varieties resistant to both cold and drought, adapting to both cold and warm climates, and they rotate among the three types of pastures.

The impact of climate change on traditional animal husbandry knowledge is evident from the statistics on the number of these seven cattle breeds raised in six research villages over the past six years (Table 13).

From the data in the table above, it is evident that the gradual warming of the climate, particularly the prolonged drought lasting five years, has impacted the breed composition and quantity of cattle in local animal husbandry. In response to these changes, herdsmen have increasingly selected and bred drought-tolerant breeds such as Yong, Ba, and Jimo, as well as Zong, Gele, and Ai, which can adapt to varying altitudes and temperature environments. Conversely, Ra, Nega, Nezu, Le, and Dele, which are adapted to low-temperature environments, are no longer bred on a large scale. These breeds are now primarily used for

Table 13 Changes in the number and proportion of traditional cattle breeds. (The information is sourced from field surveys conducted among villagers)

Types	Year											
	2012		2013		2014		2015		2016		2017	
	N	%	N	%	N	%	N	%	N	%	N	%
ra、nega	294	21.2	267	20.5	214	17.0	194	16.8	187	17.6	133	14.4
yong、ba	317	22.9	303	23.3	332	26.3	314	27.2	293	27.6	246	26.5
zong、gele	235	17.0	217	16.7	231	18.3	233	20.2	196	18.5	177	19.1
nezu、le	173	12.5	167	12.8	128	10.2	117	10.2	93	8.8	84	9.1
ai	121	8.7	107	8.2	127	10.1	115	10.0	122	11.5	131	14.1
dele	141	10.2	133	10.2	107	8.5	96	8.3	73	6.9	54	5.8
jimo	103	7.4	108	8.3	122	9.7	84	7.3	98	9.2	102	11.0
Total	1384	100.0	1302	100.0	1261	100.0	1153	100.0	1062	100.00	927	100.00

Note N represents the number of traditional cattle breeds;% is the percentage of the total

breeding and grazing in high-altitude pastures year-round.

4.3.3 Participatory Technology and Climate Change Adaptation

Ethno-veterinary medicine, also known as traditional veterinary medicine, encompasses the knowledge, skills, methods, practices, and beliefs of communities involved in animal health management. In rural areas, ethno-veterinary medicine often employs community-based approaches to enhance animal health and provide basic veterinary services. This knowledge system, which is locally rooted and innovative, plays a crucial role in mitigating the adverse effects of climate change on animal diseases in livestock management.

In Deqin, ethno-veterinary medicine is vital for the livelihood of local animal husbandry due to its low cost, abundant medicinal resources, and convenience. However, recent increases in animal diseases and pests attributed to climate change have exposed the limitations of traditional practices, which may not fully address new and emerging diseases. Therefore, there is a need to develop participatory technology that includes capacity training for local experts and encourages the experimentation with national veterinary technologies to address livestock health issues more effectively.

Under the participatory technology development model, local experts have begun to collect and analyze samples of veterinary plant medicine, catalog traditional utilization methods, and select certain ethno-veterinary plants for semi-wild conservation planting, considering the impact of climate change on livestock diseases. Ethno-veterinary medicine offers valuable solutions for managing livestock diseases and pests affected by climate change. Investigations by local experts have identified useful plants and documented their treatment methods (see Appendix A).

Through participatory technology, local experts integrate ethno-veterinary knowledge with modern veterinary practices to adapt to local climate changes. This approach includes several key activities:

1. Training Local Experts: Local experts organize and catalog ethno-veterinary medicine technologies and medicinal plant resources. They acquire modern knowledge of animal epidemic prevention and disease control. They conduct experiments with traditional veterinary technologies tailored to local diseases and pests,

form experimental groups among villagers, and innovate traditional veterinary practices.

2. Establishing Local Veterinary Clinics: Local experts are involved in setting up veterinary clinics, equipping them with necessary tools and appliances, and protecting some medicinal plant resources through conservation efforts. This initiative aims to enhance veterinary services and includes training villagers and younger generations to promote the inheritance and development of ethno-veterinary knowledge.
3. Creating a Climate Change Fund: The fund supports technological innovation by providing small loans for experimental projects proposed by local experts and villagers. After collective discussion and approval, this financial support helps improve livelihood innovation and self-development capabilities.
4. Training in Breeding Technology: Agricultural technicians are invited to train local experts and villagers in yak breeding, walnut cultivation, and other relevant technologies. This training helps in selecting appropriate livestock, crops, and economic forest varieties, diversifying livelihood strategies, and enhancing economic income sources.

Climate change negatively impacts animal husbandry by increasing the frequency and severity of climate-related diseases and pests. Over time, local villagers have developed and refined their use of local biodiversity resources as veterinary medicine, accumulating valuable traditional knowledge and experience.

The participatory technology model enables local experts to organize, inherit, develop, and innovate ethno-veterinary knowledge, addressing the challenges posed by climate change to animal husbandry and livestock health. Through research, analysis, and practical application, this approach aims to identify methods for coping with climate change in animal husbandry and other livelihood areas, offering potential directions for future research in the field of ethno-veterinary science.

5 Conclusion

Frequent climate change and extreme climate events significantly increase the vulnerability and pressure on the traditional livelihoods of Indigenous Peoples and Local Communities (IPLCs) worldwide, including the Tibetan people in the eastern Himalayas. Climate change, combined with environmental degradation, intensive livestock production, population growth, urbanization, and other factors, is a critical element impacting traditional livelihoods.

For IPLCs, traditional livelihoods are not only essential for sustainable living but also integral to their cultural identity. In the case of Tibetan people in the eastern Himalayas, climate change has led to shifts in species, environments, and traditional knowledge, which in turn affect cultural practices and social values.

1. Challenge to Ethnic Identity: For Tibetans, highland barley and yaks are more than just agricultural products; they are potent symbols of their ethnic identity. Elders often express that "there will be no Tibetan people if highland barley is not planted and yaks are not raised." Therefore, the decline in highland barley cultivation and yak herding threatens Tibetan cultural identity.
2. Changes in Customs and Traditions: Climate change affects agricultural cycles and, consequently, traditional customs. For example, the timing of the wheat germination period used to coincide with the spring festival's arrow shooting match. As the germination period has shortened, this tradition can no longer be observed without damaging the wheat seedlings.
3. Shifts in Food and Drink Culture: Traditionally, highland barley is used to make "Zanba" (a staple food) and yak milk is used to produce butter, which are considered vital for health and strength. Recently, wheat and rice are increasingly replacing Zanba and yak butter as staple foods in the Tibetan diet.

4. Impact on Beliefs and Religion: Tibetan Buddhism incorporates highland barley and Zanba into religious festivals and ceremonies, where these items are used in rituals and prayers. The transition to using rice or wheat in place of highland barley and Zanba in some households represents a significant shift from traditional practices.

Traditional livelihoods are crucial for sustaining local communities and preserving biodiversity. They embody a holistic understanding of knowledge, culture, and cosmology. For IPLCs, adapting to and mitigating climate change is essential. While climate change is a global issue, localized efforts and adaptations based on traditional knowledge are vital. Additionally, innovating and evolving traditional knowledge can help communities better cope with the impacts of climate change.

Appendix A

Catalogue of medicinal plants for ethno-veterinary medicine. (The information is sourced from field surveys conducted among villagers).

Latin name of plant	Medicinal part	Curing disease
Aconitum carmichaeli	Root	Loss of appetite, refusal to eat
Acorus calamus	Root	Diarrhea, spasms, tremors, dyspnea
Agrimonia pilosa	Root	Diarrhea, no afterbirth
Allium sativum	Stem	Spasm, tremor, dyspnea, parasitosis, poisoning, fever, cold, diarrhea
Anemone hupehensis	Root and stem	Diarrhea, unable to stand for a long time
Anemone rivularis	Root	Parasitosis, unable to stand for a long time
Asparagus setaceus	Root	Fever, cold
Berberis wilsonae	Whole plant	Diarrhea
Bupleurum yunnanense	Root and fruit	Fever, cold
Caltha palustris	Whole plant	Skin disease
Cardamine hirsuta	Whole plant	Diarrhea, flatulence
Clematis yunnanensis	Whole plant	Fever, cold
Clerodendrumbungei	Root and stem	Sores
Coptischinensis	Root and stem	Diarrhea, poisoning
Davidia involucrata	Fruit	Diarrhea, fever, cold
Epilobium brevifolium	Root	Sores, no afterbirth
Euphorbia hirta	Whole plant	Skin disease, diarrhea
Foeniculum vulgare	Whole plant	Constipation
Gentiana rhodantha	Whole plant	Fever, cold
Hemiphragmaheterophyllum	Whole plant	No afterbirth
Heracleum scabridum	Root	Fever, cold
Impatiens lecomtei	Whole plant	Sores, no afterbirth
Kadsura interior	Root and stem	No afterbirth
Leonurus artemisia	Whole plant	No afterbirth
Magnolia rostrata	Stem	Spasm, tremor, dyspnea
Mahonia microphylla	Stem	Poisoning, diarrhea
Nothopanaxdelavayi	Root, stem, and skin	Loss of appetite, refusal to eat, spasm, tremor, dyspnea
Picrorhizascrophulariiflora	Root and stem	Fever, cold, parasitosis, sores
Plantago depressa	Whole plant	Flatulence, poisoning, constipation

(continued)

(continued)

Latin name of plant	Medicinal part	Curing disease
Polygonatumkingianum	Root and stem	Flatulence
Polygonum paleaceum	Root and stem	Fever, cold, poisoning
Potentilla fulgens	Root	Diarrhea, flatulence, constipation
Ranunculus chinensis	Whole plant	Diarrhea, parasitosis
Rubus corchorifolius	Whole plant	Diarrhea, parasitosis
Rumex nepalensis	Root and leave	Diarrhea, parasitosis, parasitosis
Sambucus williamsii	Root and leave	Fracture, unable to stand for a long time
Saussureacostus	Whole plant	Fever, cold
Schisandra sphaerandra	Fruit	Diarrhea, Fever, cold, Sores
Senecio scandens	Whole plant	Constipation, poisoning, diarrhea, dermatosis
Sinodielsiayunnanensis	Root	No afterbirth
Spinacia oleracea	Whole plant	No afterbirth, Sores
Vaccinium fragile	Root	Parasitosis, unable to stand for a long time
Verbena officinalis	Whole plant	Fracture, no afterbirth
Viola yezoensis	Whole plant	Sores, poisoning
Zea mays	Seed	Sores

References

Adams, R. M., McCarl, B. A., Segerson, K., Rosenzweig, C., Bryant, K. J., Dixon, B. L., Conner, R., Evenson, R. E. & Ojima, D. (1998). The economic effects of climate change on US agriculture. In R. Mendelsohn, & J. Neumann (Eds.), *Chapter 2, The Economics of Climate Change*, Cambridge University Press, Cambridge, MA.

Archer, D., & Dodman, D. (2015). Making capacity building critical: Power and justice in building urban climate resilience in Indonesia and Thailand. *Urban Climate. Available:* https://doi.org/10.1016/j.uclim.2015.06.007

Balgah, R. A., Mbue, I. N. & Buchenrieder, G. (2015). When nature frowns: A comprehensive impact assessment of the 2012 Babessi floods on people's livelihoods in rural Cameroon. *Jamba: Journal of Disaster Risk Studies* 7(1), 1–8. https://doi.org/10.4102/jamba.v7i1.197

Balgah, R. A., Kimengsi, J. N., & Buchenrieder, G. (2016). Understanding and building on indigenous agro-pastoral adaptation strategies for climate change in Sub-Saharan Africa: Experiences from rural Cameroon. *J Adv Agri, 6*(1), 833–840.

Carter, J. G., Cavan, G., Connelly, A., Guy, S., Handley, J., & Kazmierczak, A. (2015). Climate change and the city: Building capacity for urban adaptation. Progress in Planning 95, 1–66. http://linkinghub.elsevier.com/retrieve/pii/S0305900614000397

Cruz, R. V., Harasawa, H., Lal, M., Wu, S., Anokhin, Y., Punsalmaa, B., & Huu Ninh, N. (2007). Asia. Climate change 2007: impacts, adaptation and vulnerability. In M.L. Parry, O.F. Canziani, J.P. Palutikof, P.J. van der Linden, & C.E. Hanson (Eds.),*Contribution of Working Group II to the Fourth Assessment Report of the Intergovernmental Panel on Climate Change* (pp. 469–506). Cambridge, UK.

FAOSTAT. (2015). Food and agricultural organization of the UnitedNations, statistical division. Addis Ababa, Ethiopia. http://faostat3.fao.org/browse/Q/QA/

Freeman, H. A., Kaitibie, S., Moyo, S., & Perry, B. D. (2008). Livestock, livelihoods and vulnerability in Lesotho, Malawi and Zambia: designing livestock interventions for emergency situations. ILRI Research Report 8. ILRI (International Livestock Research Institute), Nairobi. http://mahider.ilri.org/bitstream/10568/236/1/FAO_ILRI_RRNo8.pdf

Goldman, M. J., & Riosmena, F. (2013). Adapting to Variabilityand Drought: The Shifting Relationship between InstitutionsandLandcapes in East Africa Savanna land. *Global Environmental Change, 23*, 588–597.

Hunn, E. (1993). What is traditional ecological knowledge? In N. Williams, G. Baines (Eds.) Traditional ecological knowledge: wisdom for sustainable development. ANU Canberra: Centre for Resource and Environmental Studies, (pp. 13–15).

IPCC. (2001). Climate change 2001: the scientific basis. Contribution of working group 1 to the third assessment report of the intergovernmental panel on climate change. In J.T. Houghton, Y. Ding, DJ. Griggs, M. Noguer, P.J. van der Linden, X. Dai, K. Maskell, C.A. Johnson (Eds.) Cambridge University Press, Cambridge.

IPCC. (2007). Climate change 2007: synthesis report. Contribution of working groups I, II and III to the fourth assessment report of the intergovernmental panel on climate change. In: Core Writing Team, Pachauri RK, Reisinger A (eds) IPCC, Geneva.

IPCC (Intergovernmental Panel on Climate Change). (2007).Climate change: impacts, adaptation and vulnerability. Contribution of Working Group II to the Fourth AssessmentReport of the Intergovernmental Panel on Climate Change. In I. Annex, edited by M. L. Parry, O. F. Canziani, J. P. Palutikof, van der Linden, and C. E. Hanson). Cambridge: Cambridge University Press.

IPCC. (2011a). Climate change: The physical science basis. Contribution of working group I to the fourthassessment report of the IPCC. Cambridge. Cambridge University Press.

Jones, P. G., & Thornton, P. K. (2009). Croppers to Livestock Keepers: Livelihood Transitions to 2050 in Africa Due to Climate Change. *Environmental Science & Policy, 12*(4), 427–437.

Kangalawe, R. Y. M., Majule, A. E., & Shishira, E. K. (2005). *An Analysis of Land Use Dynamics and Land Degradation Process In the Great Rift Valley, Central Tanzania: A Case of Iramba District*. Book Published by OSSREA. ISBN: 1904855628

Kangalawe, R. Y. M., & Liwenga, E. T. (2005). Livelihoods in the Wetland of Kilombero Valley in Tanzania: Opportunities and challenges to integrated Water Resour. *Management. Physics Chem. Earth, 30*, 968–975.

Krisna, A., Kristjanson, P., Radeny, M., & Nindo, W. (2004). Escaping poverty and becoming poor in 20 Kenyan villages. *Journal of Human Development, 5*(2), 211–226.

Lambi, C. M. (2001). Environmental constraints and indigenous agricultural intensification in NdopPlain (Upper Nun Valley of Cameroon), In C.M. Lambi., & B.E. Eze (Eds.) *Readings in Geography, Unique Printers*, Bamenda, Cameroon, 179–190.

Lema, M., & Majule, A. E. (2009). Impacts of climate change, variability and adaptation strategies on agriculture in semi arid areas of Tanzania: The case of Manyoni District in Singida Region, Tanzania. *African Journal of Environmental Science and Technology, 3*(8), 206–218. https://doi.org/10.5897/AJEST09.099

Luseno, W. K., Mcpeak, J. G., Barrett, C. B., Little, D., & Gebru, G. (2003). Assessing the value of climate forecast information for pastoralists: Evidence from Southern Ethiopia and Northern Kenya. *World Development, 31*(9), 1477–1494.

McPeak, J. (2006). Confronting the risk of asset loss: What role do livestock transfers in northern Kenya play? *Journal of Development Studies, 81*, 415–437.

Miller, D. (1995). Herds on the move: Winds of Change AmongPastoralists in the Himalayas and on the Tibetan Plateau. Discussion paper MNR 95/2. ICIMOD, Kathmandu.

Mulder, M. B., Fazzio, I., Irons, W., McElreath, R. L., Bowles, S., Bell, A., & Hazzah, L. (2010). Pastoralism and Wealth Inequality. *Current Anthropology, 51*(1), 35–48.

Muntifering, R. B., Chappelka, A. H., Lin, J. C., Karnosky, D. F., & Somers, G. L. (2006). Chemical composition and digestibility of trifolium exposed to elevated ozone and carbon dioxide in a free air (FACE) fumigation system. *Functional Ecology, 20*(2), 269–275.

Nelson, G.C. (Ed.) (2009). Agriculture and climate change: An agenda for negotiation in Copenhagen. No. 16. May 2009. www.ifpri.org/2020/focus/focus16.asp

Ngwa, K. A., & Balgah, R. A. (2016). Determinants of livelihood security among disaster victims in rural Cameroon. *International Journal of Recent Scientific Research, 7*(1), 8328–8334.

Nori, M., & Davies, J. (2007). Change of wind or wind of change? Climate Change, Adaptation and Pastoralism. WISP, IUCN, Nairobi.

Oerke, E. C. (2006). Crop losses to pests. *The Journal of Agricultural Science, 144*(1), 31–43.

Oppenheimer, M., Campos, M., Warren, R., Birkmann, J., Luber, G., O'Neill, B. & Takahashi, K. (2014). Emergent risks and key vulnerabilities. In C.B. Field, V.R. Barros, D.J. Dokken, K.J. Mach, M.D. Mastrandrea, T.E. Bilir, M. Chatterjee, K.L. Ebi, Y.O. Estrada, R.C. Genova, B. Girma, E.S. Kissel, A.N. Levy, S. MacCracken, P.R. Mastrandrea, and L.L. White (Eds.) *Climate Change 2014: Impacts, Adaptation, and Vulnerability. Part A: Global and Sectoral Aspects. Contribution of Working Group II to the FifthAssessment Report of the Intergovernmental Panel on Climate Change* (pp. 1039–1099). Cambridge University Press, Cambridge, United Kingdom and New York, NY, USA.

Oyekale, A. S. (2009). Climatic variability and its impacts on agricultural income and households' welfare in Southern and Northern Nigeria. *Electronic Journal of Environmental, Agricultural and Food Chemistry, 8*(1), 13–34.

Reyes-García, V., Fernández-Llamazares, A., Guèze, M., Garcés, A., Mallo, M., Vila-Gomez, M., & Vilaseca, M. (2015). Local indicators of climate change: The potential contribution of local knowledge to climate research. *Wiley Interdisciplinary Reviews: Climate Change, 7*(1), 109–124. https://doi.org/10.1002/wcc.374

Robinson, J., & Herbert, D. (2001). Integrating climate change and sustainable development. *International Journal of Global Environmental Issues, 1*(2), 130–148.

Scoones, I, & Wolmer, W. (2000). Pathways of change: Crop-livestock integration in Africa. In I. Scoones & W. Wolmer (Eds.), *Pathways of change: Crops, livestock and livelihoods in Africa* (pp. 1–28).

Seo, S. N. (2010). Is an integrated farm more resilient against climate change? *A Micro-Econometric Analysis of Portfolio Diversification in African Agriculture, Food Policy, 35*(1), 32–40.

Sewando, P. T., Mutabazi, K. D., & Mdoe, N. Y. S. (2016). Vulnerability of agro-pastoral farmers to climate risks in northern and central Tanzania. *Development Studies Research, 3*(1), 11–24. https://doi.org/10.1080/21665095.2016.1238311

Silvestri, S., Bryan, E., Ringler, C., Herrero, M., & Okoba, B. (2012). Climate change perception and adaptation of agro-pastoral communities in Kenya. *Regional Environmental Change, 12*(4), 791–802.

Swart, R., Robinson, J., & Cohen, S. (2003). Climate change and sustainable development: Expanding the options. *Climate Policy, 3*(1), 19–40.

Thornton, P. K., Herrero, M., Freeman, H. A., Mwai, A. O., Rege, E., Jones, P. G., & McDermott, J. (2007). Vulnerability, climate change and livestock-opportunities and challenges for the poor. *Journal of SAT Agricultural Research, 4*(1), 1–23.

Thornton, P. K., van de Steeg, J., Notenbaert, A., & Herrero, M. (2009). The impacts of climate change on livestock and livestock systems in developing countries: A review of what we know and what we need to know. *Agricultural Systems, 101*(3), 113–127.

UNFCC. (2016). UN: Managing disaster risk vital for sustainable development. http://bonnsustainabilityportal.de/?p=42326

UNISDR. (2015b). Disaster risk reduction and resilience in the 2030 agenda for sustainable development. http://www.unisdr.org/files/46052_disasterriskreductioninthe2030agend.pdf

Van Aalst, M. K., Cannon, T., & Burton, I. (2008). Community level adaptation to climate change: The potential role of participatory community risk assessment. *Global Environmental Change, 18*, 165–179.

Wane, W., Touré, A. I., & Ancey, V. (2009). Assets of the market, assets of the rural world: pastoral market income distribution in the Senegalese Sahel (Ferlo). *Journal of Income Distribution, 18*(3–4), 232–248.

Watson, R. T., & Albritton, D. L. (Eds.) (2001) Climate change 2001: Synthesis report: Third assessment report of the Intergovernmental Panel on Climate Change. Cambridge University Press.

Watson, C. (2010). Gender issues and pastoral economic growth and development in Ethiopia background note for a study of "Pastoral economic growth and development in Ethiopia". DFID &Government of Ethiopia, Adis Ababa.

White, P. C. L., Jennings, N. V., Renwick, A. R., & Baker, N. H. L. (2005). Questionnaires in ecology: A reviewof past use and recommendations for best practice. *Journal of Applied Ecology, 42*, 421–430.

Wolmer, W. (1997). Crop-livestock integration: The dynamics of intensification in Contrasting Agro-ecological Zones: A Review. Sussex: Institute of Development Studies.

Yanda, P. Z., Majule, A. E., & Mwakaje, A. G. (2005). Wetland utilization, poverty alleviation and environ. Conservation in Semi Arid Areas of Tanzania, the Case of Singida Region, Tanzania. In *Proceedings of East Africa "Integrated River Basin Management.* In*Conference Held on 7th-9th 2005 at ICE Hall*, Sokoine University of Agric. Morogoro, Tanzania.

Yin, L., Zachary, M., Zheng, Y., & Zhang, X. (2020). The impacts of climate change on the traditional agriculture of ethnic minority in China. *J. Environ. Eng. Sci 9*, 43–55. https://doi.org/10.17265/2162-5298/2020.02.001

Yin, L., & Zachary, M. (2018). Traditional Knowledge and Community-Based Research to Support Climate Change Adaptation in Eastern Himalayas. In S. J. P. Tume & V. I. Tanyanyiwa (Eds.), *Climate Change Perception and Changing Agents in Africa & South Asia* (pp. 159–170). Vernon Press.

Yin, L., Zachary, M., & Zheng, Y. (2018). The traditional knowledge associated to biodiversity in an age of climate change. *Earth Sciences, 7*(5), 209–215. https://doi.org/10.11648/j.earth.20180705.12

6 Traditional Knowledge in Forest Resource Management for the Climate Change Adaptation

1 Introduction

The global climate is changing, and the frequency and intensity of extreme weather events are projected to increase in future (IPCC, 2007). Climate change now represents a significant threat to the environment, natural resources, and ecosystems, including forests. Approximately 350 million people worldwide who live near dense forests depend on them for both subsistence and income (World Bank, 2004). Additionally, forest-dominated mountainous landscapes offer a broad range of ecosystem services to people living in the mountains, as well as to those in lower hills and plains (Grêt-Regamey et al., 2013). For many indigenous peoples and local communities, their livelihoods are closely tied to their surrounding environment, particularly forest resources and ecosystems. These populations are often among the first to experience the direct or indirect impacts of climate change, which increasingly poses diverse and complex threats to their way of life (Zvobgo et al., 2023).

Indigenous peoples and local communities possess rich and diverse traditional knowledge, technologies, and cultural practices that have been utilized for generations in the management of natural resources, climate change adaptation, and disaster risk reduction (UNFCCC, 2013). The global importance of traditional knowledge in climate change adaptation, mitigation, and disaster risk reduction is widely acknowledged (IPCC, 2014). Indigenous peoples and local communities rely on traditional knowledge for local decision-making concerning forest resource management. This knowledge allows them to assess available forest ecosystem resources, improve their understanding of climatic risks, and identify gaps in adaptive capacity within both individuals and institutions. Furthermore, it helps reduce adaptation costs by leveraging local knowledge and physical resources (Dahal, 2013). Drawing on their traditional knowledge, indigenous peoples and local communities have developed numerous strategies to cope with the impacts of climate change. While they may lack specific experience with modern climate change, they have a long history of adapting to environmental changes. Traditional knowledge reduces their vulnerability to the effects of climate change and enhances their resilience in increasingly challenging environments (Karki & Adhikari, 2015; Ahmed & Haq, 2017).

The Himalayas, often referred to as the "third pole" of the planet, possess ecosystems that are particularly fragile compared to other regions of the world. The effects of climate change are more pronounced here, with the region warming at a rate faster than the global average (Liu & Chen, 2000; Shrestha et al., 1999). Temperature increases are more significant during the winter and autumn seasons, and these increases are more pronounced at higher altitudes (Liu & Chen, 2000). In this context, Himalayan forests

Y. Lun, *Climate Change and Traditional Knowledge*, Sustainable Development Goals Series,
https://doi.org/10.1007/978-3-032-12015-1_6

are especially vulnerable to the impacts of climate change and face severe ecological degradation due to anthropogenic pressures (Ma et al., 2012). The effects of climate change are already evident in the forest systems of the Himalayas, contributing directly and indirectly to vegetation degradation through the invasion of alien species, increased occurrences of forest fires, and a reduction in forest productivity, including both timber and non-timber forest products. Himalayan forests are known for their multifunctionality, providing a range of ecosystem services that support the livelihoods of local communities (Rasul, 2014). These communities embody the long-standing legacy of human-nature interactions, which have shaped cultural landscapes and traditional forest resource management systems across many regions (Chakraborty et al., 2018).

In the Eastern Himalaya region of southwest China, climate change is increasingly affecting forest resources and posing significant threats to the traditional livelihoods of the local Tibetan people, whose survival relies heavily on these resources. Over generations, the local Tibetans have developed a rich body of traditional knowledge in forest resource management, encompassing observation, classification, utilization, protection, and spiritual beliefs related to the forest. This traditional knowledge is now undergoing dynamic changes as it adapts to the evolving climate (Hopping et al., 2016). The present research examines how, in the context of climate change, traditional knowledge helps local Tibetans manage their forest resources and develop climate adaptation strategies at the community level. Thus, in the Eastern Himalayas, Tibetan traditional knowledge of forest management and its adaptation to climate change offer valuable local insights for understanding the global implications of climate change.

2 Background

Due to the unique landforms and climate of Deqin, the region exhibits distinct vertical forest belts. The traditional livelihood of the local Tibetan people revolves around agropastoralism and the collection of non-timber forest products (NTFPs). These forest belts at varying altitudes are closely tied to their traditional way of life, providing essential natural and material resources.

2.1 Alpine Shrub, Meadow, and Animal Husbandry

The alpine shrub and meadow zones are located between 4,200 and 4,600 m above sea level. This area has a cold climate, expansive terrain, and frequent cold air convection. Common shrubs in these areas include azalea, cypress, and willow. This zone serves as the summer pasture for the local Tibetans, providing grazing grounds for their livestock. During the summer months, herders graze yaks, buffaloes, and yellow cattle here (Choki, 2021).

2.2 Temperate Evergreen Coniferous Forest and Cold Temperate Coniferous Forest Belts, and the Collection of Non-Timber Forest Products

The temperate evergreen coniferous forest and cold temperate coniferous forest belts are situated between 3,000 and 4,200 m above sea level. These areas experience abundant rainfall and a cold climate, with common tree species including spruce, fir, hemlock, pine, and cypress. Local Tibetans primarily collect NTFPs in these forests, such as the economically valuable Tricholoma matsutake and other wild fungi. The collection of NTFPs is a primary source of income for many villagers.

2.3 Subtropical Broad-Leaved Forest Belt and Agriculture

The subtropical broad-leaved forest belt ranges from 2,200 to 3,000 m above sea level and is characterized by arbors and shrubs such as linden, oak, hazel, yellow fir, and pine. In this zone, local

Tibetans traditionally cut down trees and branches for firewood and farmyard manure, providing vital support for agricultural activities. Historically, villagers also cleared forests in this area to create farmland. However, due to national policies promoting the return of farmland to forest, much of this farmland has since been abandoned and restored to forest.

3 Methodology

3.1 Ethno-Ecology Methods

Through semi-structured interviews and questionnaires with villagers, we investigated the relationships between climate change, different forest belts, and traditional livelihoods. Simultaneously, we conducted surveys to assess the role of forest environments and resources in supporting animal husbandry, foraging, and agriculture.

To facilitate this, we divided the surveyed villagers into three groups based on the forest belt and their traditional livelihoods. Additionally, we prioritized a gender-balanced approach, with each group consisting of 50 men and 50 women, for a total of 300 villagers across the three groups.

3.2 Community-Based Research

Community-based research methods were employed to explore the villagers' traditional knowledge of forest biodiversity, including traditional technologies, customary laws, and cultural practices. We also encouraged villagers to participate actively in research, documenting the impacts of climate change on forests.

The villagers were organized into three research groups according to their traditional livelihoods. Each group was led by local experts with traditional knowledge, who guided the villagers in conducting community research. Here again, a gender perspective was emphasized.

The first group focused on animal husbandry, which is traditionally male-dominated; thus, most members of this group were men with extensive knowledge of livestock management.

The second group studied the collection of non-timber forest products (NTFPs), a practice commonly undertaken by women. Consequently, this group was predominantly composed of women with deep knowledge of foraging and collection practices.

The third group investigated agriculture, which involves both men and women. As a result, the composition of this group was more evenly balanced in terms of gender (Moulton, 2024).

3.3 Traditional Knowledge-Based Forest Management and Adaptation Planning

Based on traditional knowledge and forest management practices, we collaborated with the villagers to discuss and establish a climate change adaptation plan. This plan addresses the negative impacts of climate change on forest resources and traditional livelihoods. It incorporates traditional knowledge, customary laws, traditional organizations, and cultural beliefs to respond to the various threats and challenges posed by climate change.

4 Results and Discussion

4.1 The Perspective of Forest in Climate Changes

With the increasing impacts of climate change, local villagers have both observed and experienced its effects on forests, often in ways closely tied to their traditional livelihoods.

4.1.1 Expansion of Alpine Shrublands and Shrinking Meadows

In the high-altitude alpine shrub and meadow regions, villagers have noticed a significant rise in temperatures over recent years, with the area becoming noticeably warmer than before. As a result, they have observed an expansion of alpine shrubs encroaching on the high-altitude meadows, causing these meadows to shrink year by year.

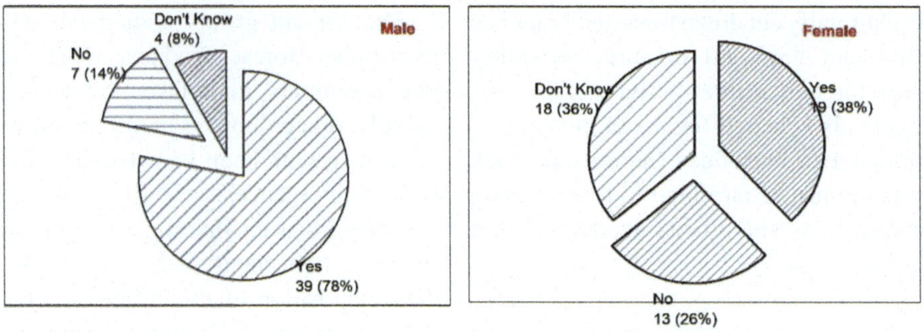

Fig. 1 Shows the expansion of alpine shrub and atrophy of meadow for men and women. (The information is sourced from a survey conducted among the villagers)

This reduction in meadowland has had a direct negative impact on animal husbandry, as the shrinking pastureland limits the availability of grazing areas for livestock.

To better understand this phenomenon, we conducted a questionnaire survey with 50 men and 50 women from the community. The results of this survey are presented below:

From Fig. 1, we can observe that compared to women, more men have reported noticing the effects of climate change on alpine shrubs and meadows. This is largely because men are primarily engaged in animal husbandry in this region. During grazing, they directly experience how rising temperatures are causing alpine shrubs to gradually encroach upon high-altitude meadows, leading to a reduction in meadow area. In contrast, women are generally not involved in animal husbandry or do not frequent the alpine pastures, so their observations of climate change's impact on forest vegetation and livestock are less direct.

4.1.2 Instability in NTFP Yields

In the temperate evergreen coniferous forests and cold temperate coniferous forest zones, villagers have become increasingly aware of the impacts of climate instability and extreme weather events, such as droughts and heavy rainfall. These climatic fluctuations directly affect the collection of non-timber forest products (NTFPs), resulting in significant instability in their yields.

To further explore this phenomenon, we conducted a questionnaire survey with 50 men and 50 women, and the results are summarized below:

From the above Fig. 2, we can see that women are primarily engaged in the collection of non-timber forest products (NTFPs). Consequently, women are more attuned to the impacts of climate instability and frequent extreme weather events in this region, which increase the uncertainty of their NTFP collection. Men, on the other hand, are generally not involved in NTFP collection and thus have less direct observation of its impacts. However, since NTFP collection is a major source of income for local families, men also experience indirect effects of these changes.

4.1.3 Slowdown in Vegetation Restoration

In the subtropical broad-leaved forest area, villagers have observed that rising temperatures are slowing down the restoration of vegetation following deforestation. Compared to previous years, the regeneration of new seedlings has noticeably diminished.

To investigate this phenomenon further, we conducted a questionnaire survey with 50 men and 50 women, and the results are presented below:

From Fig. 3, we can observe that both men and women are involved in agriculture, but their roles differ. Men primarily engage in firewood cutting in this forest belt, while women focus on cutting

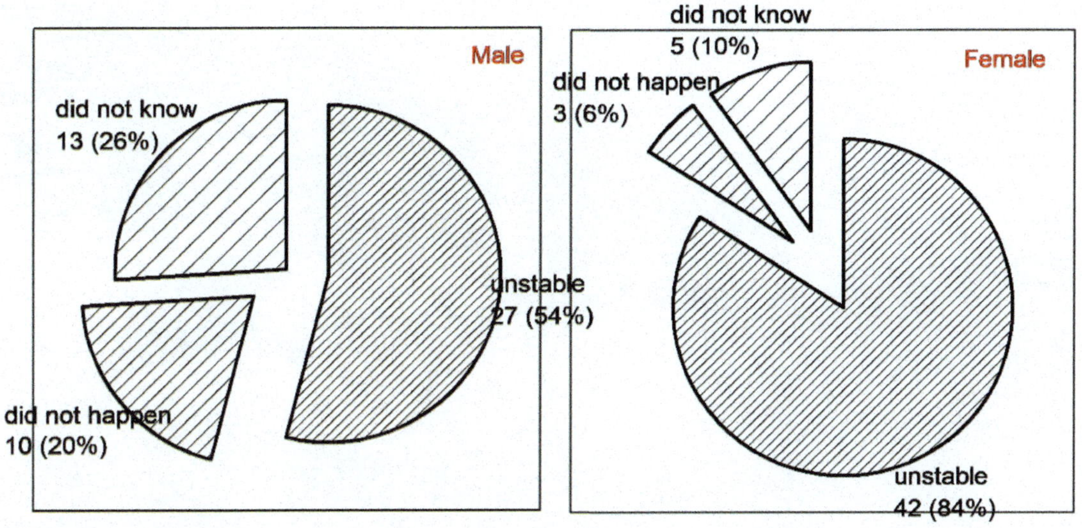

Fig. 2 Shows the instability of the yield of NTFP for both men and women. (The information is sourced from a survey conducted among the villagers)

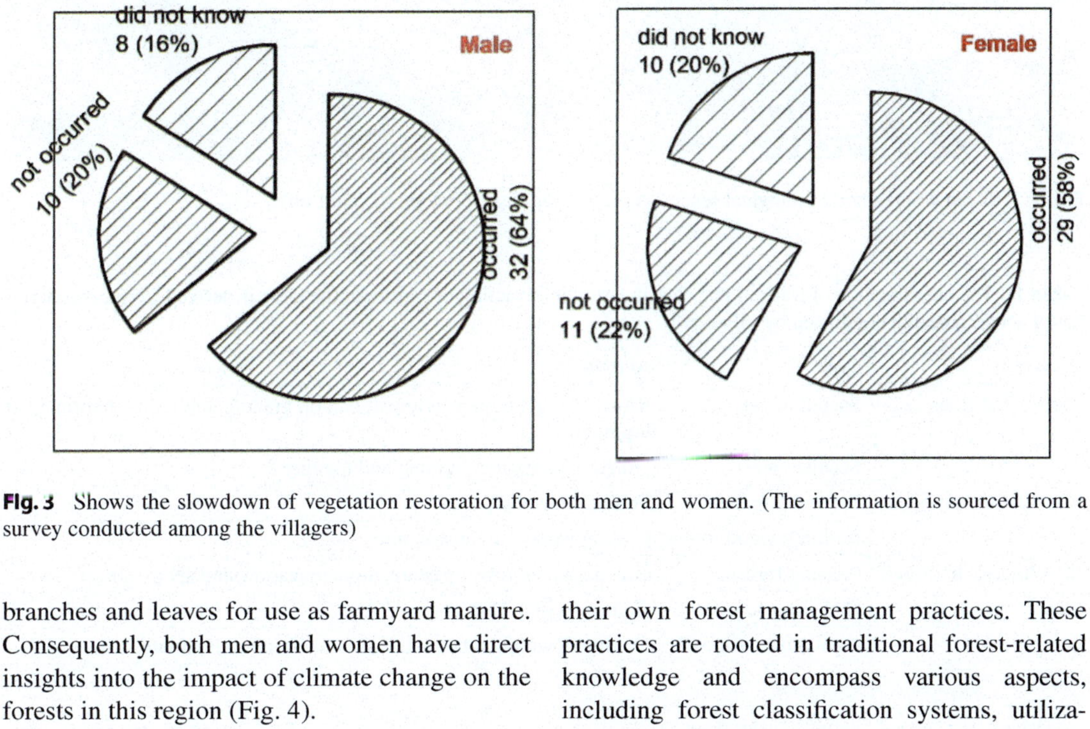

Fig. 3 Shows the slowdown of vegetation restoration for both men and women. (The information is sourced from a survey conducted among the villagers)

branches and leaves for use as farmyard manure. Consequently, both men and women have direct insights into the impact of climate change on the forests in this region (Fig. 4).

4.2 Traditional Knowledge-Based Forest Management

Through generations of experience and knowledge accumulation, local villagers have developed their own forest management practices. These practices are rooted in traditional forest-related knowledge and encompass various aspects, including forest classification systems, utilization and protection measures, and cultural beliefs associated with forests.

4.2.1 Classification and Utilization of Forests

Based on traditional knowledge, villagers classify forests according to several criteria (Table 1).

Fig. 4 Tibetan women build biological fences to protect saplings (photoed by Yin Lun)

Table 1 The classification of forest. (The classification information of forests in the table is derived from community-based action research conducted by villagers)

Criterions	Types	Activities
Belief traditions	Sacred forest	Rituals for the deities in forest; ban on grazing, gathering, hunting and logging
	Secular forest	Grazing, gathering, hunting and logging
Utilization traditions	Firewood forest	Felling firewood
	Building wood forest	Cutting wood for construction
Livelihoods traditions	Grazing forest	Grazing in the forest pasture, alpine meadow and sparse forest
	Gathering forest	Gathering of NTFP
	Agricultural forest	Cutting down branches and leaves for farmyard manure

First, according to traditional beliefs, forests are classified into two categories: sacred forests and secular forests. Sacred forests are associated with sacred mountains, and all flora and fauna within these areas are considered part of the sacred mountain. Various deities are believed to reside in these sacred forests, leading local customary laws to prohibit activities such as logging, hunting, plant gathering, and livestock grazing within these areas. Additionally, white pagodas are constructed near sacred forests, and rituals are regularly performed. Small ceremonies are conducted by elders, while larger ceremonies are led by monks, who pray for the blessings of the sacred forest and mountain

Fig. 5 Tibetan women carry out afforestation on traditional sacred mountains (photoed by Yin Lun)

for the village and its people. In contrast, secular forests are traditionally utilized by villagers for various resources, including animals, trees, and plants. However, recent national regulations have imposed strict prohibitions on hunting and logging in these areas as well (Fig. 5).

Second, based on their use, forests are categorized into two types: firewood forests and building wood forests. Firewood forests are generally found at lower altitudes and contain trees that are less dense, making them suitable for household firewood. Building wood forests, located at higher altitudes, feature denser and more robust trees ideal for construction.

Third, in terms of traditional livelihoods, forests are divided into three categories: grazing forests, gathering forests, and agricultural forests. Grazing forests are situated in high-altitude areas and consist of forest pastures, sparse forests, and alpine meadows. These areas are cold, and herders migrate here in the summer, setting up temporary wooden huts or tents to live in for about three months while grazing their livestock. Gathering forests are found at medium altitudes and are typically characterized by dense, humid environments with rich biodiversity. Villagers collect various non-timber forest products (NTFPs) from these forests during the summer and autumn. Agricultural forests are located at lower altitudes and historically were cleared through slash-and-burn techniques to create farmland (Pedroso-Junior et al., 2009). Although such practices are now prohibited by national laws and agricultural methods have evolved, these areas have been restored to forest zones. Today, villagers primarily cut branches and leaves for use as farmyard manure in these forests (Fig. 6).

4.2.2 Customary Law for Use of Forest

In addition to the customary laws prohibiting the logging of sacred forests, there are also traditional regulations governing the use of forest resources.

In grazing forest areas, herdsmen implement a system of rotational grazing across different forest pastures and alpine meadows. This practice aims

Fig. 6 Tibetan women use traditional knowledge to maintain forest ecosystems trees (photoed by Yin Lun)

to protect pasture resources and forest vegetation by preventing overgrazing in specific areas and reducing vegetation degradation. By rotating grazing locations, villagers effectively mitigate the risk of overgrazing and preserve the health of the vegetation.

In firewood and agricultural forests, villagers divide the forest into five to twelve sections and cut wood in these sections sequentially, creating a cycle that allows for forest vegetation restoration. In villages with limited forest resources, villagers distinguish between dry and wet wood. Dry wood refers to naturally fallen trees, while wet wood pertains to living trees. Villagers follow a traditional practice of using dry wood for four days and wet wood for three days each week to manage their resources sustainably.

In building wood forests, the use of wood is strictly regulated. Only villagers who are constructing houses are permitted to cut wood, and they must apply for and obtain prior approval from the village authorities. Both national laws and local customs prohibit the sale of cut wood.

In gathering forests, where non-timber forest products (NTFPs) are directly linked to household income, customary laws are particularly stringent. Among these products, Tricholoma matsutake is the most valuable, and its collection is governed by the strictest regulations. The collection of Tricholoma matsutake seedlings is prohibited to protect the habitat and ensure future growth. Each village has designated forest areas, and villagers from other villages are forbidden from collecting matsutake without permission. Unauthorized collectors must pay an "entry fee," and those caught collecting without authorization face significant fines. Additionally, a designated individual supervises the collection and market sales of Tricholoma matsutake seedlings, imposing fines on those who violate the regulations.

4.2.3 Traditional Cultures Related to the Forest

Forests hold significant cultural and spiritual value for the local Tibetans, intertwining deeply with their belief systems and practices concerning forest conservation.

(1) Sacred Forest Belief and Ecosystem Protection

For local Tibetans, forests on sacred mountains are considered sacred and belong to the mountain deity. The belief that these forests are divine dictates that humans should not exploit their resources. This sacred belief effectively safeguards the ecosystems and biodiversity of these forests. As a monk explained:

> The worship of the sacred forest is a fundamental aspect of our tradition. We believe that the sacred forest, its animals, plants, rivers, and lakes are the property of the sacred mountain. People are forbidden from logging, hunting, or polluting these sacred areas. Violations are believed to incur divine retribution. For instance, cutting down trees or killing animals in the sacred forest results in illness for the perpetrator or their family, while unexplained deaths of livestock are seen as punishment from the sacred mountain and its deities.

(2) Divine Tree Belief and Protection of Ancient Trees

In both forests and villages, many ancient trees are revered as divine. These trees, such as walnut, pine, cypress, and fir, are considered sacred and are treated with special reverence. White silk and prayer flags are often hung on these trees to honor their sanctity, and stone fences are constructed around them for protection. According to an elderly villager:

> There is a monumental walnut tree in our village, its age estimated to be several centuries based on its size. It is believed to have been planted by our earliest ancestors. Although historical records are lacking, this tree is regarded as a symbol of our village's history and continuity, connecting us to our ancestors. Thus, it is considered a divine tree and a guardian of our village.

A herdsman who frequently grazes in the mountains remarked:

> There is a massive fir tree in the alpine forest of our village, estimated to be several hundred years old. We always offer white silk and prayer flags when passing this tree to seek its protection. This act is believed to keep wild animals away from our livestock and ensure a bountiful harvest of matsutake mushrooms and other fungi.

These traditional beliefs and cultural practices contribute to the effective conservation of forest ecosystems, flora, fauna, and biodiversity. Many ancient trees in the region have been preserved due to these spiritual and cultural values.

4.3 Community-Based Adaptation for Climate Change

In recent years, local Tibetans have become increasingly aware of climate change and its impacts on forest ecosystems, which, in turn, directly affect their traditional livelihoods. In response to these challenges, villagers need to develop a community-based adaptation plan that leverages their traditional knowledge.

This adaptation plan is primarily community-driven, with villagers at its core, allowing them to fully utilize their traditional knowledge. Additionally, it employs an integrated approach that involves various stakeholders, including government agencies, researchers, NGO workers, and other interested parties. This collaborative strategy aims to harness the strengths of all involved to enhance the effectiveness of the adaptation plan (Fig. 7).

Specifically, the adaptation plan addresses the effects of climate change on forest ecosystems and the resulting challenges to local livelihoods. It encompasses the following three main components.

4.3.1 Traditional Knowledge of Burning Pasture and Adaptive Measures of Animal Husbandry

Historically, local herdsmen have relied on traditional knowledge of burning pastures to manage alpine shrub and bramble invasions. This practice was aimed at maintaining pasture areas, promoting the growth of forage grass, and improving soil fertility, thereby enhancing the quality of the pasture for future grazing. The process was meticulously planned, involving detailed strategies for fire location and area,

Fig. 7 Tibetan women establish ‘carbon neutral forest’ on the basis of traditional sacred forest (photoed by Yin Lun)

wind direction, climate conditions, division of labor, creation of firebreaks, firefighting measures, and emergency protocols. For thousands of years, these methods effectively minimized the risk of forest fires, improved pasture quality, and supported sustainable animal husbandry (Makondo & Thomas, 2018).

However, the implementation of the National Forest Protection Law has led to a ban on traditional burning practices to prevent forest fires. Concurrently, climate change has accelerated the spread of alpine shrubs and brambles, leading to more severe pasture degradation. Many pastures are now completely overtaken by these shrubs, making them unsuitable for grazing. The restrictions imposed by national laws, such as the ban on burning and policies like returning farmland to forest, have exacerbated the spatial imbalance of grasslands, increasing pressure on pastures and hindering the sustainable development of animal husbandry.

In response, villagers advocate for the continued use of their traditional burning methods to combat pasture degradation caused by climate change. They hope that local governments and forestry authorities will permit controlled burning in alpine pastures. Recognizing the threat to animal husbandry, local governments have begun collaborating with villagers, the forestry department, and fire departments to pilot controlled burns in severely degraded areas. This collaborative approach is gradually validating the value of traditional knowledge as a viable local strategy for addressing grassland degradation and adapting to climate change.

4.3.2 Traditional Institution of Women's Organizations and Adaptive Measures of Gathering

Traditionally, women have been primarily responsible for gathering non-timber forest products (NTFPs), including mushrooms such as Tricholoma matsutake. Recently, climate variability and extreme weather events like droughts and heavy rainstorms have significantly impacted forest environments and mushroom growth. Consequently, the traditional knowledge and management mechanisms that women have developed over years of collecting Tricholoma matsutake are being challenged. For instance, changes in climate have led to earlier or later harvesting times, reduced production in traditional areas, and altered growth conditions, complicating the prediction of Tricholoma matsutake's growth and affecting collection activities.

In response, women have adapted their traditional management mechanisms to address these challenges. A local women's organization known as the "sisterhood" plays a central role in this adaptation. Traditionally involved in organizing women's activities, the sisterhood has taken on the responsibility of managing Tricholoma matsutake collection. After discussions within the village, it was decided that the sisterhood would oversee the collection efforts. The sisterhood has implemented several measures to adapt to climate change:

1. **Environmental Protection**: To mitigate the effects of droughts and rainstorms on Tricholoma matsutake growth, the sisters instruct each woman to protect their designated collection areas. This includes backfilling with humus soil or covering the ground with leaves after collecting Tricholoma matsutake to ensure continued growth in subsequent years.
2. **Collection Regulation**: The sisterhood has established a strict collection system that specifies the size and maturity of Tricholoma matsutake to be harvested. Only mushrooms larger than 4 cm are permitted for collection. They also restrict access to the collection areas for outsiders and impose fines for unauthorized collection, thereby safeguarding resources and economic interests.

A local woman commented:

> It is more effective for women to manage forests and collect Tricholoma matsutake. Women are more familiar with the growth environment and climate conditions of Tricholoma matsutake and know how to protect its resources. The sisterhood can create and enforce strict rules for forest protection and matsutake management in our village. Men might cause conflicts or disputes, making rule implementation challenging. With women in charge, there are fewer conflicts, and the rules are more effectively enforced.

The actions of the sisterhood underscore the critical role of traditional organizations and management mechanisms in climate change adaptation. Their strategies for forest protection and matsutake resource management are exemplary local adaptations to the impacts of climate change.

4.3.3 Traditional Ways of Forest Utilization and Agricultural Adaptive Measures

Locally, forests provide farmyard manure for agriculture. As mentioned above, villagers cut down branches and leaves and stock them up for farmyard manure. In recent years, villagers have observed the impact of climate change on this part of the forest. They believe that climate warming makes the recovery and regeneration of trees in the forest slow down. Therefore, the villagers can no longer cut down as casually as before.

In order to adapt to this situation, the villagers divide the agricultural forest into different areas according to the tradition, and formulate rules and regulations to regulate the cutting of trees. The specific regulations include: first, after dividing the forest into several areas, the first area that can be cut down is determined every year. The villagers should cut down trees in this area in accordance with the regulations. In the following years, they should cut down trees in other areas in turn, and finally return to the first area. Secondly, in the same area, dry wood and wet wood should be distinguished, and dry wood should be used as much as possible.

The villagers believe that the above methods can maximize the use of forest resources to provide farmyard manure for agriculture and adapt to the impact of climate warming.

5 Conclusion

This research explores the traditional forest knowledge of Tibetans in the eastern Himalayas, encompassing their understanding, customary laws, and cultural practices related to forests. It posits that traditional knowledge is both the core and foundation of community-based forest management for local Tibetans. In the context of climate change, the study highlights the significant role and value of this traditional knowledge and community-based forest management in enhancing the effectiveness of local climate change adaptation plans in the eastern Himalayas.

Consequently, the research aims to illustrate the local Tibetan people's observations of climate change, their forest resource management practices, and their adoption of coping strategies to adapt to climate change and its impacts on forest resources. It reveals that relying solely on traditional knowledge is insufficient. To effectively adapt to climate change, it is essential to complement community-based forest management with cooperation mechanisms among multiple stakeholders, including villagers, local governments, scientific researchers, and NGOs. This collaborative approach should integrate traditional knowledge with national policies, scientific research, and technological advancements to develop a comprehensive climate change adaptation plan (Meena et al., 2019).

Additionally, the study offers a gender perspective on climate change adaptation. Given women's crucial role in forest management and livelihoods, their traditional knowledge possesses significant value and contributes to unique traditional institutions. This positions women as key players in the adaptation process.

Finally, although the climate change adaptation strategies based on traditional knowledge and forest management methods are specific to the Tibetan communities in the eastern Himalayas, they also offer a global perspective (Chauhan et al., 2023). The insights gained regarding the value of traditional knowledge, the mechanisms of multi-stakeholder cooperation, and the pivotal role of women can serve as references for indigenous communities worldwide in formulating their own climate change adaptation strategies.

References

Ahmed, M. N. Q., & Haq, S. M. A. (2017). Indigenous people's perceptions about climate change, forest resource management, and coping strategies: A comparative study in Bangladesh. *Environment Development and Sustainability*. https://doi.org/10.1007/s10668-017-0055-1

Chakraborty, A., Saha, S., Sachdeva, K., & Joshi, P. K. (2018). Vulnerability of forests in the Himalayan region to climate change impacts and anthropogenic disturbances: A systematic review. *Regional Environmental Change, 18*(6), 1783–1799. https://doi.org/10.1007/s10113-018-1309-7

Chauhan, H. K., Gallacher, D., Bhatt, A., & Bisht, A. K. (2023). The Himalayas: A climate change laboratory. *Environmental Development, 45*, Article 100814.

Choki, K. (2021). Cordyceps, climate change and cosmological imbalance in the Bhutan highlands. In *Environmental Humanities in the New Himalayas* (pp. 152–166). Routledge.

Dahal, S. (2013). *Indigenous knowledge in climate change adaptation*.

Grêt-Regamey, A., Brunner, S. H., & Kienast, F. (2013). Mountain ecosystem services: Who cares? *Mountain Research and Development, 32*, S23–S34. https://doi.org/10.1659/MRD-JOURNAL-D-10-00115.S1

Hopping, K. A., Yangzong, C., & Klein, J. A. (2016). Local knowledge production, transmission, and the importance of village leaders in a network of Tibetan pastoralists coping with environmental change. *Ecology and Society, 21*(1).

IPCC (Intergovernmental Panel on Climate Change). (2007). *Climate change: The third IPCC assessment report*. Cambridge University Press.

IPCC. (2014). Climate change 2014: Impacts, adaptation, and vulnerability. Part A: Global and sectoral aspects. Contribution of Working Group II to the Fifth Assessment Report of the Intergovernmental Panel on Climate Change.

Karki, M., & Adhikari, J. R. (2015). Integrating indigenous, local and modern knowledge for sustainable conservation and management of forest eco-systems in Nepal.

Liu, X., & Chen, B. (2000). Climate warming in the Tibetan Plateau during recent decades. *International Journal of Climatology, 20*, 1729–1742. https://doi.org/10.1002/1097-0088(20001130)20:14%3c1729::AID-JOC556%3e3.0.CO;2-Y

Ma, M., Singh, R. B., & Hietala, R. (2012). Human driving forces for ecosystem services in the Himalayan region. *Environmental Economics, 3*, 53–57.

Makondo, C. C., & Thomas, D. S. (2018). Climate change adaptation: Linking indigenous knowledge with western science for effective adaptation. *Environmental Science & Policy, 88*, 83–91.

Meena, R. K., Verma, T. P., Yadav, R. P., Mahapatra, S. K., Surya, J. N., Singh, D., & Singh, S. K. (2019). Local perceptions and adaptation of indigenous communities to climate change: Evidences from High Mountain Pangi valley of Indian Himalayas. *Indian Journal of Traditional Knowledge, 18*(1).

Moulton, H. (2024). Indigenous women are the "guardians of Pachamama": Territorial sovereignty is indispensable for just climate change adaptations in Peru. *Global Environmental Change, 89*, Article 102934.

Pedroso-Junior, N. N., Adams, C., & Murrieta, R. S. (2009). Slash-and-burn agriculture: A system in transformation. *Current Trends in Human Ecology, 12*(34), 12–34.

Rasul, G. (2014). Food, water, and energy security in South Asia: A nexus perspective from the Hindu Kush Himalayan region. *Environmental Science and Pollution, 39*, 35–48. https://doi.org/10.1016/j.envsci.2014.01.010

Salick, J., Yang, Y., & Amend, A. (2005). Tibetan land use and change near Khawa Karpo Eastern Himalayas. *Economic Botany, 59*, 312–325. https://doi.org/10.1663/0013-0001(2005)059[0312:TLUACN]2.0.CO;2

Shrestha, A. B., Wake, C. P., Mayewski, P. A., & Dibb, J. E. (1999). Maximum temperature and trends in the Himalaya and its vicinity: An analysis based on temperature records from Nepal from period 1971–94. *Journal of Climate, 12*, 2775–2787. https://doi.org/10.1175/1520-0442(1999)012%3c2775:MTTITH%3e2.0.CO;2

UNFCCC. (2013). Best practices and available tools for the use of indigenous and traditional knowledge and practices for adaptation, and the application of gender-sensitive approaches and tools for understanding and assessing impacts, vulnerability and adaptation to climate change; Technical paper No., FCCC/TP/2013/11. United Nations Framework Convention on Climate Change (UNFCCC), pp. 1–62.

World Bank. (2004). *Sustaining forests: A development strategy*. Report no. 29704. Washington, DC: World Bank. https://doi.org/10.1596/0-8213-5755-7

Zvobgo, L., Johnston, P., Olagbegi, O. M., Simpson, N. P., & Trisos, C. H. (2023). Role of Indigenous and local knowledge in seasonal forecasts and climate adaptation: A case study of smallholder farmers in Chiredzi, Zimbabwe. *Environmental Science & Policy, 145*, 13–28.

Traditional Knowledge in Use of Plant Species and Genetic Resources for the Climate Change Adaptation

7

1 Introduction

Agro-biodiversity is a crucial component of biodiversity, significantly contributing to food and agriculture production. It is considered a subset of biodiversity, encompassing the diversity and variability of plants, animals, microorganisms, and the in situ and ex situ conservation of genetic resources associated with agriculture (Partap and Sthapit, 1998).

Plant Genetic Resources (PGRs) represent the plant component of agro-biodiversity. Globally, the total number of higher plant species is estimated to be 270,000, with approximately 43% of these being crops, cultivated plants, and their wild relatives, currently classified as PGRs for food and agriculture (Hilger et al., 2015). PGRs for food and agriculture include the genetic diversity found in landraces, traditional varieties, improved cultivars, and the wild relatives of crops, as well as other wild plant species that could be utilized for crop improvement now or in the future (Javaid and Steve, 2016).

Indigenous and local communities are often situated in regions rich in genetic resources but located in fragile and vulnerable environments. These communities have cultivated and utilized biological diversity since the advent of agriculture, and their practices have been recognized for their role in enhancing and maintaining global biodiversity, particularly in domesticated plants and animals (Blackburn and Anderson, 1993; Harlan, 1995; Nabhan, 1997). For generations, these communities have relied on traditional knowledge and practices to conserve, use, and improve PGRs, ensuring on-farm diversity as a strategy for securing livelihoods (FAO, 2011).

The impact of climate change on crop production and global food security is a widely discussed issue today (Reynolds, 2010). Climate change encompasses record-breaking temperatures, shifts in precipitation patterns, increased frequency of droughts and floods, and rising atmospheric greenhouse gas concentrations. These factors can adversely affect crop yield (both quality and quantity), growth rates, photosynthesis, transpiration rates, and can also lead to more frequent disease and pest epidemics (Mahato, 2014). Climate change has already had a profound impact on plant genetic resources and the livelihoods of people, particularly smallholders living in marginal environments (FAO, 2011).

At the same time, plant genetic resources (PGRs) and associated traditional knowledge (TK) offer valuable genetic traits to address the challenges posed by climate change. PGRs are essential for global food security and agriculture, as they provide the genetic diversity needed for species to adapt to selective pressures such as abiotic stresses (e.g., drought, heat, frost) and biotic stresses (e.g., new diseases and pests) as well as changing climatic conditions (Hammer and Teklu, 2008). Genetic resources for food and agriculture play a crucial role in overcoming

Y. Lun, *Climate Change and Traditional Knowledge*, Sustainable Development Goals Series,
https://doi.org/10.1007/978-3-032-12015-1_7

climate change challenges, enhancing food security and nutrition, and improving agricultural productivity, rural livelihoods, sustainability, and resilience (FAO, 2015; Asfaw and Lipper, 2011).

Traditional knowledge related to plant genetic resources is also vital for climate change adaptation. Farmers have long domesticated plant species, creating a diverse range of crop varieties tailored to specific needs and environmental conditions (GIZ, 2009). Additionally, the traditional knowledge of indigenous and farming communities is instrumental in the conservation and development of genetic resources. This indigenous knowledge provides a foundational basis for developing effective adaptation and natural resource management strategies. It offers cost-effective, participatory, and sustainable approaches to adaptation (Nakashima et al., 2012).

Mountains generally support rich biological diversity due to varied physiography, ecological conditions, and resultant habitat diversity (Badola and Aitken, 2003; Price, 2004). The Himalayas, with their diverse climate types (ranging from subtropical to alpine), altitude variations, topography, soil types, river valleys, and forest resources, are hotspots for biodiversity, including agro-biodiversity and a wide array of plant genetic resources. Several studies have explored agricultural biodiversity, plant genetic resources, and climate change in the Himalayan region (Pande et al., 2016; Radh et al., 2013; Sundriyal et al., 2004). These studies highlight the significance of plant genetic resources for the sustainable development of mountain agriculture and their critical role in adapting to adverse climate conditions and promoting food security. However, there remains a gap in research and discussion concerning traditional knowledge related to plant genetic resources, particularly regarding adaptation plans at the community level based on this knowledge.

The Eastern Himalayas in northwestern Yunnan are renowned for their rich biodiversity and cultural diversity. Recently, the effects of climate change, including natural disasters, have increasingly threatened the livelihoods of local Tibetan communities, impacting their agro-pastoral activities, water resources, and biodiversity, such as mountain pastures, valleys, and forests (Klein et al., 2014). Given the fragile natural environment and the deep dependence of these communities on these resources, these challenges have significantly jeopardized their sustainable development and well-being. Agro-pastoralism is crucial for local Tibetans, serving not only as a livelihood strategy but also as a means of biodiversity management and cultural expression. In this system, agriculture and herding are mutually dependent.

Local Tibetans have developed traditional knowledge related to the use of plant species and genetic resources within this agro-pastoral system. This traditional knowledge encompasses practices for domesticating, cultivating, and utilizing cultivated plants or varieties, as well as managing other biological resources. It has been created and refined by the local community through their production and daily living processes. This knowledge primarily focuses on the exploitation and preservation of biological and genetic resources, including local crop and pasture varieties.

This paper examines the impacts of climate change and extreme weather events on plant species and genetic resources. It also explores traditional knowledge related to crop and pasture varieties among local Tibetan communities. Through action research, this paper aims to evaluate the value of traditional knowledge in developing community-level adaptation plans for managing plant resources.

2 Methodology

In this study, we mainly use two methods. For plant genetic resources, we use the ethno-ecology methods; for traditional knowledge related to plant genetic resources, we use the community-based action research.

2.1 Ethno-Ecology Methods

This study was carried out in collaboration with traditional farmers, herdsmen, knowledgeable local peoples and with reference to extensive agricultural and botanical literature. After obtaining prior informed consent, two investigation groups were formed, one for farmers and the other group for herdsmen; thereafter the information on crop genetic resources, pasture genetic resources, and climate change adaptation were recorded.

2.1.1 Investigation of Crop Genetic Resources

Among the local Tibetan villagers, some possess extensive traditional agricultural knowledge. These experts, with their deep understanding of crop plant distribution, conduct investigations into crop plants and genetic resources. This study involved examining crop plants and genetic resources across various farmlands, considering different elevations and slopes.

2.1.2 Investigation of Pasture Genetic Resources

Certain local Tibetan villagers, primarily herdsmen, spend the entire year grazing livestock in alpine pastures and possess considerable traditional knowledge about pasture plants. This knowledge includes information on forage distribution and the climatic adaptation of these plants. Consequently, these herdsmen carried out investigations into pasture plant and genetic resources across different altitudes and grassland types.

2.2 Community-Based Research and Adaptation Plan

Building on the survey using ethno-ecological methods and previous investigation data, we initiated community-based research. This research focuses on plant genetic resources and incorporates traditional knowledge related to these resources. It includes the classification, use, and management of plant genetic resources by local Tibetan villagers, as well as the customary laws and cultural practices associated with them (Ayers, 2009).

The research was divided into two groups: the first group focused on traditional knowledge related to crop genetic resources, while the second group concentrated on traditional knowledge related to pasture genetic resources. For both groups, the community-based research was structured into three parts:

1. Traditional classification knowledge of genetic resources.
2. Traditional utilization knowledge of genetic resources.
3. Observations and understanding of the impact of climate change on genetic resources.

Based on this traditional knowledge, we discussed how to support local Tibetan villagers in developing community-based strategies for adapting to climate change.

3 Results and Discussion

In agro-pastoralism, genetic resources for food and agriculture encompass the diversity and variability of animals, plants, and microorganisms utilized by farmers and pastoralists. These resources are crucial for producing both food and non-food agricultural products, as well as for maintaining the ecosystem structures, functions, and processes within and around production systems. In this context, plant genetic resources include both crop and pasture varieties.

Climate change impacts not only plant genetic resources but also the associated traditional knowledge (Islam et al., 2024). Thus, the traditional knowledge of Tibetan villagers regarding plant genetic resources offers valuable local insights for analyzing the effects of climate change.

In Deqin, the plant genetic resources within agro-pastoral systems are abundant and closely tied to the way local Tibetans manage their farmland and grasslands. For these communities, plant genetic resources are categorized into two main areas: crop genetic resources and pasture genetic resources. Correspondingly, traditional knowledge about plant genetic resources is also divided into two segments: agricultural farmers possess extensive traditional knowledge about crop genetic resources, while herdsmen engaged in animal husbandry have rich traditional knowledge about pasture genetic resources.

In high-altitude villages, where animal husbandry is prevalent and the proximity to grasslands is significant, villagers tend to possess more extensive knowledge of pasture genetic resources. Conversely, in low-altitude villages, where agriculture is the primary occupation and the climate supports the cultivation of a variety of crops, villagers generally have more knowledge about crop genetic resources.

Traditional knowledge also varies by age and gender (Owusu, 2017). For instance, men over the age of 50, who are predominantly involved in animal husbandry, often have extensive knowledge of pasture genetic resources. On the other hand, women, especially older women, who are more frequently engaged in agricultural activities, possess greater knowledge about crop genetic resources.

3.1 Traditional Classification Knowledge of Plant Genetic Resources

3.1.1 Traditional Classification Knowledge of Crop Genetic Resources

In local Tibetan agriculture, farmland is categorized into two types: Qingwan (irrigated land) and Rixing (dry land). Qingwan refers to flat terrain where water can be channeled for irrigation, allowing for intensive cultivation and the planting of crops up to twice a year. In contrast, Rixing is situated on higher, less accessible mountain slopes where irrigation is impractical, and crops can only be planted once annually. Dry land cultivation relies on ancient methods that depend on weather conditions and include shifting cultivation. Common crops grown on dry land include various varieties of highland barley, maize, buckwheat, bitter buckwheat, potatoes, and nettle.

Irrigated land, located around the villages at lower altitudes with flat terrain, facilitates easy access and transportation. The high productivity of this land is due to intensive cultivation practices, including regular watering, fertilizing, soil loosening, and timed planting. Main crops include different varieties of wheat, maize, highland barley, beans, and various vegetables. Crops on irrigated land are grown twice annually: maize from June to October and wheat and highland barley from November to the following June.

Highland barley is a representative crop of Deqin, well-suited to high-altitude conditions. It has a long cultivation history in the region, with different varieties grown in both dry and irrigated lands. The local Tibetans possess extensive traditional knowledge about highland barley, which holds significant cultural value. Highland barley is not only used as a sacred offering in religious ceremonies but also symbolizes local identity. For this reason, highland barley has been selected as the representative crop genetic resource for our research into traditional knowledge.

Through field surveys using ethno-ecological methods, we recorded 17 varieties of highland barley (see Appendix A). The local Tibetan people classify these varieties into three categories based on climatic and altitude characteristics: low-altitude varieties adapted to warmer climates, mid-altitude varieties suited to cooler climates, and high-altitude varieties adapted to cold climates (Table 1).

Table 1 Highland barley varieties categories. (The information is sourced from field surveys conducted among the villagers)

Land type	Climatic condition	Highland barley varieties categories (local Tibetan name)
Qingwan (irrigated land in low altitude)	Warm	Cuwa, Jiezhigeluna, Ge nie, Ge jiang, Ni wo, Jiu ge
Qingwan (irrigated land in middle altitude)	Cool	Ge na, Gecai, Naina, Malu, Ge si, he li
Rixing (dry land in high altitude)	Cold	Jiongnai, Rige, Ge le, Helugeme, Shi na

Source Survey data (2017)

3.1.2 Traditional Classification Knowledge of Pasture Genetic Resources

Due to variations in topography, sea level, temperature, and soil, the grasslands in Deqin can be classified into four main categories: alpine meadow grassland, shrub meadow grassland, sparse forest grassland, and mountain shrub grassland.

1. Alpine Meadow Grassland: Found at elevations above 3800 meters, this grassland exhibits distinct zonal and seasonal characteristics. Pasture resources in this type include Anemarrhena, Fritillaria, Saussurea involucrata, Cordyceps sinensis, Rhododendron microphylla, Platycladus orientalis, alpine willow, Carex, dragon's blood, primrose, rice pocket, floating finger grass, deer antler Artemisia, wild onion, and wild leek. (Fig.1)

2. Shrub Meadow Grassland: Also situated at elevations above 3800 meters, this type of grassland is characterized by high irrigation coverage but low utilization rates. Key forages include dragon's blood, Kentucky bluegrass, Carex, Festuca arundinacea, Stipa, turning white leaves, toothed fruit acid mold, Gentiana calyx, wild leek, and horseshoe grass. (Fig. 2)

3. Sparse Forest Grassland: Found between 2500 and 3600 meters above sea level, this grassland has a cold, cool, and humid climate. It consists of grasses such as Poa pratensis, sedge, horseshoe grass, and Breviscapus.

4. Mountain Shrub Grassland: Characterized by its mountainous terrain, this grassland is located at elevations between 1900 and 3200 meters. It features grasses such as cattail grass, little thrush grass, horse bean grass, annual wormwood, plantain grass, and rhubarb. It is noted for its long grass growth period, year-round grazing, and significant slopes in some areas. (Fig. 3)

Through a field survey using ethno-ecological methods, we recorded 51 pasture species (see Appendix B). The local Tibetan herdsmen

Fig. 1 Distribution of plant resources on high mountain flowing stone beach (photoed by Yin Lun)

Fig. 2 Local experts investigate the distribution of plant resources on alpine meadows (photoed by Yin Lun)

Fig. 3 Local experts investigate the distribution of plant resources in forests (photoed by Yin Lun)

possess detailed classification knowledge of these pastures. Based on the survey, they have conducted community-based research on traditional classification systems. For instance:

- By Season: Pastures are categorized into three types: spring, summer, and autumn.
- By Temperature: Pastures are divided into cold-resistant and drought-resistant types.
- By Growing Environment: Pastures are classified according to their adaptability to arid or humid environments.
- By Elevation and Location: Pastures are divided into five types: valley grass dam, near the field, in the field, on the hillside, and on snow mountain pastures (Table 2)

3.2 Traditional Utilization Knowledge Related to Genetic Resources

3.2.1 Traditional Utilization Knowledge of Crop Genetic Resources

Tibetan farmers possess extensive traditional knowledge regarding the utilization of highland barley seed varieties. Based on their experience and expertise, these farmers are well-acquainted with the specific characteristics of various highland barley seeds (Zhou, 2024). They classify the seeds of 17 different varieties according to their distinct traits. Through participatory assessment and grading, the farmers' traditional knowledge

Table 2 Pasture varieties categories. (The information is sourced from field surveys conducted among the villagers)

Classification basis	Classification standard	Pasture varieties categories (local Tibetan name)
According to the growing place	In the plain	Pa nazua, gi an jia ma, pa zua
	Beside the field	Wabo ma, wajiu ma, yage ma, a zongxuzong
	In the field	Huohua, hu
	On hillside	Jia hu la, rong mu rong gong, shuoshu
	Beside the snow	Ye zi, lu mu ya sha, dong zha, nahua
According to the altitude	Summer alpine	Ye zi, lu mu ya sha, dong zha, nahua
	Winter valley	Wabo ma, wajiu ma, yage ma, a zongxuzong
According to the growing environment	In dry places	Rongqu, gi an jia ma, shuoshu
	In wet places	Pa zua, pa nazua
According to the mode of reproduction	Flowering and seed bearing	Ye zi, huohua, yage ma, lu mu ya sha
	Not flowering and bearing seeds	Na wo, rongba, ga mu geishi
According to the collection method	Collected by excavation	Wabo ma, wajiu ma, pa nazua, gi an jiama, dasimu, azongxuzong, dong xi mu
	Collected by cutting	Hu, huohua, pazua
According to plant size	Tall	da gu, gu hu
	Medium sized	Gang de shin
	Short	Sizha
According to the preference of livestock	Very like	Gin, gululu, mi lu
	Average	Shilu pa nu
	Dislike	a duo ri duo, jionongna, lijia pong
According to the nutritional content	Good nutrition	di baqiongjie
	Medium nutrition	Cha bulu
	Poor nutrition	Jiubanaju
According to feeding effect	Grass that can fatten livestock	Baginri ma, chuoboga
	Grass that can cure livestock	Gu duo
For plants with similar morphology, use the principal and the secondary to classify	Principal	Jiona, runuo
	Secondary	Jio gai, runuoru gai

Source Survey data (2017)

about the adaptive characteristics of these seed varieties is summarized in Table 3.

Over the past 20 years, rising temperatures and increased drought occurrences have led farmers to preserve drought-resistant varieties such as Cu wa, He li, Ge le, and Ge nie, while less resilient varieties have gradually declined. Simultaneously, traditional seed exchange practices persist among farmers within the same village and across different villages. These practices involve displaying traditional seed varieties from various households or villages, which are then exchanged or traded based on needs. Additionally, farmers share knowledge about seeds and the germplasm itself. Through these exchanges, farmers can identify and obtain seed varieties better suited to adapting to climate change.

Table 3 Assessment and grading about the adaptive characteristics of the different kind of seeds. (The information is sourced from field surveys conducted among the villagers)

	Drought resistance	Barren resistance	High-temperature resistance	Cold resistance	Moisture resistance
Cuwa	9	8	6	4	3
Jiezhigeluna	8	7	8	2	1
Ge nie	6	7	9	3	3
Ge jiang	3	8	6	5	4
Ni wo	6	7	4	7	9
Jiu ge	5	4	6	10	9
Ge na	7	5	4	7	7
Gecai	7	3	1	7	10
Naina	2	6	8	4	5
Ma lu	4	4	7	3	5
Ge si	7	8	3	7	5
He li	9	9	6	4	2
Jiongnai	8	7	1	4	5
Rige	3	1	2	6	7
Ge le	7	3	5	6	8
Helugeme	4	6	4	7	6
Shi na	6	2	6	3	6

Source Survey data (2017). The full score is 10 points, from 1 to 10, the higher of the adaptive characteristic has the higher score, the lower the scores of weaker adaptabilities

3.2.2 Traditional Utilization Knowledge of Pasture Genetic Resources

Through long-term grazing and the accumulation of generational knowledge, Tibetan herdsmen in Deqin have developed a rich understanding of pasture utilization, particularly regarding its nutritional value. For instance, they believe that forage from high-altitude areas is more nutritious than that from low-altitude areas. Similarly, within the same altitude, forage from wetlands is deemed more nutritious than that from drylands, and forage from sunny slopes is considered more nutritious than that from shady slopes.

Based on a survey of 158 herdsmen, the nutritional value of 51 types of pastures was evaluated and scored. Ultimately, 9 types of pastures were identified as having the highest nutritional value according to local herdsmen. Detailed resource investigations and community-based research were then conducted on these 9 selected pastures, as shown in Table 4.

To optimize the use of forage resources, local Tibetan herdsmen possess extensive traditional knowledge about the growing times and feeding periods of pastures, including the 9 high-quality types identified. This knowledge is closely linked to local climate conditions and phenological events.

Growing Time Knowledge:

Traditional knowledge regarding the growing times of these 9 pastures is informed by both climatic and phenological factors. Specifically:

- For four of the pastures, growing times are determined based on climatic factors such as precipitation and temperature.
- For the remaining five pastures, growing times are assessed through phenological indicators, such as bird behavior, snowmelt patterns, and wetland conditions.

Table 4 Traditional knowledge of 9 kinds of fine pastures. (The information is sourced from field surveys conducted among the villagers)

No.	Latin name	Local Tibetan name	Growing time	Growing place	Feeding period	Nutritive value
1	Chenopodium album	Hui	Growing in April with little rain	Farmland	There are black birds flying and singing in the farmland	Increasing weight of livestock.
2	Brassica campestris	Huohua	Growing in April after twice rains	Farmland	Grass grows higher that chickens can hide in the grass	Increasing weight of livestock.
3	Taraxacummongolicum	Wang bao mu	Growing in march after the snow melts	Grassland	After three rains	Increasing milk production of dairy cows.
4	Senecio oryzetorum	Wang ju mu	Growing in April when the birds are singing	Forest	The smell of the forest has changed	Increasing milk production of dairy cows; increasing weight of livestock
5	Plantago depressa	Gai jia mu	Growing in April when birds fly from north to south	Grassland	After three to four rains	Increasing milk production of dairy cows; increasing weight of livestock; herbal medicine for diarrhea of livestock.
6	Fagopyrum esculentum	Xie ye ma	Growing in April after two or three rains	Forest	The air in the forest becomes moist	Increasing weight of livestock.
7	Phlomis atropurpurea	Pa nazhua	Growing in may when the grassland starts to warm	Grassland	The direction of the wind has changed and in the early morning, there was a lot of fog in the grassland	Increasing weight of livestock.
8	Juncus allioides	Gong zhua	Growing in April when the wetland begins to overflow	The wetland of alpine meadow	At the time of seed setting	Increasing milk production of dairy cows; increasing weight of livestock; herbal medicine for diarrhea of livestock.
9	Cyananthus flavus var. flavus	Yizi	Growing in may after the snow melts	Alpine screes	Snow lotus begins to bloom and blue sheep began to appear in groups	Increasing milk production of dairy cows; increasing weight of livestock; herbal medicine for diarrhea of livestock.

Source Survey data (2017)

Feeding Period Knowledge:

The optimal feeding periods for livestock vary for each type of pasture. Herdsmen use their observations and experience to determine when each type of forage reaches its ideal feeding stage. For the 9 pastures mentioned:

- Three are assessed based on climatic indicators, such as rainfall and fog.
- Six are evaluated using phenological cues, including bird and blue sheep behavior, forest odors, humidity levels, and observations of grass growth, flowering, and seed setting.

It is notable that the traditional utilization knowledge of pastures encompasses not only visual observations but also auditory and olfactory cues.

3.3 Local Perspective on Impacts of Climate Change on Plant Genetic Resources

3.3.1 Impacts of Climate Change on Crop Resources

Based on surveys conducted with 120 farmers, and their assessments of the severity of impacts, the results indicate that climate change and associated disasters, including secondary effects such as landslides, increased pest and disease incidences related to warming, and declining soil fertility, are the most significant factors affecting crop resources (see Table 5). The impacts, ranked in order of their significance, are as follows: drought, increased pest and disease occurrences linked to warming, and changes in precipitation.

Among the climatic factors perceived by farmers, drought is the most severe threat to crop resources. Historical data from the past twenty years, as recalled by farmers, indicate that the frequency of pest and disease outbreaks associated with warming is increasing. Additionally, there have been significant changes in precipitation patterns, which critically affect crop growth. Farmers report uncertainty regarding precipitation trends: while they observe a decrease in regular rainfall, they also experience an increase in intense, short-term rainstorms. As a result, many farmers believe that precipitation is becoming increasingly unpredictable.

3.3.2 Impacts of Climate Change on Pasture Resources

In parallel with the investigation of farmers, we also surveyed 120 herdsmen to rank the various factors impacting pasture resources. These factors included climate changes (temperature and precipitation), extreme weather events, and secondary disasters (see Table 6). According to their rankings, the most significant impacts are: 1. temperature warming; 2. changes in precipitation; 3. drought.

Herdsmen have observed the following key impacts of climate change on pasture resources:

1. Temperature Warming: Increased temperatures lead to earlier growth of grass, which is the most significant climate change impact on pasture resources.

Table 5 Impacts of climatic factors to crop resources by proportion (%). (The information is sourced from field surveys conducted among the villagers)

Impacts of climatic factors to crop resources	Farmers (n = 120)
Drought	27.5
Increased pests and diseases linked to warming	22.5
Precipitation changes	18.3
Temperature changes	15.8
Humidity changes	7.5
Low soil fertility	2.5
Rainstorm	2.5
Landslides	1.6
Flood	1.6
Total	100

Source Survey data (2017)

Table 6 Impacts of climatic factors to pasture resources by proportion (%). (The information is sourced from field surveys conducted among the villagers)

Impacts of climatic factors to pasture resources	Herdsmen (n = 120)
Temperature warming	30.8
Precipitation changes	25.8
Drought	19.2
Snowfall changes	9.2
Snowstorm	6.7
Humidity changes	3.3
Rainstorm	1.7
Landslides	1.7
Flood	1.7
Total	100

Source Survey data (2017)

2. Precipitation Changes: Many types of pastures rely on rainfall to initiate growth, so variations in precipitation significantly affect pasture resources.
3. Drought: Continuous drought conditions, characterized by a lack of rain and high temperatures, negatively impact both the growth and quality of pastures.
4. Snowfall Changes: Variations in snowfall and snowstorms affect high-altitude grasslands, which are crucial for livestock nutrition.

3.4 Community-Based Adaptation Plan to Climate Change

Based on traditional knowledge and local perspectives regarding the impacts of climate change on plant genetic resources, we assisted Tibetan villagers in developing a community-level adaptation plan for climate change. This plan is divided into two parts based on crop and pasture resources. We worked separately with farmers and herdsmen to integrate their traditional knowledge into these adaptation strategies.

3.4.1 Traditional Knowledge of Crop Genetic Resources and Climate Change Adaptation

Through community-based action research on crop resources and associated traditional knowledge, as well as risk assessments of climate change impact factors, local Tibetan farmers have begun to develop a community-level adaptation plan for climate change. The core components of this plan are variety selection and seed exchange.

It is important to note that traditional knowledge related to variety selection and the mechanism of seed exchange have long been practiced in local communities, independent of current climate change concerns (Garai and Ku, 2024). However, within the context of climate change, these traditional practices have become more targeted. This demonstrates that traditional knowledge is not static but rather a dynamic process that evolves to address new challenges.

First, variety selection plays a crucial role in adapting to climate change. Villagers select crops based on the characteristics of different varieties, choosing those that are more resistant to higher temperatures, drought, diseases, and pests in response to recent climate changes.

Second, seed exchange is a traditional mechanism that facilitates the improvement of crop varieties. Farmers from different villages exchange seeds and knowledge, helping to identify varieties that are better suited to the local climate. This exchange has proven to be an effective strategy for adapting to climate change.

Third, planting experiments with high-quality highland barley are underway. With support from local government and scientific researchers, farmers have established experimental fields to test and select traditional and superior highland barley varieties. The goal is to develop the best variety through crossbreeding traditional highland barley strains, integrating traditional knowledge with modern planting technologies. This approach not only utilizes traditional knowledge but also creates a platform for collaboration between traditional practices and scientific advancements, facilitating effective cooperation between these two knowledge systems.

3.4.2 Traditional Knowledge of Pasture Genetic Resources and Climate Change Adaptation

Based on traditional knowledge related to pasture resources, local Tibetan herdsmen have developed an adaptation plan to address the most severe impacts of climate change. This plan involves collaboration with local government and pasture researchers, creating a multi-stakeholder approach that integrates community-based and government-supported strategies. A key component of this adaptation plan is the conservation planting of pasture.

First, the exchange and development of traditional knowledge are emphasized. Local experts and herdsmen have classified and identified pasture resources, conducted field investigations in alpine pastures, and documented growth patterns of various pastures across different months. They have also transferred this knowledge to younger generations to ensure its continuity and dissemination among herdsmen.

Second, conservation planting of pasture is a major focus. Based on the identification of high-quality pastures by herdsmen, and with support from local government and researchers, nine types of excellent pastures were selected for artificial grassland construction and conservation planting in alpine and other grasslands (Sun et al., 2016). Additionally, experiments in grassland ecological restoration, informed by traditional knowledge, are being conducted to protect pasture genetic resources.

Third, grassland co-management is being developed. Local government and herdsmen are collaborating to build biological fences to safeguard pasture resources, establishing a cooperative model that enhances grassland productivity. Incorporating traditional knowledge into grassland co-management practices promotes the protection of pasture biodiversity and the sustainable development of local livelihoods. The integration of pasture conservation planting, technological innovation in pasture management, and effective co-management strategies aims to mitigate pasture resource degradation, improve productivity, and support sustainable grassland utilization, thereby reducing the risks and vulnerabilities associated with climate change and enhancing the resilience of animal husbandry.

4 Conclusion

Global climate change, marked by rising temperatures and altered precipitation patterns, has significantly impacted plant genetic resources, posing unprecedented challenges to indigenous peoples and local communities worldwide who rely on traditional livelihoods. This paper explores the potential role of traditional knowledge in addressing these climate change challenges by examining Tibetan traditional knowledge related to plant resources. We aim to highlight how traditional knowledge can aid in adapting to climate change and reveal the value of plant genetic resources in this context. Existing traditional knowledge and mechanisms within communities have the potential to be revitalized and innovated to meet climate change challenges, demonstrating that traditional knowledge is dynamic and adaptable rather than static and unchanging.

The traditional agro-pastoralism practiced by Tibetan people in the Eastern Himalayas offers a valuable case study for understanding plant genetic resources. This traditional livelihood encompasses both crop and pasture genetic resources, broadening our research perspective (Baul and McDonald, 2015). While crops have been domesticated for thousands of years, pastures have experienced minimal human interference. Consequently, adaptation strategies for these two types of plant genetic resources differ, and the role of traditional knowledge in each process varies.

For Tibetans and other indigenous peoples facing unprecedented climate change impacts, it is essential to recognize and harness the value and vitality of traditional knowledge, alongside integrating it with modern science and technology. Climate change presents an opportunity to bridge these two knowledge systems, fostering mutual understanding and collaboration. In this regard, community-based adaptation plans should involve not only the actions of indigenous peoples and local communities but also the support and cooperation of government, scientific research institutions, NGOs, and other stakeholders. Such multi-stakeholder collaboration will help create more effective and integrated climate change adaptation strategies.

Appendix A

Highland barley species. (The information is sourced from field surveys conducted among the villagers).

No.	Latin name	Local Tibetan name	Grade[1]	Trend[2]
1	Hordeum vulgare	Naina	A	c
2		Ma lu	B	b
3		Ge si	B	a
4		Jiezhigeluna	C	a
5		Ge na	A	c
6		Ge le	A	b
7		Ge cai	C	c
8		Helugeme	B	c
9		Ge nie	D	c
10		Ge jiang	D	c
11		He li	B	b
12		Ni wo	A	b
13		Cu Wa	B	a
14		Jiu ge	A	b
15		Shi na	D	a
16		Jiongnai	A	a
17		Ri ge	A	b

Appendix B

Pasture species. (The information is sourced from field surveys conducted among the villagers).

No.	Families	Genera	Latin name	Local Tibetan name	Locality	Grade[3]	Trend[4]
1	Osmundaceae	Osmunda	*Osmunda claytonianum*	Jia hu la	Damp valley	B	a
2	Ranunculus	Anemone	*Anemone rivularis*	Za dan gu song	Hillside grassland	C	b
3	Cruciferous	Brassica	*Brassica campestris*	Huohua	Farmland	B	c

(continued)

[1] The four grades of A \ B \ C \ D indicate the quality of highland barley in order of high quality, good quality, medium quality, and low quality.

[2] The trend of highland barley decrease, change, and increase under the influence of climate change are expressed in three grades of a \ b \ c.

[3] The four grades of A \ B \ C \ D indicate the quality of pasture in order of high quality, good quality, medium quality, and low quality.

[4] The trend of pasture decrease, change, and increase under the influence of climate change are expressed in three grades of a \ b \ c.

(continued)

No.	Families	Genera	Latin name	Local Tibetan name	Locality	Grade	Trend
4	Saxifragaceae	Rodgersia	*Rodgersia pninata*	Rong men rong gong	In the forest with an altitude of 2500-3500 m	B	b
5	Caryophyllaceae	Stellaria	*Stellaria media*	Gai mu gai xia	Farmland, stream side grassland	D	b
6	Polygonaceae	Fagopyrum	*Fagopyrum esculentum*	Xie ye ma	Hillside grassland	A	b
7			*F. Leptopodum*	Gong me	Farmland	D	c
8		Polygonum	*Polygonum nepalense*	Dong me	Farmland, stream side grassland, damp valley	B	a
9			*P. Macrophyllum*	Rong qu	Farmland, stream side grassland, damp valley	C	a
10			*P. Sinomontanum*	Gai dougu duo	Farmland, stream side grassland, damp valley	C	a
11			*P. Glaciale*	Dong me	Farmland, stream side grassland, damp valley	B	a
12		Rumex	*Rumex dentatus*	A bengke le	Stream side grassland, wetland	B	a
13	Chenopodiaceae	Chenopodium	*Chenopodium album*	Hui	Farmland	A	c
14	Geraniaceae	Geranium	*Geranium yunnanensis*	Dong zhua	Grassland, forest	D	b
15	Balsaminaceae	Impatiens	*Impatiens chinensis*	Dong riyue	Farmland, wetland	C	b
16	Rosaceae	Sorbus	*Sorbus rehderiana*	Meng lou	Grassland with altitude of 2500-4000 m	B	a
17		Potentilla	*Potentilla macrophylla*	Da gai	Grassland with altitude of 3000-4000 m	B	b
18			*P. Fulgens*	Da gai	Grassland with altitude of 2500-4000 m	B	b
19			*P. Freyninna*	Si ren	Grassland with altitude of 2500-4000 m	B	b
20			*P. Anserina*	Rong Moya sha	Grassland with altitude of 2500-4000 m	B	b

(continued)

(continued)

No.	Families	Genera	Latin name	Local Tibetan name	Locality	Grade	Trend
21	Papilionaceae	Astragalus	*Astragalus tataricus*	Sai long mu	Hillside grassland	D	c
22	Fagaceae	Quercus	*Quercus spinosa*	Shu bu	Grassland with altitude of 2400-4000 m	B	a
23	Aceraceae	Acer	*Acer davidii*	La Ba	Grassland with altitude of 2000-3500 m	B	a
24	Umbelliferae	Carum	*Carum carvi*	Dong xia mu	Hillside grassland	C	b
25			*P. Nanum*	Mu xia	Grassland with altitude of 3000-4500 m, alpine screes	B	a
26	Caprifoliaceae	Sambucus	*Sambucus adnata*	Shuoshu	Under the forest or in the shrub	B	b
27	Dipsacaceae	Dipsacus	*Dipsacus asperoides*	Ca xia mu	Grassland, forest, farmland	C	c
28	Compositae	Cirsium	*Cirsium eriophoroides*	Da si mu	Hillside grassland, shrub	D	c
29		Ligularia	*Ligularia latihastata*	Na guo	Grassland, forest, hillside grassland	B	c
30			*L. Cremanthodioides*	Mu tong	Hillside grassland	C	b
31		Taraxacum	*Taraxacum mongolicum*	Wang bao mu	Grassland, forest, farmland	A	b
32		Arctium	*Arctium lappa*	A zong xi za	Grassland, forest, farmland	C	c
33		Senecio	*Senecio oryzetoru*	Wang qu mu	Grassland, forest, farmland	A	c
34	Campanulacea	Cyananthus	*Cyananthus flavus*	Yi zi	Grassland with altitude of 2800-4500 m, alpine screes	A	a
35	Borraginaceae	Microporous	*Microula sikkimensis*	Ye ge mu	Grassland with altitude of 2800-4500 m	B	b
36	Scrophuhrhceae	Pedicularis	*Pedicularis siphonantha*	Di dizhuazhua	Grassland with altitude of 2800-4500 m	D	a
37	Plantaginaceae	Plantago	*Plantago depressa*	Gai jia mu	Grassland, forest, farmland	A	b
38	Rubiaceae	Galium	*Galium elegans*	Qiongjia	Grassland, forest, farmland	C	b
39	Malvaceae	Mallow	*Malva verticillata*	Zhong ba	Grassland, forest, farmland	B	b
40	Labiatae	Salvia	*Salvia prezewalskii*	Rong zui ma	Grassland with altitude of 2800-4000 m	B	a

(continued)

(continued)

No.	Families	Genera	Latin name	Local Tibetan name	Locality	Grade	Trend
41		Phlomis	*Phlomis atropurpurea*	Pa nazhua	Grassland with altitude of 2800-4000 m	A	a
42			*P. Pratemsis SP*	Shua gai	Grassland with altitude of 2800-4000 m	B	b
43	Iridaceae	Iris	*Iris bulleyana*	Bie bi	Hillside grassland	B	c
44	Juncaceae	Juncus	*Juncus allioides*	Gong zhua	Grassland with altitude of 2800-4500 m	A	a
45	Cyberaeae	Kobresia	*Kobresia parva*	Ba zhua	Grassland with altitude of 2800-4500 m	B	a
46		Carex	*Carex limosa*	Zhua sha	Grassland with altitude of 2800-4500 m	C	a
47			*C. Thomsonii*	Zhuajia	Grassland with altitude of 2800-4500 m	C	a
48	Gramineae	Agrostis	*Agrostis limprichtii*	Zhua chu	Grassland with altitude of 2500-3500 m	B	b
49		Elymus	*Elymus tangutorum*	Rang Ba	Grassland with altitude of 2500-3500 m	B	a
50			*E. Nutans*	Rang gu	Grassland with altitude of 2500-3500 m	C	b
51	Gramineae	Fargesia	*Fargesia cuticontracta*	He ya	Grassland with altitude of 2000-3500 m, forest	B	b

References

Asfaw, S., & Lipper, L. (2011). *Economics of PGRFA management for adaptation to climate change: A review of selected literature*. Commission on genetic resources for food and agriculture. Background Study Paper No. 60.: FAO. (available at http://www.fao.org/docrep/meeting/023/mb695e.pdf).

Ayers, J., & Forsyth, T. (2009). Community-based adaptation to climate change. *Environment: Science and Policy for Sustainable Development, 51*(4), 22–31.

Badola, H. K., & Aitken, S. (2003). The Himalayas of India: A treasury of medical plants under siege. *Biodiversity, 4*, 3–13. https://doi.org/10.1080/14888386.2003.9712694

Baul, T. K., & McDonald, M. (2015). Integration of indigenous knowledge in addressing climate change. *Indian Journal of Traditional Knowledge, 1*(1), 20–27.

Bhat, J. A., & Taiwo, S. (2016). Plant genetic resources and traditional knowledge in light of recent policy developments. In R. K. Salgotra & B. B. Gupta (Eds.), *Plant genetic resources and traditional knowledge for food security*. Springer Science+Business Media Singapore. https://doi.org/10.1007/978-981-10-0060-7_10

Blackburn, T. C., & Anderson, K. (1993). *Before the wilderness: Environmental management by native Californians*. Ballena Press.

Deutsche Gesellschaft fürInternationaleZusammenarbeit (GIZ). (2009). *Farmers as breeders—Participatory plant breeding. Issue papers—People and biodiversity*. Working Paper.: GTZ Imprint. star-www.giz.de/dokumente/bib/04-5104a1.pdf

FAO. (2011). Draft updated global plan of action for the conservation and sustainable utilization of plant genetic resources for food and agriculture.,: Food and agriculture Organization of the United Nations (FAO), United Nations. Document CGRFA/WG-PGR-5/11/2 Rev 1. Available at http://www.fao.org/fileadmin/templates/agphome/documents/PGR/ITWG/ITWG5/CGRFA_WG_PGR5112__REV1finalnumbX.pdf

FAO. (2015). *Coping with climate change—The roles of genetic resources for food and agriculture. Rome* (available at http://www.fao.org/3/a-i3866e.pdf).

Garai, J., & Ku, H. B. (2024). Cultural adaptation in the era of climate change: An ethnographic study on the resilience of indigenous people at Chittagong hill tracts area in Bangladesh. *Natural Hazards Research, 4*(4), 643–652.

Hammer, K., & Teklu, T. (2008). Plant genetic resource: Selected issues from genetic erosion to genetic engineering. *Journal of Agriculture and Rural Development in the Tropics and Subtropics, 109*, 15–50.

Hao, Z. H. O. U., & Lun, Y. I. N. (2024). Using traditional knowledge to reduce disaster risk-a case of Tibetans in Deqen County, Yunnan Province. *International Journal of Disaster Risk Reduction, 108*, 104492.

Harlan, J. (1995). *The living fields: Our agricultural heritage*. Cambridge University Press.

Hilger, T., Lewandowski, I., Winkler, B., Ramsperger, B., Kageyama, P., & Colombo, C. (2015). Seeds of change: Plant genetic resources and people's livelihoods (INTECH). https://doi.org/10.5772/59984.

Islam, F. B., Naylor, L., Bryan, J. E., & Coker, D. J. (2024). Climate coloniality and settler colonialism: Adaptation and indigenous futurities. *Political Geography, 114*, 103164.

Klein, J. A., Hopping, K. A., Yeh, E. T., Nyima, Y., Boone, R. B., & Galvin, K. A. (2014). Unexpected climate impacts on the Tibetan plateau: Local and scientific knowledge in findings of delayed summer. *Global Environmental Change, 28*, 141–152.

Mahato, A. (2014). Climate change and its impact on agriculture. *International Journal of Scientific and Research Publications, 4*(4).

Nabhan, G. P. (1997). *Cultures of habitat: On nature, culture, and story*. Counterpoint.

Nakashima, D. J., Galloway, M. K., Thulstrup, H. D., Ramos, C. A., & Rubis, J. T. (2012). *Weathering uncertainty: Traditional knowledge for climate change assessment and adaptation* (p. 100). UNESCO/UNU.

Owusu, M. (2017). Gender vulnerability to climate change and livelihood security in urban slum communities in Accra, Ghana (doctoral dissertation).

Pande, P. C., Awasthi, P., Bargali, K., & Bargali, S. S. (2016). Agro-biodiversity of Kumaun Himalaya, India: A review. *Current Agriculture Research Journal, 40*(1). https://doi.org/10.12944/CARJ.4.1.02

Partap, T., & Sthapit, B. (1998). *Managing agrobiodiversity, farmers' changing perspectives and institutional resources in the Hindi Kush Himalayan region* (pp. 1–30). International centre for integrated mountain development.

Price, M. (Ed.). (2004). *Conservation and sustainable development in mountain areas*. IUCN.

Radh, B., Rawat, D. S., & Tiwari, J. K. (2013). Wild edible plant resources of the lohba range of Kedarnath forest division (KFD), Garhwal Himalaya, India. *International research journal of biological sciences, 2*, 65 73.

Reynolds, M. P. (2010). *Climate change and crop production*. CABI International. https://doi.org/10.1079/978 1845936334.0000

Salick, J., Yang, Y., & Amend, A. (2005). Tibetan land use and change near KhawaKarpo eastern Himalayas. *Economic Botany, 59*, 312–325. https://doi.org/10.1663/0013-0001 (2005)059 [0312, TLUACN] 2.0. CO; 2

Sun, J., Qin, X., & Yang, J. (2016). The response of vegetation dynamics of the different alpine grassland types to temperature and precipitation on the Tibetan plateau. *Environmental Monitoring and Assessment, 188*(1), 20.

Sundriyal, M., Sundriyal, R. C., & Sharma, E. (2004). Dietary use of wild plant resources in the Sikkim Himalaya, India. *Economic Botany, 58*, 626–638. https://doi.org/10.1663/0013-0001 (2004)058 [0626: DUOWPR] 2.0. CO; 2

8 Gender Perspective on Traditional Knowledge in Irrigation Management and Climate Change Adaptation

1 Introduction

1.1 Women and Climate Change

Climate change impacts men and women differently due to variations in traditional roles, societal expectations, and livelihoods (Schalatek, 2009). Differences in gender roles globally have significant implications for understanding whether the impacts of climate change will be similar or divergent for men and women (Enarson, 2012). Variations in economic responsibilities, family duties, political rights, education, and health can render men and women more or less vulnerable to the effects of climate change (Joan Nagel, 2016). The gender dimension of climate change operates on two levels: first, women, particularly those in developing countries, are generally more vulnerable than men due to their roles as caregivers and dependents, which compounds their exposure to climate-induced poverty and livelihood precarity, limiting their ability to build resilience (Bhagat, 2017; Bhatta et al., 2015). Second, men and women undertake different roles in responding to climate change, with women playing crucial roles in both mitigating and adapting to its effects. Women are often key agents of change, actively engaging with and adapting to both climatic and non-climatic stressors (Sandra Bäthge, 2010; Tschakert & Machado, 2012).

Women, who constitute the majority of the developing world's 1.4 billion poor, generally have lower incomes, limited access to credit and decision-making authority, and restricted control over resources, thereby increasing their vulnerability to climate impacts. Due to unequal divisions of labor and social discrimination, women and men experience climate change differently (Arora-Jonsson, 2011a, 2011b; Dimitrov, 2019; Jackson, 1993). Limited access to information and resources, combined with restricted participation in decision-making, exacerbates the negative effects of climate change on women, amplifying existing social inequalities (Brody et al., 2008). Gender inequality thus contributes to more pronounced adverse impacts of climate change, often making women more vulnerable than men (Joan Nagel, 2016). Additionally, women frequently lack secure and effective means to access and control land, water, forests, and other resources. As a result, when their survival and livelihoods are threatened by climate change, they must cope with extremely limited resources, with indigenous and local women facing particularly severe challenges (Demetriades et al., 2008). Their livelihoods, often reliant on natural resources, subsistence agriculture, and informal sectors, are especially vulnerable to climate change (Resurreccion, 2013; Lambrou & Piana, 2006; Dankelman et al., 2008). In other words, there is a bidirectional relationship between climate change and gender: climate change exacerbates existing gender inequalities, and gender inequalities, in turn, lead to more severe negative impacts on women compared to men (Wendy, 2011).

Y. Lun, *Climate Change and Traditional Knowledge*, Sustainable Development Goals Series,
https://doi.org/10.1007/978-3-032-12015-1_8

Gender not only affects vulnerability but also influences responses to climate change. Women play a crucial role in climate change mitigation and adaptation (IUCN, 2007; UNDP, 2009; UNFPA/WEDO, 2009). They are not merely passive victims but also active agents in mitigating and adapting to climate change (Raty & Carlsson-Kanyama, 2009). Women worldwide—especially at the community level—have been instrumental in resisting, mitigating, and even reversing the effects of climate change. They possess unique knowledge and skills that significantly contribute to effective climate change adaptation and mitigation (Masum, 2009).

In indigenous peoples and local communities, men and women have distinct roles and responsibilities, leading to varying levels of vulnerability and consequences in the face of climate change (CDMP, 2009). For women in these communities, the impacts of climate change are not only directly felt in the environment but also affect their economic conditions, livelihoods, health, and social well-being (Masum, 2009). Their traditional knowledge plays a crucial role in adapting to climate change (Cannon, 2002; Rivero, 2002; Ahmed & Fajber, 2009; Ahmed et al., 2013). Women are not only the primary supporters of their families and implementers of agricultural production but their labor is closely tied to natural resources such as water, land, and forests, which are directly affected by climate change (Agarwal, 2002; Boyd, 2002; Yin et al., 2018). They possess extensive knowledge of their natural environment and prioritize the preservation and selection of crop varieties to ensure food security amidst changing climatic conditions (IUCN/UNDP/GGCA, 2009). Moreover, women can leverage their knowledge to develop and promote innovative farming methods to adapt to climate change (Mitchell et al., 2007; Bridge, 2008).

However, at both the community and international levels, women's issues related to climate change often receive inadequate attention. There are limited channels for women to voice their unique needs regarding the different impacts of climate change on men and women (Sandra Bäthge, 2010). Women's potential in climate change mitigation and adaptation remains underutilized, and their practical knowledge about the environment, livelihoods, and resources is frequently unrecognized (Agarwal, 2002). Social and cultural constraints often hinder women's participation in public affairs and their involvement in decision-making processes related to climate change at various levels (Boyd, 2002; Rodenberg, 2009).

Climate change may exacerbate existing water shortages, with extreme and prolonged droughts having cascading effects on food production, and on drinking and irrigation water supplies. In indigenous peoples and local communities, women are primarily responsible for procuring water for their families and for irrigating their farmland (Yin, 2014; Boateng, 2018). Given their central role in water irrigation, women are particularly sensitive to changes in seasons and climatic conditions that impact water quantity and accessibility, making irrigation even more time-consuming (Tandon, 2007). Women's responsibilities encompass various aspects of water use, including household consumption and agricultural irrigation, which means their understanding of climate change extends beyond scientific and physical changes to encompass socio-economic dimensions (Masum, 2009; Yin et al., 2020). However, women often remain excluded from water and irrigation decision-making processes, as paternalistic water governance systems frequently fail to address their needs, especially in the context of climate change (Ahmed, 2013).

1.2 Theory and Development of Gender Studies on Climate Change

Gender research on climate change examines the relationship between climate change and women through the lens of gender analysis. This research encompasses two main aspects: First, due to gendered divisions of labor and inequality, the impacts of climate change differ between men and women. Women are generally more vulnerable to climate change than men, experiencing more pronounced effects on their lives, livelihoods, and health. Climate change exacerbates existing gender inequalities. Second, men and women play different roles in responding to climate change. Women contribute unique perspectives, technologies, and knowledge in adaptation efforts, providing invaluable contributions that men cannot easily replace.

In recent years, gender research on climate change has gained attention across various disciplines. This research not only explores the relationship between climate change and individuals, particularly women, but also examines its intersection with environmental, social, and cultural contexts.

Gender research on climate change can be categorized into seven areas: First, the impact of climate change is more severe on women due to social and labor divisions. Second, the effects of market-oriented climate change mitigation policies. Third, differing perspectives and coping strategies between men and women regarding climate change. Fourth, varying degrees of climate change impacts on men and women. Fifth, increased health problems for women caused by climate change. Sixth, heightened risks for women and children due to natural disasters and competition for resources. Finally, the lack of gender perspectives in climate policy-making due to low female participation (Brody & Demetriades et al., 2008; Esplen& Demetriades, 2008; Seema, 2011; Sandra, 2010; Rodenberg, 2009; Chinwe, 2011).

Gender analysis offers a research perspective and methodology for understanding the differential impacts of climate change on different genders, highlighting the disparities between how women and men are affected. It focuses on the relationships between men and women, gender dynamics within households, and the empowerment, knowledge, control, and participation of both genders in decision-making at all levels (Iipinge, 2000). Gender relationships refer to the social structure and power dynamics between women and men, which influence how resources are distributed and utilized (Watson, 2006). Different societal roles and responsibilities lead to varied impacts of climate change on men and women (Banda, 2005). Thus, understanding the connection between gender and climate change necessitates an examination of social roles, responsibilities, resource access, and productive activities, and how these factors shape the different impacts and risks of climate change for men and women.

Due to their distinct societal roles and statuses, men and women exhibit different attitudes, perspectives, knowledge, and methods for addressing climate change. Gender analysis of climate change explores how men and women use their varying resources, knowledge, and experiences to cope with its impacts. It also examines how their social status, mechanisms, and processes either constrain or enhance their capacity to manage these impacts, and how such constraints can be mitigated (Chinwe, 2011). Gender research on climate change not only provides theoretical and practical guidance for women's adaptation strategies but also holds broader theoretical and practical significance for human responses to climate change.

Figure 1 above illustrates the three aspects and structure of gender research on climate change.

By outlining the theoretical structure, we can define the scope of gender research in this field. It is evident that gender studies on climate change exhibit different characteristics depending on the environment, country, nationality, and cultural background. Consequently, gender research on climate change has the following attributes:

First, the impact of climate conditions varies across different living environments, influencing how women access and utilize sustainable resources or enhance their survival skills in response to climate change. This diversity in climatic conditions leads to varied gender research findings, reflecting distinct environmental characteristics. For instance, women living in plains, plateaus, islands, oases, or deserts experience different impacts and responses to climate change. Similarly, gender studies in urban versus rural areas also reveal significant differences in climate change impacts and adaptations.

Second, women in developed and developing countries are affected differently by climate change. Their attitudes, status, roles, resources, and coping strategies vary between these contexts.

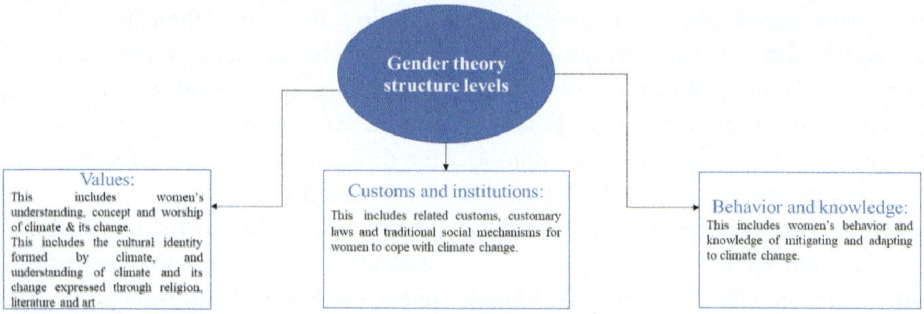

Fig. 1 Theoretical structure and attributes on gender research on climate change

Fig. 2 Tibetan women prepare to make traditional wooden water channels (photoed by Yin Lun)

For example, women in developing countries often face greater challenges from climate change and disasters compared to men, while women in developed countries tend to have a lower carbon footprint than men.

Third, women from different ethnic groups have diverse perspectives, understandings, and responses to climate change. Women from various cultural backgrounds, such as Native American, Australian Aboriginal, African tribal, and Chinese ethnic groups, exhibit significant differences in their traditional knowledge and coping strategies regarding climate change.

Lastly, gender studies on climate change are influenced by cultural and religious contexts. Women from different cultural backgrounds, including Indian, Arab, European, and Chinese cultures, have distinct perceptions of climate

Fig. 3 Traditional wooden water channels (photoed by Yin Lun)

change. In China, ethnic cultures such as Tibetan, Mongolian, Han, and Uygur have unique interpretations of climate and its changes. For example, Tibetan culture views sacred mountains as central to climate change, Mongolian beliefs link longevity to climate, Han philosophy integrates climate into the concept of harmony between heaven and humans, and Uygur perspectives are influenced by Islam.

In summary, gender research on climate change encompasses three theoretical structures: values, customs and institutions, and behavior and knowledge. It also reflects the attributes of environmental, national, ethnic, and cultural diversity, positioning it as an interdisciplinary field. This diversity provides a comprehensive framework for future gender research on climate change.

1.3 Focus and Contents of Gender Studies on Climate Change

1.3.1 The Impacts of Climate Change on Women

Due to differing social divisions of labor and gender discrimination, men and women experience climate change impacts unequally. For instance, the abnormal climate patterns have intensified the "feminization of poverty" with higher female mortality rates in climate-related disasters compared to men. In many regions, prolonged droughts and floods have led to increasingly scarce natural resources, placing greater survival pressure on women than on men. Consequently, women are more vulnerable to the effects of climate change (Hemmati, 2007). Additionally, social and cultural constraints often

Fig. 4 Tibetan women cleaning water channels (photoed by Yin Lun)

hinder women's participation in public affairs. During extreme climate events, their access to weather forecasts and disaster warnings is typically limited. This inequality in economic and social spheres exacerbates the vulnerability of women to natural disasters. In such disasters, the mortality rate for women and children is 14 times higher than for men (Brody and Demetriades et al., 2008). For example, in the 2004 tsunami in Indonesia, women accounted for 65% of the fatalities (Demetriades & Esplen, 2008; Rohr, 2006).

1.3.2 The Relationship Between Women and Climate Change

Compared to men, women generally exhibit a more environmentally friendly lifestyle, characterized by lower carbon emissions and greater environmental protection. This perspective aligns with eco-feminism within gender studies on climate change. Some research suggests that women, as a group, maintain a more harmonious relationship with the environment, including the climate, than men. This view is particularly prevalent in studies from developed countries, reflecting discussions in gender, development, and eco-feminism. Research indicates that women are often more sensitive to environmental risks, more likely to adapt their behaviors in response to climate crises, and more supportive of stringent climate policies and measures.

According to this perspective, men are often seen as the primary contributors to environmental pollution and climate change, irrespective of their socio-economic status. For instance, gender studies on transportation reveal that men tend to drive more frequently and over longer distances, resulting in higher carbon emissions. In contrast, women are more likely to use public transport, bicycles, or walk, making more sustainable

Fig. 5 Tibetan women building reservoirs (photoed by Yin Lun)

and environmentally conscious choices. Similarly, studies on meat consumption show that men consume more meat than women, with livestock accounting for 18% of global greenhouse gas emissions. This research also indicates that women's consumption patterns are generally more sustainable, resulting in a smaller carbon footprint and reduced climate impact (Oldrup & Breengaard, 2009).

Despite these insights, the current climate change response is often dominated by technical teams composed mainly of men, who are perceived as the primary problem-solvers. This lack of female participation and gender perspective in these expert teams means that women's viewpoints and contributions are insufficiently represented (Johnsson, 2007).

1.3.3 Women's Response to Climate Change

Different genders play distinct roles in coping with climate change. Women's social roles and traditional knowledge are crucial in both mitigating and adapting to climate change (Rodenberg, 2009; IUCN, 2007; UNDP, 2009; UNFPA, 2009). In indigenous and local communities, women often take on unique roles in adaptation strategies. They are adept at responding to the consequences of climate change and understand the needs of their families (Brody and Demetriades et al., 2008). For example, women are key to ensuring family food security. They are primarily responsible for livelihood activities, hold significant decision-making power in family affairs, and possess extensive traditional

Fig. 6 Vegetable greenhouse cultivation carried out by Tibetan women (photoed by Yin Lun)

knowledge about their local environment. This expertise enables them to select and preserve crop varieties in response to climate and environmental changes (IUCN, 2009). Such practices can boost agricultural production, secure food supplies, and create surplus products for market trade, ultimately contributing to household income and enhancing community resilience to climate change.

Women's roles in mitigation strategies are equally important. For instance, providing household energy, such as using efficient cooking stoves, can reduce emissions at the family level. Women also engage in natural resource conservation and forest management, including tree planting and vegetation restoration, which helps to prevent emissions from deforestation and significantly reduces greenhouse gases in the atmosphere. Given their critical contributions to both mitigation and adaptation efforts, women must be actively involved in related decision-making processes (Sandra, 2010).

2 Background

In Deqin, the local Tibetans have long relied on a traditional irrigation system based on their agro-pastoral lifestyle. However, this system, along with its customary laws and traditional knowledge, is increasingly threatened by climate change (Yin and Xue, 2012). The challenges in effectively utilizing water resources are creating risks for many livelihoods. Additionally, as most men leave for work, agriculture has become increasingly feminized, with women emerging as the primary workforce. Despite this, traditional irrigation practices and decision-making processes remain dominated by men, leading to gender conflicts and contradictions (Yin, 2014).

To address these issues, a community-based irrigation management committee was established, leveraging the traditional women's organization known as "sisterhood." This committee has been tasked with discussing, formulating, and implementing new irrigation rules and regulations. Through this mechanism, women have been actively involved in the decision-making process for irrigation management. This article explores how community-based irrigation management in Tibetan villages, rooted in traditional organizations and knowledge, can effectively adapt to climate change by integrating gender mainstreaming.

The traditional irrigation system comprises customary law, technology, and infrastructure.

3 1Traditionalcustomary Law

In the traditional irrigation system, customary law plays a crucial role. In Deqin, Tibetan villages are often divided into upper and lower parts based on kinship and geographical location, and the irrigation customary law is structured around this division.

First, villagers construct irrigation facilities based on the village's topography and the social dynamics between the upper and lower parts. This arrangement aims to maximize the efficient use of water resources and ensure fair distribution. Prior to construction, the village head convenes a meeting with family representatives to discuss the repair and development of irrigation facilities. Subsequently, they assess the fields to determine the number of irrigation lines needed to adequately serve every household.

Second, the villagers establish rules for water usage according to the division between the upper and lower parts of the village. These rules are designed to ensure fair water allocation and prevent conflicts during dry seasons when water is scarce. Since local irrigation relies primarily on river water and rainfall, the village head gathers household representatives to deliberate and decide on the fair distribution of river water resources for both dry and rainy seasons. The irrigation rules are based on river flow and temperature variations. If the river flow is sufficient and there is a significant temperature difference between morning and afternoon, irrigation will alternate between the upper and lower parts of the village daily. If the river flow is low and the temperature difference is minimal, irrigation will be conducted in both morning and afternoon sessions.

3.1 Traditional Technology and Knowledge

The traditional irrigation technology and knowledge include building irrigation equipment, controlling water flow speed and increasing water temperature.

Farmers constructed a comprehensive irrigation system by building reservoirs, digging canals, and erecting ditches, which eventually directed river water into their fields. The traditional reservoirs are made from rammed earth and stone, with the size of each reservoir determined by the number of farmers using the irrigation line. Canals are dug along the hillsides, with stones strategically placed at corners to redirect water flow. In cases where deep gullies make it impractical to dig canals, villagers construct ditches. They dig a deep channel in a whole log, position the log across the gully, and secure both ends with stones. Once the reservoirs, canals, and ditches are built, a complete irrigation system is established. When all the irrigation lines are in place, the village forms a fully functioning irrigation network.

Since many of the fields in these villages are located on hillsides, the soil in the fields risks erosion if the irrigation water flows too quickly. Therefore, villagers make every effort to construct canals in flatter areas to slow the flow of water. When dealing with steep channels, they install a series of adjustable wooden gates along the canal and use stones or cloth to further control the flow, ultimately reducing water speed.

Additionally, the local rivers are primarily fed by melting glaciers and snow, resulting in very cold water. If the irrigation water is too cold, it can hinder crop growth. To address this issue, villagers construct additional reservoirs along the irrigation lines and design the canals to be as winding as possible. This increases the time the water remains on the surface, allowing sunlight to warm the water before it reaches the fields.

4 Methodology

The research employed a combination of Participatory Rural Appraisal (PRA) and gender-based action research. Additionally, structured interviews, open-ended interviews with key informants, and focus group discussions were conducted. We also reviewed local background information, including data on farmland irrigation, meteorology, disasters, and river hydrology.

4.1 Participatory Rural Appraisal (PRA)

The PRA with village women centered on traditional irrigation systems and the primary risks posed by climate change. These risks included the depletion of irrigation water resources due to drought, the breakdown of traditional irrigation practices caused by heavy rainfall, and the damage to irrigation infrastructure from debris flows and landslides. The PRA involved 20 women from each village, representing different age groups: older women over 60, middle-aged women between 30 and 50, and young women under 30. The PRA followed these steps: first, the women collectively identified all the irrigation risks and challenges related to climate change; second, they ranked and prioritized these issues, selecting the five most critical concerns.

After the main risks were identified, group discussions were held in the village. These discussions included 30 men and women of all ages, ensuring a diverse representation of traditional irrigation knowledge. The purpose of the discussions was to confirm the importance and urgency of the five key risks identified by the women, while also considering the perspectives and suggestions of the men.

Structured interviews and open-ended interviews with key informants focused primarily on women, particularly older women with extensive experience in irrigation. The aim of these interviews was to understand how women mitigate and adapt to irrigation-related risks in the face of climate change. Interviews with key informants also helped contextualize women's roles within their traditional organizations, knowledge systems, and coping mechanisms.

4.2 Gender-Based Irrigation Plan (GBIP)

The Gender-based Irrigation Plan (GBIP) is a methodology designed to empower women, women's groups, and local institutions to actively participate in identifying and addressing irrigation-related water resource and social issues. This approach emphasizes the leadership of local women in managing irrigation systems, with local priorities, knowledge, needs, and capacities serving as central elements in the planning process.

To implement GBIP more effectively, three models have been developed based on data from the PRA surveys and interview results in the villages:

I. Gender Mainstreaming Model: This model emphasizes the cultural practices of women related to the irrigation system.
II. Gender Participation Model: This model focuses on the role of traditional women's organizations in managing irrigation systems.
III. Gender Management Model: This model highlights the use of traditional knowledge held by women in relation to the irrigation system.

5 Results and Discussion

The results were structured to demonstrate the impact of climate change on traditional irrigation systems and to present the gender-based irrigation plan for climate change adaptation. Our analysis examined the research results through two primary climate change indicators—rising temperatures and increased instability, as well as extreme weather and disasters—and explored how these factors affect traditional irrigation systems.

The gender-based irrigation plan, developed during this action research project, addresses three key aspects: irrigation management, protection of irrigation water sources, and irrigation technology development. Women play a central role in implementing adaptive strategies to respond to climate change in these areas.

5.1 Impact of Climate Change on Traditional Irrigation Systems

Through PRA, group discussions, and interviews, our survey gathered information on climate indicators and villagers'—particularly women's—perspectives on how climate change is affecting traditional irrigation systems. Women were then asked to score and rank these indicators based on the severity of their impact.

A total of 60 women participated in the participatory rural appraisal process. After extensive discussions, they identified 10 key climate indicators that affect traditional irrigation systems. Each woman was allowed to vote for up to five indicators, resulting in a total of 300 votes. For example, 21% of the votes indicated that prolonged droughts had the most significant impact on traditional irrigation systems due to reduced water resources. Another 18% of the votes highlighted rising temperatures as a major concern, mainly because of accelerated glacier and snowmelt, leading to secondary disasters such as floods and landslides. Furthermore, 14% of the votes pointed to unstable rainfall, which undermines traditional irrigation practices and knowledge, as another critical impact (see Table 1).

According to the scoring and ranking of women, there are two main types of climate indicators affecting traditional irrigation systems. One is extreme weather phenomena and disasters represented by persistent drought, and the other is climate change represented by temperature rise and rainfall instability.

5.1.1 Extreme Weather and Disasters

Extreme weather events and disasters have had a profound impact on traditional irrigation systems, particularly the persistent droughts of the past decade. In Deqin, irrigation relies heavily on river water and rainfall, but continuous drought has caused a significant reduction in both. Cili Lamu, an elderly local woman, shared her experience:

> In the past ten years, *the drought* has become increasingly severe, something I never witnessed in my youth. During a drought, because there is little or no rain, we are forced to rely on river water for irrigation, even though we prefer rainwater. Traditionally, we believe that rainwater is better for crop growth and requires less labor. Additionally, river water is too cold to promote healthy crop growth.
>
> However, when there is no rain, we have no choice but to use the river for irrigation. The drought also greatly affects the river itself. Without the replenishment of rainfall, the river's flow decreases significantly. Furthermore, some rivers are fed by springs from the high mountains, and during droughts, these springs dry up or diminish, reducing the river's flow even more. This has created severe water shortages for irrigation. As a result, when we irrigate the farmland in the village, the downstream part of the river often stops flowing—a situation that was rare when I was younger.

At the same time, floods, landslides, and debris flows triggered by intense rainfall over short periods also pose significant threats to traditional irrigation systems. Danzhen Lachu, a local woman, explained:

> In recent years, rainstorms have become increasingly frequent. These heavy downpours often lead to floods and, in severe cases, disasters such as debris flows and landslides. Since most of the irrigation ditches and canals are built on hillsides, where vegetation is sometimes sparse, these floods, landslides, and debris flows often destroy the pools, canals, and ditches. After such disasters, we are forced to repair or rebuild the entire irrigation system.

The water shortages caused by drought frequently lead to conflicts over water use among farmers during peak farming seasons, particularly for those with fields downstream, where river water often runs dry. Additionally, floods, landslides, and debris flows not only damage irrigation infrastructure but also render the entire system unusable during critical farming periods. The constant need for repairs and rebuilding further increases the labor burden on villagers, especially

Table 1 Women's responses about impact of climate indicators to irrigation. (The information is sourced from a participatory rural appraisal involving 60 Tibetan women)

No	Indicator	Impact	Importance (N = 300)
1	Lingering drought	Reducing irrigation water resources	62(21%)
2	Temperature rise	Glaciers and snow melt rapidly, causing secondary disasters such as floods and landslides	53(18%)
3	Erratic rains	Unstable rainfall makes irrigation experience and traditional knowledge invalid	42(14%)
4	Heavy rainfall	Causing secondary disasters such as floods and landslides	37(12%)
5	Flood	Destroy irrigation pools, canals and ditches	31(10%)
6	Landslide	Destroy irrigation pools, canals and ditches	20(7%)
7	Snowstorm	Destroy irrigation pools, canals and ditches, reducing water temperature	17(6%)
8	Shift in therein timings	Makes irrigation experience and traditional knowledge invalid	15(5%)
9	Change in the duration of rains	Makes irrigation experience and traditional knowledge invalid	13(4%)
10	Rains late	Makes irrigation experience and traditional knowledge invalid	10(3%)

women, who are primarily responsible for maintaining traditional irrigation systems due to the growing feminization of agriculture.

6 2Temperature Rise and Rainfall Instability

The impact of rising temperatures on traditional irrigation systems is evident, as the melting of glaciers and snow feeds most of the local rivers. As temperatures rise, this melting accelerates, causing the flow rate and volume of river water to increase during the summer months. LurongZhuoma, an elderly local woman, shared her insights:

> We used to judge the flow and velocity of the river by listening to the sound of the water, which helped us determine the optimal time for irrigation. We could also adjust and control the flow speed and temperature of the irrigation water accordingly. However, in recent years, the rivers have had much higher flow rates during the summer, which I believe is due to the increasing melt of glaciers and snow at the rivers' sources. This has made it difficult for me to rely on my experience of listening to the river's sound, as it no longer provides the same information.
>
> Additionally, the traditional methods we use to regulate and control the flow rate and temperature of the irrigation water have become insufficient. With faster water flow, we need to construct more gates to slow it down. At the same time, as the river water becomes colder, we are forced to build more winding irrigation canals to raise the water temperature before it reaches the fields.

Local people believe that rain irrigation is more beneficial for crop growth compared to river water, as rain promotes healthier crops. However, the increasingly erratic rainfall in recent years—including shifts in timing, changes in duration, and delays—has sometimes rendered traditional irrigation knowledge ineffective. Nima Meiduo, a local woman, shared her experience:

> In the past, the rainy season was fairly predictable, so we would wait for it to irrigate because crops irrigated by rain grew better and tasted better. We could predict the onset of the rainy season by observing changes in wind direction, the scent in the air, and the calls of birds, and we estimated the duration of the rains based on the number of birds in the fields. We also prepared in advance by releasing river water from pools to store the rainwater.

> However, the timing of the rainy season has become unpredictable. Sometimes it arrives early, sometimes very late, or it may not come at all. Moreover, we can no longer rely on our experience with wind direction and birdcalls to accurately predict its arrival. Similarly, the duration of the rainy season is now difficult to foresee—sometimes we have continuous rain, and other times we experience short, intense downpours.

For local women, irrigation-related knowledge and experience is not only visual but also relies on auditory and olfactory cues. They traditionally judged the flow and speed of rivers and predicted the timing and duration of the rainy season through these senses to determine when and how to begin irrigation. However, due to climate change and the resulting shifts in natural phenomena, this traditional knowledge has become partially ineffective. Women can no longer rely solely on their past experiences and knowledge for managing irrigation.

6.1 Gender-Based Irrigation Management Plan for Climate Change Adaptation

Traditional knowledge is a process of continuous development and innovation. Despite the challenges posed by climate change, it remains the most effective tool for local women to navigate these changes. Drawing upon their traditional knowledge, local women have initiated a series of actions to help irrigation systems adapt to the changing climate as effectively as possible.

6.1.1 Management of Irrigation Water Sources

Traditionally, the management of irrigation systems is primarily decided by the village head (typically male) and male representatives from each household. However, the women, who are the primary users of these irrigation systems, are often excluded from both management and decision-making processes. Although they may voice their opinions and suggestions through male representatives in their families, their contributions are frequently overlooked, resulting in limited influence. This issue has become even more pronounced in recent years, as many men have left to seek work elsewhere, leading to the feminization of agriculture. Women have become the primary labor force and, consequently, the main users of the irrigation systems. Given this shift, their voices and perspectives are increasingly critical. It is essential that women be included in the management and decision-making processes related to irrigation, and that they are empowered to take on these roles.

In the local villages of Deqin, there exists a traditional organization known as "The Sisterhood." This group was spontaneously established by female villagers and is responsible for organizing various festivals and religious activities, such as singing and dancing performances, dinner parties, sacred mountain prayers, and Buddhist worship. While all women in the village are members, active participation generally includes those between the ages of 12 and 65. "The Sisterhood" operates with a concerted management approach: every year during the Spring Festival, six women from six different households are elected as leaders for that year, with new elections held annually.

In recent years, as the impacts of climate change on the irrigation system have intensified, "The Sisterhood" has adapted its role to address these challenges. The organization has begun to coordinate collective and spontaneous responses from women to manage and mitigate the effects of climate change. Through "TheSisterhood,"women have taken on responsibilities such as formulating and implementing customary laws for irrigation management, establishing water resource allocation rules, compiling and preserving traditional irrigation knowledge, protecting vegetation and forests along riverbanks, and conducting regular inspections and repairs of irrigation facilities like pools, ditches, and canals.

These actions demonstrate how women are increasingly involved in irrigation management and decision-making processes. They reflect the capacity of local women to respond proactively to climate change and its associated challenges, highlighting their critical role in managing and adapting to environmental changes and disasters.

6.1.2 Conservation of Irrigation Water Source

In the local villages, irrigation is not only a vital livelihood activity but also deeply embedded in the local culture, which intertwines with beliefs related to sacred natural features such as mountains, lakes, and rivers. The water resources used for irrigation, derived from both rainwater and river water, are closely linked to these sacred sites. Sacred mountains and lakes, associated with rainwater, and sacred rivers, associated with river water, hold significant positions in the local irrigation culture, leading to the development of associated religious rituals.

Notably, in local beliefs, these sacred natural features are often gender-specific, with many associated with female deities. These sacred elements are also seen as protectors of local women, who predominantly organize and participate in the related worship ceremonies. According to tradition, before initiating irrigation, villagers conduct rituals for both rainwater and river water. The rain ceremony involves women performing a sacred mountain ritual to pray for adequate rainfall. During persistent droughts, a special rain prayer ceremony is held at the sacred lake. Similarly, for river water, women conduct a sacred river ceremony to ensure sufficient and appropriate water flow. Additionally, they hold disaster prevention ceremonies at vulnerable riverbank areas prone to landslides and mudslides to request protection from these risks.

As the adverse effects of climate change on traditional irrigation systems become increasingly evident, women recognize that merely performing rituals is insufficient. They have begun to integrate traditional beliefs with practical actions to address and adapt to these changes. Under this approach, women have combined cultural practices with modern ecological concepts to protect irrigation water resources and watershed environments.

Firstly, they designated a specific day for traditional belief ceremonies as "Irrigation Water Resources Protection Day." On this day, alongside performing traditional ceremonies, they engage in educational activities to raise awareness about the impacts of climate change on irrigation and advocate for water conservation.

Secondly, women have expanded traditional rituals to include practical environmental actions. In addition to traditional practices like chanting sutras and burning incense, they now undertake activities such as cleaning river garbage, maintaining irrigation pools, ditches, and canals, and reinforcing riverbanks to mitigate flood risks.

Thirdly, women focus on preserving original vegetation and trees in landslide-prone areas along riverbanks. This includes pruning trees to prevent lodging, fencing to protect trees from livestock, and removing parasitic plants to prevent strangulation. They also engage in afforestation with local tree species, particularly cedar, which is not only well-suited to the local environment but also holds a special place in traditional beliefs. Cedars are burned during irrigation ceremonies to fulfill ritualistic purposes and promote rain.

6.1.3 Distribution Institution of Irrigation Water Resources

As previously mentioned, the traditional distribution of irrigation water resources in the village is based on a division between the upper and lower parts of the village. However, with the escalating impacts of climate change, this traditional system has become insufficient to meet the villagers' needs and to resolve conflicts arising from water shortages between the upper and lower parts of the village. Consequently, there is a pressing need to establish a new system for allocating irrigation water resources. In response, women have undertaken the task of improving and innovating the traditional water distribution system.

Rather than completely overhauling the traditional system, women have enhanced and reinforced it by maintaining the division between the upper and lower village areas while introducing improvements. They have developed distinct irrigation lines for each area, ensuring that water resources are allocated to each household's fields. The new system includes more refined rules and regulations governing these irrigation lines, specifying the rights and obligations of each household. This includes detailed schedules and sequences for irrigation, as well as responsibilities for the maintenance of irrigation lines and facilities. The new system aims to optimize water resource use, ensure fair distribution, minimize water shortages, and prevent conflicts caused by inequitable resource allocation.

In addition to revising the distribution system, women have also addressed the materials used in constructing irrigation facilities. Traditionally, irrigation infrastructure was built with rammed earth, wood, and stone—materials that are prone to water leakage and susceptible to damage from climate-related disasters such as heavy rainfall, floods, debris flows, and landslides. The high cost of rebuilding and maintaining these facilities after disasters places a significant burden on women. To tackle these challenges, women have begun to incorporate modern materials, such as cement, iron pipes, steel bars, and bricks, alongside traditional materials. This combination enhances the durability and resilience of irrigation facilities, reducing the frequency and cost of repairs and rebuilding.

7 Conclusion

The gender-based irrigation management practiced by Tibetan women illustrates how traditional knowledge and culture can enhance adaptation to the impacts of climate change. This approach offers valuable insights for developing equitable and effective irrigation management plans. For Tibetan women, traditional knowledge encompasses three key elements: the "sisterhood" organization, the sacred beliefs surrounding mountains and rivers, and the traditional mechanisms for water distribution between the upper and lower parts of the village. These elements form the foundation for their involvement in water resource management, foster a belief system that supports the protection of irrigation resources, and drive innovations in water resource allocation. This demonstrates that traditional knowledge is not static but a dynamic process of continuous development and innovation (Berkes, 1999).

In the face of climate change, adaptation is crucial for communities whose lives are directly affected by environmental variability and extremes (Adger et al., 2003). Tibetan women rely on water resources for irrigation, and their management strategies, rooted in traditional knowledge and culture, represent a form of adaptation to climate change. This offers valuable lessons for indigenous peoples and local communities worldwide, who often base their natural resource management decisions on cultural values (Wolf, 2011). Traditional knowledge and culture thus play a critical role in building awareness of climate change and developing local adaptation strategies.

While climate change poses significant challenges to Tibetan women regarding irrigation water resources, it also presents an opportunity for greater involvement in decision-making processes. Tibetan women have used their traditional knowledge to manage irrigation systems, establish customary laws, and formulate regulations. Although this is a localized example, it underscores a broader principle: women from indigenous and local communities globally are engaged with climate change issues, such as temperature increases, extreme weather events, and resource governance, in ways that resonate with policymakers and scientists (Berkes, 2009; Ford et al., 2007). This highlights the importance of including indigenous women in policy and decision-making processes. Policymakers and scientists should recognize the value of women's traditional knowledge, consider their local experiences, and integrate these perspectives with scientific knowledge to create more equitable and effective climate change adaptation policies and plans.

References

Adger, N. W., Huq, S., Brown, K., Conway, D., & Hulme, M. (2003). Adaptation to climate change in the developing world'. *Progress in Development Studies, 3*(3), 179–195. https://doi.org/10.1191/1464993403ps060oa

Agarwal, B. (2002). Participatory exclusions, community forestry, and gender: An analysis for south Asia and a conceptual framework. *World Development, 29*(10), 1623–1648. https://doi.org/10.1016/S0305-750X(01)00066-3

Ahmed, S. I. (2013). *Water for Poor Women: Quest for an Alternative Paradigm*. Lexington Books.

Ahmed, S., & Fajber, E. (2009). Engendering adaptation to climate variability in Gujarat, India. *Gender and Development, 17*, 33–50. https://doi.org/10.1080/13552070802696896

Ahmed, A., Lawson, E. T., Mensah, A., et al. (2016). Adaptation to climate change or non-climatic stressors in semi-arid regions? Evidence of gender differentiation in three agrarian districts of Ghana. *Environmental Development, 20*, 45–58. https://doi.org/10.1016/j.envdev.2016.08.002

Annecke, W. (2011). Gender and Climate change adaptation. Adaptation and beyond: published by Indigo development & change as a contribution towards effective and participatory adaptation to climate change. Indigo development & change. http://www.indigo-dc.org

Arora-Jonsson, S. (2011a). Virtue and vulnerability: Discourses on women, gender and climate change. *Global Environmental Change, 21*, 744–751.

Arora-Jonsson, S. (2011b). Virtue and vulnerability: Discourses on women, gender and climate change. *Global Environmental Change, 21*, 744–751. https://doi.org/10.1016/j.gloenvcha.2011.01.005

Banda, K. (2005). *Climate change, gender and livelihoods in Limpopo Province: Assessing impact of climate change, gender and biodiversity in Bahlabela District, Limpopo Province*. University of Witwatersrand.

Bäthge, S. (2010). Climate change and gender: economic empowerment of women through climate mitigation and adaptation? *Programme Promoting Gender Equality and Women's Rights. Deutsche Gesellschaft für Technische Zusammenarbeit (GTZ) GmbH*. http://www.gtz.de/gender

Berkes, F. (1999). *Sacred Ecology: Traditional Ecological Knowledge and Management Systems*. Taylor & Francis.

Berkes, F. (2009). Indigenous ways of knowing and the study of environmental change. *Journal of the Royal Society of New Zealand, 39*, 151–156. https://doi.org/10.1080/03014220909510568

Bhagat, R. (2017). Migration, gender and right to the city. *Economic and Political Weekly*. https://doi.org/10.1093/acprof:oso/9780190241490.003.0002

Bhatta, G. D., Aggarwal, P. K., Poudel, S., & Belgrave, D. A. (2015). Climate-induced migration in South Asia: Migration decisions and the gender dimensions of adverse climatic events. *Journal of Rural Community Development, 10*, 1–23.

Tandon, N. (2007). Biopolitics climate change and water security: impact, vulnerability and adaptation issuesfor women. *Agenda, 21*, 4±17. https://doi.org/10.1080/10130950.2007.9676064

Boateng, G. O., Collins, S. M., MbulloP, W. P., Onono, M., Neilands, T. B., et al. (2018). Anovel household water insecurity scale: Procedures and psychometric analysis among postpartum women in western Kenya. *PLoS ONE, 13*(6), Article e0198591. https://doi.org/10.1371/journal.pone.0198591

Boyd, E. (2002). The Noel Kempff project in Bolivia: Gender, power, and decision-making in climate mitigation. *Gender and Development, 10*(2), 70–77. https://doi.org/10.1080/13552070215905

Bridge. (2008). Gender and climate change: mapping the linkages. A scoping study on knowledge and gaps.

Brody, A., Demetriades, J., & Esplen, E. (2008). Gender and climate change: Mapping the linkages. A scoping study on knowledge and gaps. BRIDGE, Institute of Development Studies (IDS), Brighton.

Cannon, T. (2002). Gender and climate hazards in Bangladesh. *Gender Dev, 10*(2), 45–50. https://doi.org/10.1080/13552070215906

CDMP. (2009). Climate change, gender and vulnerable groups in Bangladesh. Climate Change Cell, DoE, MoEF; Component 4b, MoFDM. Dhaka.

Dankelman, I., Alam, K., Ahmed, W. B., Gueye, Y. D., Fatema, N., & Kutin, R. M. (2008). Gender, climate change and human security: lessons from Bangladesh, Ghana and Senegal, ELIAMEP: WEDO, 1–71.

Demetriades, J., & Esplen, E. (2008). The Gender Dimensions of Poverty and Climate Change Adaptation. *IDS Bulletin, Institute of Development Studies*. https://doi.org/10.1111/j.1759-5436.2008.tb00473.x

Dimitrov, B. E. G. (2019). Effects of climate change on women. *Research Review, International Journal of Multidisciplinary, 4*(5), 210–215.

Elaine, E. (2012). Women confronting natural disaster: from vulnerability to resilience. Boulder, CO: Lynne Rienner Publisher; Enarson, Elaine, and P.G. Dhar Chakrabarti (eds.). 2009. Women, Gender, and Disaster: Global Issues and Initiatives. Thousand Oaks, CA: Sage Publications.

Ford, J. D., Pearce, T., Smit, B., Wandel, J., Allururt, M., & Shappa, K. (2007). Reducing vulnerability to climate change in the Arctic: the case of Nunavut, Canada. Arctic, 60, 150–166. https://doi.org/10.14430/arctic240

Hemmati, M., & Rohr, U. (2007). *A Huge Challenge and a Narrow Discourse*. Women and Gender Studies Institute, New College, University of Toronto, Toronto.

Iipinge, E. M., & Williams, M. (2000). Gender and development. Windhoek: John Meinert Printing.

IUCN. (2007). Gender and climate change. Women as agents of change. http://cmsdata.iucn.org/downloads/climate_change_gender.pdf

IUCN/UNDP/GGCA. (2009). Training manual on gender climate change. http://data.iucn.org/dbtw-wpd/edocs/2009-012.pdf

Jackson, C. (1993). Doing what comes naturally. Women and environment in development. *World Development, 21*, 1947–1963. https://doi.org/10.1016/0305-750X(93)90068-K

Lambrou, Y., & Piana, G. (2006). Gender: the missing component of the response to climate change. *Food and Agriculture Organization of the United Nations (FAO)*

Lun, Y. (2014). Theory and development of gender research on climate change. *Journal of Yunnan University of Ethnology, 31*(6), 73–77.

Lun, Y., & Dayuan, X. (2012). Gender perspective on climate change and disaster–case of Hongpo village in Deqin County. *Yunnan, Journal of Yunnan Normal University, 44*(5), 65–72.

Lun, Y., & Xiaohan, Z. (2020). Traditional knowledge is the light of wisdom for conserving biodiversity and adapting to climate change. *DasQuestões, 10*(1), 104–108.

Lun, Y., Zachary, M., Zheng, Y., & Zhang, X. (2020). The Impacts of Climate Change on the Traditional Agriculture of Ethnic Minority in China. *Journal of Environmental Science and Engineering, 9*, 43–55. https://doi.org/10.17265/2162-5298/2020.02.001

Yin, L., Zachary, M., & Zheng, Y. (2018). The traditional knowledge associated to biodiversity in an age of climate change. *Earth Sciences 7*(5), 209–215. https://doi.org/10.11648/j.earth.20180705.12

Mitchell, T., Tanner, T., & Lussier, K. (2007). We know what we need!" South Asian women speak out on climate change adaptation, London: Action Aid International and the Institute of Development Studies (IDS).

Masum Momaya. (2009). Women Address Climate Change by Connecting the Dots. AWID.

Nagel, J. (2016). *Gender and Climate Change: Impacts, Science, Policy*. Routledge. https://doi.org/10.4324/9781315675275

Oldrup, H., & Breengaard, M. H. (2009). Desk study on gender, gender equality, and climate change. Nordic Council of Ministers.

Raty, R., & Carlsson-Kanya-ma, A. (2009). Comparing energy use by gender, age and income in some European countries. Swedish Defense Research Agency, Stockholm.

Resurrección, B. P. (2013). Persistent women and environment linkages in climate change and sustainable development agendas. *Women's Stud Int Forum, 40*, 33–43. https://doi.org/10.1016/j.wsif.2013.03.011

Rivero, R. (2002). Gendering responses to El Niño in rural Peru. *Gender; Development, 10*(2), 60–69. https://doi.org/10.1080/13552070215907

Rodenberg, B. (2009). Climate change adaptation from a gender perspective: Across-cutting analysis of development-policy instruments. German Development Institute. Discussion Paper 24/2009.

Rohr, U. (2006). Gender and climate change. Tiempo: A bulletin on climate and development. http://www.tiempocyberclimate.org.

Schalatek, L. (2009). Es sind die Machtverhältnisse, die Frauen für den Klimawandelverwundbarermachen. *Böll THEMA, Heft, 2*, 14.

Chinwe Ifejika Speranza. (2011). Promoting gender equality in responses to climate change--The case of Kenya.

Tschakert, P., & Machado, M. (2012). Gender justice and rights in climate change adaptation: Opportunities and pitfalls. *Ethics and Social Welfare, 6*, 275–289. https://doi.org/10.1080/17496535.2012.704929

UNDP. (2009). Resource Guide on Gender and Climate Change.

UNFPA/WEDO. (2009). Climate change connections. Gender and Population.

Watson, E. (2006). *Gender and natural resources management*. University of Cambridge.

Wolf, J., & Moser, S. (2011). Individual understandings, perceptions, and engagement with climate change: Insights from in-depth studies across the world. *WIRES Climate Change, 2*, 547–569. https://doi.org/10.1002/wcc.120

Traditional Knowledge in Disaster for the Climate Change Adaptation of Tibetan People in Eastern Himalayas, China

9

1 Introduction

At present, global climate change is causing not only significant alterations in rainfall and temperature patterns worldwide but also an increase in the frequency and intensity of extreme weather events. Scientific research indicates that the gradual climate changes historically observed on Earth are now being replaced by more abrupt climate changes, characterized by sudden severe precipitation, extreme cold, intense heat, and other extreme weather phenomena occurring within short periods (Bharara, 1986). The Intergovernmental Panel on Climate Change (IPCC) has reported a dramatic increase in hydrological and meteorological hazards over recent decades (IPCC, 2003). Additionally, the United Nations Human Settlements Programme notes that extreme weather events related to climate change surged by 50% from the 1950s to the 1990s (IDS, 2007). Such frequent and severe climate changes can lead to catastrophic climate disasters, including droughts, floods, typhoons, heavy rainfall, and landslides, along with their secondary effects. These disasters have had a profound impact on ecosystems and human societies globally, causing significant risks and losses (Fisher, 1997).

1.1 Disasters and Culture

Human interest in the relationship between climate and disasters from a cultural perspective has a long history. As early as ancient Greek times, scholars such as Hippocrates began to explore how climate influences societal formation and how various climatic conditions and seasonal factors affect human behavior (Harris, 1968). Similarly, ancient Chinese thinkers developed rich concepts regarding the relationship between climate change, disasters, and humanity, leading to ideas such as "heaven" and the "unity of heaven and man" (Lun et al., 2020). These traditional thoughts and explanations have laid a foundation for modern social science studies of climate and related disasters. In the early twentieth century, modern social sciences began to focus on climate and its associated disasters. Human Geography, for instance, considers climate as an independent variable that influences different social structures, living styles, and behavioral patterns, ultimately affecting the economic development patterns of today's world (Huntington, 1915). Historians also recognize that climate and related disasters can impact the rise and fall of civilizations (Lambert, 1975). For example, drought is identified as a major factor

Y. Lun, *Climate Change and Traditional Knowledge*, Sustainable Development Goals Series,
https://doi.org/10.1007/978-3-032-12015-1_9

contributing to the decline of Mycenaean civilization during the late Bronze Age in ancient Greece (Carpenter, 1968).

Disaster anthropology, which emerged in the 1960s, seeks to integrate research on various types of disasters, including floods, volcanoes, earthquakes, fires, and droughts, and focuses specifically on disaster-related propositions (Torry, 1979; Oliver, 1996; Hoffman & Oliver-Smith, 2002). This field posits that anthropology, due to its comprehensive approach to observing all aspects of life affected by disasters, can offer unique insights into disaster research (Oliver, 1996). Furthermore, because disasters result from the interaction between nature and society, anthropology is well-positioned to embrace disasters as a central focus of the discipline. Cultural ecology examines disasters and natural hazards through the lenses of ecological and social systems, emphasizing individuals' adaptability to environmental changes (Vayda & McCay, 1975). Since the 1980s, anthropologists have concentrated on studying the social experiences and circumstances of disaster victims, focusing on social factors in disaster research and exploring disaster and risk within this framework (Hoffman et al., 2002). Concurrently, there has been an emphasis on concepts such as vulnerability, resilience, and adaptability to understand the social underpinnings of disasters (Oliver, 1996). Recent studies on vulnerability highlight the need to distinguish between social issues and natural factors (Oliver, 2002).

1.2 Disasters, Indigenous Peoples and Traditional Knowledge

The impacts of climate disasters vary across different ecosystems and human societies. Those with the fewest resources and the lowest adaptability, such as indigenous peoples and local communities, are particularly vulnerable to climate change and extreme weather events (Houghton, 2001). Consequently, climate disasters exacerbate the already fragile environmental, economic, and social foundations of these groups, widening the development gap between them and other regions, while also threatening their sustainable development and safety.

Environmental vulnerability theory emphasizes institutional flexibility in addressing climate change by integrating aspects of climate variability, vulnerability, uncertainty, and risk into case studies (Eakin, 2000; Reilly & Schimmelpfennig, 2000; de Loe et al., 2000). In climate change research, vulnerability is addressed through the combination of local traditional knowledge, experience, and global climate phenomena (Magistro et al., 2001; Vasquez-Leon, 2002). Cultural anthropology examines climate disasters from a cultural perspective, highlighting how the interaction between humans and their environment is a key component of global cultural diversity, which is often disrupted by climate disasters. Currently, indigenous peoples and local communities are making concerted efforts to recognize, adapt to, and respond to climate-related risks and challenges. Climate disasters not only affect traditional livelihoods but also threaten the preservation and rights of traditional cultures, potentially leading to changes or even the loss of fundamental cultural elements such as religion and belief systems (Crate & Nuttall, 2009).

Climate disasters manifest in various forms around the world. Due to differences in natural environments and geographical locations, the impacts of these disasters on indigenous peoples and local communities vary significantly. In this context, local traditional knowledge has gained increasing recognition and value. For example, studies have explored the relationship between climate disasters and religion through ethnographic fieldwork with Tibetan communities in the Himalayas (Huber & Pedersen, 1997), examined the understanding and perception of climate disasters by the Inuit in Alaska (Marino & Schweitzer, 2009), observed climate disaster impacts on indigenous peoples in Nigeria (Ishaya et al., 2008), and investigated climate disaster prediction among Indian tribal groups (Pareek & Trivedi, 2011).

These case studies from different regions underscore the growing recognition of traditional knowledge's role and value in aiding local communities in coping with climate disasters. Traditional knowledge is a system developed through long-term practice and experience by local people. Communities have historically relied on this knowledge to observe environmental changes and address natural disasters. Indigenous peoples and local communities in disaster-prone areas, in particular, have accumulated extensive knowledge on disaster prevention, early warning, preparedness, response, and post-disaster recovery. This knowledge is typically passed down through generations based on accumulated experience and observation (Louise, 1998). As such, traditional knowledge has become a crucial tool for environmental protection and disaster management (John, 2005).

Traditional knowledge can significantly reduce the risk and impact of disasters. By leveraging this knowledge, local strategies can be developed to address climate disasters (Shaw et al., 2009). Additionally, traditional knowledge plays a crucial role in disaster education, risk early warning, crisis response, and the formulation of disaster risk reduction and emergency response policies (Salick et al., 2007). When combined with modern scientific and technological knowledge, traditional knowledge can more effectively mitigate and manage climate disaster risks (Reid et al., 2009).

Based on field investigations of Tibetan villages in the eastern Himalayas, this paper examines community-based research on disasters and related traditional knowledge. It categorizes and analyzes villagers' perceptions of climate disasters and documents traditional knowledge related to disaster risk reduction within the framework of disaster risk management. For effective community-based climate disaster management, it is essential to first understand local villagers' interpretations of disaster risk and the actions they take in response. A disaster risk reduction plan should then be developed based on this traditional knowledge. Finally, integrating traditional knowledge with scientific knowledge, along with aligning practice and policy, is crucial for effectively addressing climate disaster risks.

2 Methodology and Area of Study

2.1 Methodology

2.1.1 Ethnographic Research

The primary ethnographic research methods employed were semi-structured interviews and group discussions. Narrative inquiry was utilized during the semi-structured interviews with villagers. The interview topics were organized into three levels: first, focusing directly on disasters; second, examining related environmental changes; and third, exploring traditional knowledge associated with disasters. During these interviews, villagers shared their perspectives on disasters, particularly their impacts on the local environment and social changes, and narrated historical stories, legends, and myths related to disasters.

Four focus group discussions were also conducted, differentiated by age and gender according to cultural practices. The first group comprised elderly villagers, who mainly discussed major historical disasters and related stories, myths, and legends. The second group, consisting solely of female villagers, focused on the impacts of disasters on agriculture and the collection of non-timber forest products. The third group, made up of male villagers, concentrated on the effects of disasters on animal husbandry and alpine pastures. The fourth group, a mix of male and female villagers of various ages, discussed the impact of disasters on village life and daily activities.

In both the semi-structured interviews and group discussions, not only were the villagers' narratives and conversations about disasters recorded, but also songs, dances, and paintings related to disasters were documented. This comprehensive data collection establishes a foundation for the second stage of community-based research.

2.1.2 Community-Based Research

We also encourage and support villagers in conducting community-based research on traditional knowledge related to disasters. This tradi-

Table 1 Comparison between traditional classification and scientific classification of disasters

Traditional classification of disasters	Scientific classification of disasters
Mountain disaster	Earthquake Landslide Debris flow
Water disaster	Flood Rainstorm Drought
Snow disaster	Snowstorm Avalanche
Air disaster	High temperature Severe cold
Fire disaster	Fire accident

tional knowledge encompasses various aspects, including villagers' predictions of disaster risk, classification of disaster types, beliefs about disaster prevention, and technologies and mechanisms for disaster response.

Local villagers have developed their own understanding and classification of traditional knowledge related to disasters. This knowledge is extensive and deeply intertwined with the local environment, livelihood, and culture, particularly concerning environmental factors such as climate. Consequently, villagers categorize disasters into five types: mountain disasters, water disasters, snow disasters, air disasters, and fire disasters. Some of these traditional classifications align with scientific understandings of disasters, while others diverge from them (Table 1).

Based on previous ethnographic research, we identified villagers with extensive traditional knowledge related to disasters, referred to as local experts. We then organized the villagers into five research groups according to their classification of traditional knowledge related to disasters. Each group was led by two local experts to conduct community-based research on this traditional knowledge.

2.1.3 Framework of Disaster Risk Reduction Management

Following the documentation of traditional knowledge related to disasters, we assisted the villagers in further analyzing this knowledge. The analysis was conducted within the framework of disaster risk reduction management, which categorizes disaster risk into three stages: pre-disaster, during disaster, and post-disaster. We then classified and organized all types of traditional knowledge related to disasters according to this framework (Fig. 1).

Through this framework-based analysis, we have gained a deeper understanding of the value and role of traditional knowledge in disaster risk management. This understanding lays the foundation for developing a community-based disaster risk reduction action plan.

3 Results and Discussion

3.1 Results

Based on traditional knowledge, local villagers have developed a disaster risk reduction management mechanism. According to the analysis framework, this mechanism encompasses three types of traditional knowledge related to disasters: early warning and prevention before a disaster, response and relief during a disaster, and reconstruction and recovery after a disaster.

Additionally, following the villagers' traditional classification of disasters, each category of traditional knowledge can be further divided into five types.

3.1.1 Pre-disaster Warning and Prevention

For each type of disaster, traditional knowledge is categorized into early warning and prevention. Early warning knowledge is derived from accumulated experience and represents the villagers' direct perception of pre-disaster environmental conditions, including observations through vision, hearing, smell, and taste. In contrast, prevention knowledge is based on learned practices and encompasses the decision-making and actions taken by villagers informed by their understanding of the pre-disaster environment.

Take mountain disasters as an example. In Deqin, where most Tibetan villages are situated in canyons, mountain disasters are the most frequent

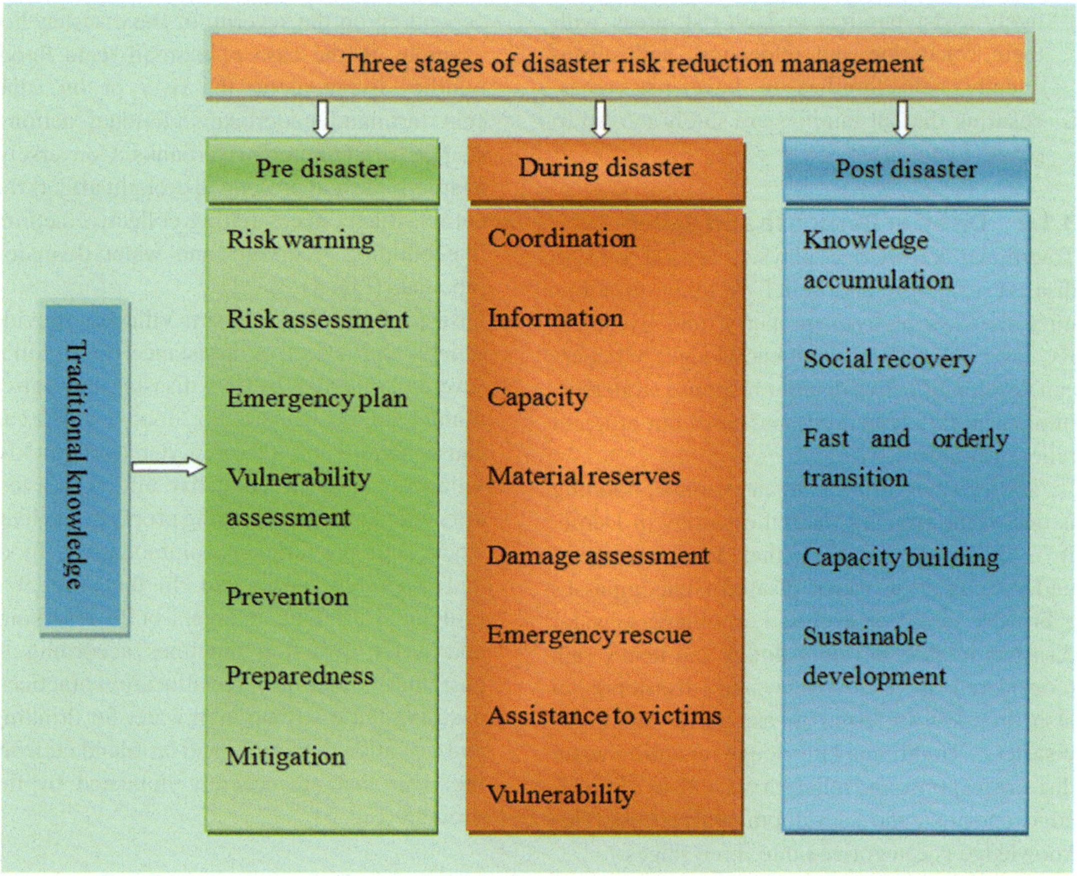

Fig. 1 Framework of Disaster Risk Reduction Management through Traditional Knowledge

and threatening. The villagers' traditional knowledge about mountain disasters is therefore quite extensive. This knowledge enables villagers to mitigate and reduce the risk of mountain disasters to a significant extent. It can be categorized into three main components:

1. Risk Warning Knowledge: This knowledge primarily relies on the villagers' auditory and visual observations. Local villagers believe that before a mountain disaster occurs, the mountain will produce a specific sound. Villagers can discern different types of sounds emanating from within the mountain, allowing them to predict whether a landslide, debris flow, or even an earthquake might occur and assess the potential scale of these disasters. Additionally, villagers observe changes in the behavior of animals and birds on the mountain to identify areas with the highest disaster risk.
2. Prevention Decision-Making and Action: Based on the warning information gathered, villagers make informed decisions and take appropriate actions. For instance, if a small landslide is anticipated, families in high-risk areas will prepare by building retaining walls and relocating their belongings and livestock to safer locations. If a large-scale landslide is predicted, all families in the affected area will prepare to evacuate, ensuring that personnel, property, and livestock are moved out of the disaster risk zone.
3. Mutual Aid Mechanism: When implementing risk prevention decisions and actions, the traditional mutual aid mechanism within local villages comes into play. Villagers will collec-

tively assist families in high-risk areas with their evacuation and relocation, and subsequently help families in lower-risk areas, ensuring that all families are safely moved to secure locations.

3.1.2 Disaster Response and Relief

Traditional knowledge plays a crucial role in disaster response and relief, helping to minimize the loss of lives, property, and livestock. By leveraging traditional knowledge, villagers can develop effective disaster response strategies, implement timely actions, and carry out efficient relief operations.

Taking water disasters as an example, flooding is one of the primary natural disasters in Deqin. It is important to note that local villagers' understanding of water disasters encompasses a broader range of events. For villagers, water disasters include not only floods and heavy rain (considered as "excessive water" disasters) but also droughts (referred to as "water shortage" disasters). Traditional knowledge regarding water disaster response and relief serves both as a knowledge repository and a developmental process. This knowledge is categorized into three stages:

1. Disaster Response Strategy: When faced with water disasters, villagers convene meetings led by village heads and local experts to develop response strategies. This strategy includes information collection, damage assessment, and material preparation. In the event of a flood, villagers evaluate the scope and intensity of the flood at the village level, and assess damage to people, livestock, homes, and farmland. Based on this assessment, they prepare materials for reinforcing riverbanks, such as stones and timber. For droughts, villagers similarly assess the drought's impact on agricultural irrigation and drinking water for both people and livestock, and then construct water storage tanks and diversion pipelines to mitigate these effects (Figs. 2 and 3).
2. Disaster Response Actions: During water disasters, villagers implement actions based on the response strategies. These actions are categorized into individual and collective efforts depending on the severity of the disaster. For example, in the case of a small-scale flood, families living along the river or on adjacent farmland undertake individual actions, such as strengthening riverbanks. Conversely, a large-scale and prolonged drought affects the entire village, necessitating collective actions like building reservoirs and water diversion pipelines (Fig. 4).
3. Disaster Relief Mechanism: Villagers provide timely and effective assistance to families severely affected by the disaster through a mutual aid mechanism. For floods, which can cause significant localized damage quickly, villagers engage in collective efforts to assist affected families by rescuing property and livestock, and providing accommodation, food, and other emergency aid. In the case of a prolonged drought, villagers build reservoirs and water diversion pipelines according to customary water resource allocation practices, ensuring fair distribution of water for drinking and irrigation to families and farmland far from the river that are severely impacted by the drought (Fig. 5).

3.1.3 Post-disaster Reconstruction and Recovery

Traditional knowledge encompasses not only applied technologies and mechanisms but also cultural and spiritual beliefs. Consequently, after a disaster, traditional knowledge contributes to both material recovery and spiritual healing, supporting the sustainable development of villages and social recovery (Fig. 6).

Taking snow disasters as an example, recent years have seen an increase in extreme weather events, with snow disasters becoming particularly impactful in Deqin. This is especially true for villages located at high altitudes, where the losses from snow disasters are significant. Additionally, in local traditional beliefs, sacred mountains, glaciers, and snow are deeply connected spiritually. Avalanches, blizzards, and other snow-related disasters are often viewed by local villagers as divine retribution from the sacred mountains due to human misconduct. Thus, for local villages, traditional knowledge plays a

Fig. 2 Tibetan villagers use traditional knowledge to consolidate river embankments and reduce the risk of flood disasters (photoed by Yin Lun)

dual role in fostering both material and spiritual recovery following snow disasters.

1. Social Recovery and Sustainable Development: Each disaster serves as a learning opportunity, enhancing the villagers' ability and experience in managing future events. For example, in response to seasonal snowstorms, villagers adjust livestock breeding structures and crop varieties for different altitudes based on traditional knowledge. Concurrently, they integrate scientific knowledge and modern technologies, such as saving livestock feed and installing greenhouses, to promote the sustainable development of agriculture and animal husbandry.
2. Spiritual Recovery: Snowstorms, avalanches, and other snow disasters are perceived by local villagers as punishments from the sacred mountains due to human transgressions. In response, villagers invite monks to perform sacrificial ceremonies to appease the sacred mountains. During these ceremonies, monks recite specific scriptures to calm the mountains' anger and pray for their continued protection of the villages. Following the ceremony, villagers believe that the spiritual bond with the sacred mountains has been restored, ensuring ongoing protection.

3.2 Discussion

The frequency and severity of climate disasters worldwide are escalating, yet despite advancements in natural and social sciences, these academic achievements have not been effectively translated into real-world applications. Consequently, the damage caused by disasters continues to rise. Current climate disaster risk reduction

Fig. 3 Tibetan villagers use traditional knowledge to fix landslide areas (photoed by Yin Lun)

plans often lack interdisciplinary approaches that combine natural and social sciences, and fail to integrate scientific and traditional knowledge, limiting their effectiveness at the community level.

In the face of global climate change, addressing disaster risks requires more than just scientific and technological solutions; it also necessitates the incorporation of traditional knowledge, particularly for indigenous peoples, ethnic minorities, local communities, and other traditional societies. Integrating science and traditional knowledge is crucial for developing a comprehensive approach to climate disaster risk management.

Integrated research on climate disaster risk encompasses several key areas: risk interpretation and action, risk prevention and disaster management, and the relationship between climate disasters and human society. The goal is to better understand and address the challenges posed by both natural and human-induced climate risks. Based on the experience in Deqin, integrated research on climate disaster risk involves four main aspects:

3.2.1 From Risk Interpretation to Action

Firstly, the villagers' interpretation of disaster risk is deeply spiritual and closely tied to local traditional beliefs. They perceive climate change and related disasters as manifestations of the emotional state of sacred mountains, influenced by human actions, especially those harming nature. This perspective aligns with modern scientific views that attribute climate change to human activities. Based on their spiritual understanding of disaster risk, villagers adopt traditional practices to mitigate and manage these risks. For instance, they engage in activities like protecting vegetation and planting trees in areas prone to landslides and floods. Additionally, villagers may request monks to conduct a "sealing the mountains" ceremony, which designates certain mountains as sacred, thereby enforcing prohibitions on deforestation and vegetation destruction. This practice helps in environmental protection and reduces disaster risk.

Fig. 4 Tibetan villagers prevent secondary disasters caused by rainstorm through vegetation restoration (photoed by Yin Lun)

Secondly, villagers' interpretation of disaster risk also encompasses environmental factors and traditional knowledge. They attribute disaster risks to environmental changes, some of which are human-induced while others are natural. Over time, villagers have developed a rich repository of traditional knowledge to address these changes. For example, in anticipation of a flood, villagers monitor changes in the river's color and flow speed, and listen to the sound of the water to gauge the flood's scale and intensity. They then implement corresponding measures to mitigate the effects of the impending disaster.

3.2.2 Community-Based Disasters Risk Reduction

Based on traditional community organization and mutual aid mechanisms, villagers are able to effectively manage climate disasters.

Firstly, villagers have established village rules and regulations that include mutual aid during disasters, post-disaster rescue, and resource allocation. They engage in long-term management and prevention efforts for climate disasters, such as afforestation, construction of reservoirs and canals, installation of water pipes, and restoration of hillside vegetation around the village.

Secondly, villagers actively promote and use renewable energy sources. Through microloans, they install greenhouses and begin planting cash crops. The adoption of solar energy and biogas reduces deforestation and greenhouse gas emissions while significantly lowering villagers' labor. The integration of new technologies, innovative farming methods, and diversified income sources enhances the community's resilience against climate disasters.

Traditional cultural response measures are effective in minimizing the negative impacts of climate disasters on indigenous peoples and local communities. These measures establish a robust system of mutual assistance, prevention, and governance for climate disasters. The sustainability and effectiveness of traditional knowledge in managing climate disasters underscore its significance for indigenous and local communi-

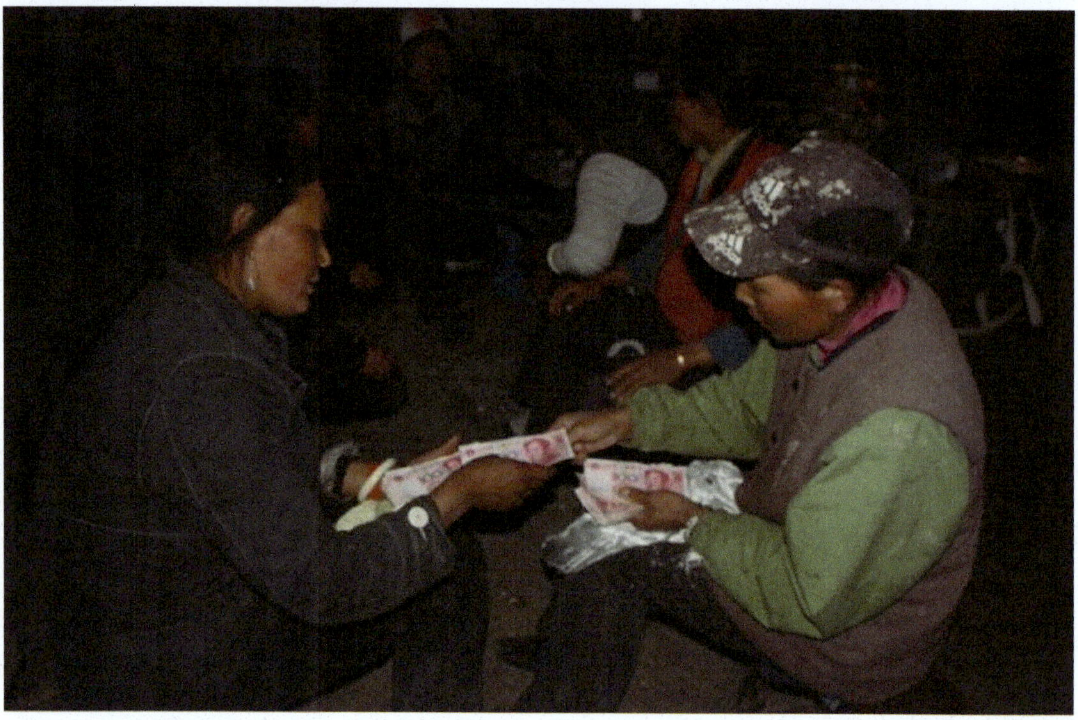

Fig. 5 Tibetan women's organizations provide small loans to support the enhancement of livelihood resilience after disasters (photoed by Yin Lun)

ties, often proving more beneficial than external interventions.

3.2.3 Integrating Traditional and Scientific Knowledge for Disaster Risk Reduction

By integrating traditional and scientific knowledge, villagers have taken effective steps to protect forage resources and mitigate the risk of climate disasters to animal husbandry.

Firstly, the villagers planted grass for protection. Utilizing traditional knowledge about herbage, they selected high-quality local grasses, created artificial grasslands, and protected and planted certain grasses in alpine pastures. They conducted ecological restoration experiments based on this traditional knowledge, which helped safeguard herbage biodiversity.

Secondly, the villagers addressed the issue of forage scarcity in winter and spring by introducing feed storage and processing technologies. They adapted and innovated these technologies to meet the local needs of livestock, improving their ability to store and process forage efficiently.

Thirdly, the local government and villagers collaborated on grassland co-management. Villagers fenced off pastures to prevent overgrazing, and the government supported the construction of pig pens, public health facilities, and other infrastructure. This collaboration established a co-management model that combined traditional forage knowledge with modern practices, enhancing pasture biodiversity and productivity.

These measures have successfully protected forage resources, improved productivity, and promoted the sustainable use of pastures. As a result, they have reduced the risk and vulnerability of animal husbandry to climate disasters, thereby improving the security of traditional livelihoods.

Fig. 6 Tibetan villagers begin post-disaster reconstruction after mudslide (photoed by Yin Lun)

3.2.4 Integrating Practices and Policies for Disaster Risk Reduction

Through the integration of local practices and national policies, villagers have implemented vegetation restoration to reduce the risk of local climate disasters.

Villagers have undertaken local vegetation restoration using traditional knowledge. In the 1970s and 1990s, the timber economy led to extensive deforestation. Villagers observed that this deforestation contributed to drier local climates and reduced precipitation. In recent years, increased climate instability and extreme weather events, such as floods, landslides, and debris flows, have underscored the need for effective vegetation restoration as a means of disaster mitigation and prevention.

The national policy of returning farmland to forest has provided a significant opportunity for restoring degraded mountain areas. Supported by the local government, villagers have combined traditional knowledge with this policy. They interviewed elders to gather information and cultivated seedlings of traditional tree species based on their memories and recommendations. Additionally, they conducted surveys of sacred mountains with well-preserved vegetation, documenting and collecting seeds and seedlings from various altitudes and slopes.

Using this information, villagers planted these seedlings in barren areas designated for reforestation under the national policy. This approach not only aligns with national objectives but also leverages local knowledge and practices. The successful integration of traditional knowledge and national policy has led to the restoration of local vegetation, mitigating the degradation caused by climate change and its associated disasters. This initiative has bolstered the ability of local communities to cope with climate-related challenges.

4 Conclusion

With the increasing severity of climate disasters, the world faces significant risks and vulnerabilities, particularly for indigenous peoples and local communities. Climate disasters impact local cultures, societies, and economies negatively. However, these communities possess valuable traditional knowledge for managing climate-related risks. Research on climate disasters and traditional knowledge underscores the importance of this knowledge and suggests it should be integrated into future disaster risk policies and regulations. Establishing community-based response strategies is crucial for influencing policy formulation and supporting the sustainable development of these communities amidst climate change.

So, there is a need of integrated research on climate disasters and traditional knowledge, we can start from the following aspects:

1. Risk Interpretation and Action: Effective risk reduction hinges on understanding human decision-making in the face of risks. The impact of natural hazards is influenced by environmental conditions and human actions, including vulnerability, decision-making, and cultural factors. Research emphasizes that understanding risk interpretations based on personal experience, perception, values, and beliefs is vital for disaster risk reduction. Thus, exploring the relationship between risk interpretation and decision-making is essential (Richard et al., 2012).
2. Traditional Knowledge in Climate Disaster Management: Traditional knowledge has gained recognition for its value in disaster management. Indigenous peoples and local communities often develop disaster response mechanisms rooted in their traditional knowledge. This knowledge plays a crucial role across all stages of disaster management: pre-disaster warning and preparation, in-disaster response, and post-disaster recovery. Despite its growing importance, further in-depth research is needed to fully understand and leverage traditional knowledge (McAdoo et al., 2009; Shaw et al., 2009).
3. Culture and Climate Disasters: Anthropology provides a framework for understanding how culture shapes responses to climate disasters. Culture influences how people perceive, understand, and respond to environmental phenomena. Anthropological research on climate disasters explores how cultural values and social practices form individual and collective coping strategies. This research includes examining cultural perspectives on climate disasters, the knowledge and concepts derived from cultural models, and the practical responses to disasters. Fieldwork is a key method for studying these dynamics (Roncoli et al., 2009).

By integrating traditional knowledge with scientific insights and considering cultural contexts, we can develop more effective disaster management strategies and policies, enhancing resilience and supporting sustainable development in the face of climate change.

Extreme weather phenomena alone do not constitute climate disasters; it is only when these phenomena impact human societies—both directly and indirectly—that they become climate disasters. Thus, the occurrence and management of climate disasters involve not just natural processes but also social, cultural, behavioral, and experiential factors. While the underlying causes of climate disasters are natural, they are deeply intertwined with social factors and the structural characteristics of human systems.

Climate disasters significantly affect societal structures and traditional lifestyles, making them a crucial focus for anthropological research. These disasters challenge the society, culture, and survival strategies of indigenous peoples and local communities. The impact of such events can severely disrupt traditional livelihoods. In response, communities often rely on traditional knowledge to manage these challenges. However, in the face of sudden and large-scale climate disasters, some traditional knowledge may become inadequate.

In these cases, innovative mechanisms within traditional knowledge frameworks can help address the negative impacts of climate disasters. Effective disaster prevention and response require a combination of policy and practical approaches, integrating both traditional and scientific knowledge. This integrated approach ensures a comprehensive response to the complexities of climate disasters.

References

Bharara, L. P. (1986). Indigenous knowledge-a coping mechanism in drought-prone areas, and changes and consequences of accelerating risks and vulnerability to desertification in Western Rajsthan. Paper presented at Seminar on control of Drought, Deserfication and Famine, New Delhi, Mimeo.

Carpenter, R. (1968). *Discontinuity in Greek civilization.* Norton.

Crate, S. A., & Nuttall, M. (2009). *Anthropology and climate change* (p. 13). Left Coast Press Inc.

de Loe, R. C., & Kreutzwiser, R. D. (2000). Climate variability, climate change and water resource management in the great lakes. *Climatic Change, 45*, 163–179.

Eakin, H. (2000). Small holder maize production and climatic risk: A case study from Mexico. *Climatic Change, 45*, 19–36.

Fisher, R. J. I. F. (1997). *Rain doesn't come: An anthropological study of drought and human ecology in Western Rajasthan.* Manohar Publishers and Distributors.

Harris, M. (1968). *Rise of anthropological theory: A history of theories of culture.* Harper and Row.

Hoffman, S. M., & Oliver-Smith, A. (2002). *Catastrophe and culture: The anthropology of disaster.* School of American Research Press.

Houghton (2001). *Climate change 2001: The scientific basis, contribution of working group I to the third assessment report of the intergovernmental panel on climate change (IPCC).* Cambridge University Press, 944.

Huntington, E. (1915). *Civilization and climate.* Yale University Press.

Huber, T., & Pedersen, P. (1997). Meteorological knowledge and environmental ideas in traditional and modern societies: The case of Tibet. *The Journal of the Royal Anthropological Institute, 3*(3), 577–597.

ICSU. (2008). *A science plan for integrated research on disaster risk: Addressing the challenge of natural and human-induced environmental hazards.* ICSU.

IDS. (2007). *Building climate change resilient cities.* IDS in Focus Issue 02.6. www.ids.ac.uk

IPCC. (2003). *Workshop on the detection and attribution of the effects of climate change.* Working Group II Workshop Report. GISS, New York. USA.

Ishaya, S., & Abaje, I. B. (2008). Indigenous people's perception on climate change and adaptation strategies in Jema's local government area of Kaduna State, Nigeria. *Journal of Geography and Regional Planning, 1*(8), 138–143.

John, B. (2005). The use of Indigenous knowledge in development: Problems and challenges. *Progress in Development Studies, 5*, 99–114.

Lambert, L. D. (1975). The role of climate in the economic development of nation. *Land Economics, 47*, 339.

Louise, G. (1998). *Working with indigenous knowledge: A guide for researcher.* International development Centre.

Lun, Y., Zachary, M., Zheng, Y., & Zhang, X. (2020). The Impacts of climate change on the traditional agriculture of ethnic minority in China. *Journal of Environmental Science and Engineering, 9*, 43–55. https://doi.org/10.17265/2162-5298/2020.02.001

Magistro, J., & Roncoli, C. (2001). Anthropological perspectives and policy implications of climate change research. *Climate Research, 19*, 91–96.

Marino, E., & Schweitzer, P. (2009). Talking and not talking about climate change in North-western Alaska. In S. A. Crate & M. Nuttall (Eds.), *Anthropology and climate change* (pp. 209–217). Left Coast Press.

McAdoo, B., Moore, A., & Baumwoll, J. (2009). Indigenous knowledge and the near field population response during the 2007 Solomon Islands tsunami. *Natural Hazards, 48*(1), 73–82.

Oliver-Smith, A. (1996). Anthropological research on hazards and disasters. *Annual Review of Anthropology, 25*, 303–328.

Oliver-Smith, A. (2002). Theorizing disaster: Nature, power and culture. In S. M. Hoffman & A. Oliver-Smith (Eds.), *Catastrophe and culture: The anthropology of disaster* (pp. 23–48). School of American Research Press.

Pareek, A., & Trivedi, P. C. (2011). Cultural values and indigenous knowledge of climate change and disaster prediction in Rajasthan, India. *Indian Journal of Traditional Knowledge, 10*(1), 183–189.

Reid, H., Cannon, T., Alam, M., Berger, R., Huq, S., & Milligan, A. (2009). *Community-based adaptation to climate change: an overview.* The International Institute for Environment and Development (IIED).

Reilly, J., & Schimmelpfennig, D. (2000). Irreversibility, uncertainty, and learning: Portraits of adaption to long-term climate change. *Climatic Change, 45*, 253–278.

Richard Eiser, J., et al. (2012). Risk interpretation and action: A conceptual framework for responses to natural hazards. *International Journal of Disaster Risk Reduction.* https://doi.org/10.1016/j.ijdrr.2012.05.002

Roncoli, C., Crane, T., & Orlove, B. (2009). Fielding climate change in cultural anthropology. In S. A. Crate & M. Nuttall (Eds.), *Anthropology and climate change* (pp. 70–86). Left Coast Press.

Salick, B., & Byg, A. (2007). *Indigenous peoples and climate change.* A Tyndall Centre Publication.

Shaw, R., Sharma, A., & Takeuchi, Y. (2009). *Indigenous knowledge and disaster risk reduction: From practice to policy*. Nova Publishers.

Torry, W. I. (1979). Anthropological studies in hazardous environments: Past trends and new horizons. *Current Anthropology, 20*, 517–540.

Vasquez-Leon, M. (2002). Assessing vulnerability to climate risk: The case of small-scale fishing in the Gulf of California, Mexico. *Investigaciones Marinas, 30*, 204–205.

Vayda, A. P., & McCay, B. J. (1975). New directions in ecology and ecological anthropology. *Annual Review of Anthropology, 4*, 293–306.

Traditional Beliefs Related to Conservation for the Climate Change Adaptation

10

1 Introduction

Climate change is a global issue with far-reaching effects on both ecosystems and human societies. Its impact on any given area is shaped not only by the climatic changes occurring locally but also by ecological, social, and economic factors (Adger, 1999). Climate change is not simply a physical or environmental phenomenon; it also encompasses social, cultural, economic, and, as demonstrated here, spiritual and moral dimensions.

1.1 Climate Change and Culture

Climate change exemplifies a "social-ecological system" (Holling, 2001), a framework that cannot be fully comprehended through scientific inquiry alone (Berkes, 2002). Within these systems, culture is often viewed as a gradually evolving element that provides stability, fostering the development of adaptive capacities (Armitage et al., 2008; Berkes, 2012). Culture not only positively influences the environment, but environmental conditions also shape cultural formation and transformation (Momaday, 1976; Izzi, 1990; Swain, 1991; Yin, 2001). The relationship between culture and the environment is deeply interconnected—understanding one necessitates an understanding of the other. Culture shapes societal-environmental interactions and mediates responses to environmental change (Adger et al., 2013).

Climate change exemplifies what is commonly referred to as a "social-ecological system" (Holling, 2001), a framework that cannot be fully grasped through scientific inquiry alone (Berkes, 2002). Within such systems, culture is often seen as a gradually evolving element that provides stability, enabling the development of adaptive capacity (Armitage et al., 2008; Berkes, 2012). Culture not only positively influences the environment, but environmental conditions likewise shape cultural formation and transformation (Momaday, 1976; Izzi, 1990; Swain, 1991; Yin, 2001). The relationship between culture and the environment is deeply interconnected—understanding one necessitates an understanding of the other. Culture shapes societal-environmental interactions and mediates responses to environmental change (Adger et al., 2013).

At the same time, cultures are dynamic and reflexive, making them susceptible to the influences of climate change. The anticipated impacts of climate change will affect cultures in diverse ways, both in the urban centers of the developed world and in resource-dependent subsistence economies. For example, as ice and snow cover diminishes in the Arctic, hunters and fishers are forced to shift their target species, resulting in the erosion of traditional knowledge and cultural practices, including traditional housing (Furgal et al., 2006). Similarly, the retreat of

Y. Lun, *Climate Change and Traditional Knowledge*, Sustainable Development Goals Series,
https://doi.org/10.1007/978-3-032-12015-1_10

glaciers at high altitudes, such as the Quelccaya Ice Cap, causes Quechua-speaking villagers in southern Peru to experience a profound sense of disconnection from their natural environment (Orlove, 2009). Cultural change is not confined to marginal societies; post-materialist values in developed nations are also argued to be significantly challenged by environmental changes (O'Brien, 2009). For instance, in England, climate change-induced disruptions to ecosystems threaten iconic, culturally significant habitats and cause shifts in phenology and seasons, leading to a sense of dislocation from familiar landscapes (Brown et al., 2011). Additionally, the loss of traditional weather patterns, such as the "soft rain" in Ireland, underscores the broader cultural impacts of climate change (Irish American Climate Project, 2008).

Culture plays a pivotal role in both mitigating and adapting to climate change, shaping societal perceptions and responses to this global phenomenon. Consequently, culture is essential for understanding not only the causes and implications of climate change but also the human responses it necessitates. The analysis of culture is central to devising and implementing adaptation strategies, as cultural contexts influence the identification of risks, decision-making processes, and methods of implementation.

1.2 Climate Change and Traditional Beliefs

Culture is defined as the symbols that convey meaning, encompassing beliefs, rituals, art, and narratives that shape collective outlooks and behaviors. These cultural elements are pivotal in formulating and implementing strategies to address various challenges (Hays, 1994; Swidler, 1986). Traditional beliefs, a fundamental component of culture, hold particular significance in indigenous and local communities. In these contexts, perspectives on and knowledge of climate change and its impacts are deeply intertwined with traditional beliefs, which frequently serve as the primary source of understanding.

Understanding the effects of climate change on indigenous peoples and local communities is essential, but equally important is recognizing how these groups perceive and interpret climate change through their underlying beliefs (Salick et al., 2012). As a fundamental cultural element, beliefs significantly influence adaptation to climate change. These beliefs encompass the spiritual philosophies and worldviews of individuals and communities, while practices serve as tangible expressions of these beliefs, manifested through rituals and actions at both individual and collective levels. Social and individual values, often shaped by religion or spirituality, play a pivotal role in shaping decisions related to adaptation strategies (Adger et al., 2013). Furthermore, religion has been positively associated with enhancing adaptive capacity by fostering social capital (Schipper, 2010; Gomez-Baggethun et al., 2012).

In the context of climate change, adaptive capacity is generally defined as the ability of a community or social system to endure environmental changes (IPCC, 2001). For indigenous peoples and local communities, traditional beliefs play a pivotal role in understanding and responding to climate change. These belief systems are often closely intertwined with Traditional Ecological Knowledge (TEK), where spirituality is deeply embedded in ancestral traditions and shaped by adaptive processes of trial and error in interacting with the environment (Murphy et al., 2016). TEK serves as a crucial resource for adaptive capacity, not only through the knowledge it provides but also via the processes of knowledge generation and continuous learning (Berkes, 2000). Within this framework, two key elements of traditional beliefs that contribute to adaptive capacity are: first, the protection of important refuges that facilitate recovery after disturbances, such as sacred groves and taboos on the use of certain plants and animals, which support adaptation during extreme events; and second, the practice of rituals that preserve institutional memory and enable collective responses to environmental pressures (Berkes & Folke, 2002).

1.3 Climate Change and Traditional Beliefs of Tibetans

The Tibetan people possess a rich array of traditional beliefs deeply intertwined with nature, particularly with biodiversity and climate change. These beliefs are expressed through the veneration of various natural elements, including deities, land, the sun, trees, mountains, water, and animals (Luo et al., 2009). Such nature worship is pivotal in both preserving biodiversity and adapting to climate change.

In Tibetan traditional beliefs, weather phenomena were intricately linked to social life through moral interpretations of the relationship between humans and the natural environment. Tibetans viewed various weather events—rain, hail, snow, clouds, fog, wind, lightning, and the sky—as manifestations of numerous deities, each endowed with distinct attributes of benevolence or malevolence. These deities were believed to control the weather and could invoke adverse conditions in response to disruptive or inappropriate human behavior. To manage the weather, priests and lamas, who were considered human intermediaries with moral authority within the traditional ritual cosmos, performed rituals to influence meteorological outcomes (Huber & Pedersen, 1997).

Tibetans often referenced spiritual concepts to elucidate climate change, interpreting observed weather phenomena as physical manifestations of their interactions with broader ecological, social, cultural, and spiritual processes. Weather conditions were intricately connected to social life and were correlated with a code of proper conduct. Consequently, a significant aspect of Tibetan understanding of weather is the perceived linkage between physical climate fluctuations and the prevailing "moral climate" influenced by human actions (Salick et al., 2012).

In Tibetan traditions, weather was associated with spiritual powers and thus perceived as susceptible to human influence. A variety of rituals aimed at controlling or influencing weather, including rain-making, hail protection, frost prevention, and snow control, were practiced. Ritual specialists, skilled in weather patterns, observed local conditions and interpreted signs such as cloud shapes and movements, wind or thunder directions, fog dispersal, and sunset colors. By interpreting these indicators, they could determine which spiritual forces might be active at any given time (Huber & Pedersen, 1997).

For the Tibetan people, the phenomenal world is thought to be inhabited by spiritual powers and deities, organized within a unified "ritual cosmos" (Samuel, 1993). Traditional beliefs include numerous rituals aimed at addressing climate change, with prayer serving as a primary method for mitigating its effects. Monks and living Buddhas, through practices such as reading texts, making offerings, and performing rituals, assert their ability to influence weather and climate. They believe that spirits and sacred mountains govern the weather, and they use rituals to communicate with these entities, either to predict weather patterns or to induce changes in the climate.

Tibetan traditional beliefs play a pivotal role in climate change adaptation. On one hand, these beliefs are deeply intertwined with Traditional Ecological Knowledge (TEK), providing specific methods for adapting to climate change and contributing directly to the adaptation process. On the other hand, the concept of climate change within Tibetan traditional beliefs aligns to some extent with modern scientific interpretations, such as the understanding that human behavior influences climate change. Consequently, integrating Tibetan traditional beliefs into contemporary climate adaptation practices and policy-making, especially at the local level, could significantly enhance adaptation efforts.

2 Method

Given the significance of traditional beliefs in the context of global climate change, the study utilized a mixed-methods approach that incorporated semi-structured surveys, in-depth interviews, focus group discussions, and participatory action research.

2.1 Semi-Structured Survey

We conducted semi-structured surveys with local villagers, monks, and living Buddhas. These surveys primarily focused on villagers' understanding of weather and climate through the lens of traditional beliefs, as well as their perceptions of the causes of climate change. Specifically, villagers frequently attribute the negative effects of climate change and climate-related disasters to "improper behavior" by individuals, which they believe incurs punishment from sacred mountains. In this context, the term "people" used by villagers includes both local inhabitants and outsiders.

In recent years, the development of tourism has significantly increased the number of visitors to Deqin. Villagers believe that the influx of vehicles driven by both locals and outsiders disturbs the sacred mountains, leading to a rise in local temperatures. Additionally, villagers perceive that some tourists' disrespectful behavior—such as making noise and engaging in other inappropriate actions at traditional sacred sites—angers the sacred mountains. This disrespect is thought to provoke the deities associated with these sites, manifesting in climate-related disasters like avalanches, snowstorms, and landslides. For example, sacred mountains, lakes, waterfalls, and other traditional sites have been transformed into tourist attractions. Due to a lack of understanding and respect for local cultural practices, tourists' behavior is seen as an offense against these sacred places, resulting in divine retribution through natural calamities.

In recent years, the development of tourism has significantly increased the number of visitors to Deqin. Villagers believe that the influx of vehicles, driven by both locals and outsiders, disrupts the sacred mountains, leading to a rise in local temperatures. Additionally, villagers perceive that certain tourists' disrespectful behaviors—such as making noise and engaging in other inappropriate actions at traditional sacred sites—provoke the sacred mountains. This perceived disrespect is thought to incite the deities associated with these sites, resulting in climate-related disasters such as avalanches, snowstorms, and landslides. For instance, sacred mountains, lakes, waterfalls, and other traditional sites have been transformed into tourist attractions. Due to a lack of understanding and respect for local cultural practices, tourists' behavior is viewed as an affront to these sacred places, leading to divine retribution through natural calamities.

Cultural studies frequently rely on qualitative methods, such as ethnography and participant observation. These qualitative approaches often stand in contrast to the quantitative methods typically employed in the social and natural sciences for studying climate change. This disparity is one reason why cultural factors have not been thoroughly integrated into climate change analyses and policies (Adger et al., 2013). To address this gap, the present study seeks to incorporate quantitative analysis by examining rituals associated with climate change and analyzing trends in climate change and disasters based on variations in the frequency of these rituals. Although this represents a preliminary effort, our goal is to advance research methodologies that bridge the study of climate change with cultural factors.

Building on the semi-structured interviews, we conducted additional in-depth interviews and focus group discussions with villagers, monks, and living Buddhas. The in-depth interviews focused on significant local climate disaster events, particularly exploring the moral and cultural impacts of incidents such as the avalanches in 1991 and 2006. To ensure the accuracy of the information, we specifically targeted elderly villagers, monks, and living Buddhas for these interviews. The focus group discussions with villagers aimed to enhance our understanding of the relationship between traditional beliefs and climate change, particularly in terms of categorizing various types of climate changes attributed to human behavior. Concurrently, the discussions with monks centered on climate change ceremonies, as well as the moral implications and conceptual frameworks related to climate change adaptation.

2.2 Participatory Action Research

We compiled the results from the semi-structured surveys, in-depth interviews, and focus group discussions, and subsequently presented these findings to the villagers, monks, and living Buddhas to validate the accuracy of our observations on traditional beliefs and climate change. Upon receiving their confirmation, we encouraged their active involvement in the analysis and research of these findings. To facilitate this, we established two research teams: the first team, composed of villagers and led by local experts, focused on classifying and studying traditional beliefs; the second team, consisting of monks and led by living Buddhas, concentrated on classifying and analyzing rituals associated with these traditional beliefs.

Building on participatory research, we facilitated action for villagers, monks, and living Buddhas to address climate change. Initially, we organized their involvement in international environmental activities to provide exposure to global perspectives on climate change, particularly those from modern scientific viewpoints. For instance, we assisted them in participating in the Climate Reality Project, initiated by former U.S. Vice President Al Gore, where they viewed Gore's presentation, An Inconvenient Truth, as part of the 24 Hours of Reality campaign.

Simultaneously, we facilitated their participation in the Paris Peace Forum, where they attended a speech by United Nations Secretary-General António Guterres on climate change and biodiversity. Additionally, we organized sessions for villagers, monks, and living Buddhas to attend meetings where officials and experts from government meteorological departments, climate change and biodiversity researchers, and environmental NGO workers delivered specialized lectures and engaged in discussions with them. Following these activities, we coordinated discussions among the villagers, monks, and living Buddhas to explore the relationship between traditional beliefs and contemporary perspectives on climate change. These discussions aimed to integrate traditional beliefs with modern ideas to develop collaborative strategies for addressing climate change.

3 Results

3.1 Conception of Climate in Sacred Mountain Belief

3.1.1 Sacred Mountain Belief System

In Deqin, villagers exhibit profound reverence for the sacred mountains surrounding their communities, reflecting a prevalent form of traditional Tibetan beliefs with distinctive local characteristics. Deqin County is home to over 300 sacred mountains, which together constitute a complex belief system categorized into macro, meso, and micro levels. Macro sacred mountains, such as Kawagebo, are revered across the Tibetan region and exert a broad influence on the populace. Meso sacred mountains hold significance for specific regions or several villages, while micro sacred mountains are typically associated with individual villages or even households. Among the local sacred mountain beliefs, the most revered are the 13 prominent snow peaks, including Kawagebo, Miancimu, and Buqiongsongjiewu, which are believed to bestow blessings upon the entire Deqin area.

From the villagers' perspective, these sacred mountains bestow blessings upon all living beings. In their traditional worldview, humans are not separate from nature but are governed by the sacred mountains. Each level of sacred mountain encompasses different areas of influence, believed to impact various aspects of human life, including health, livelihood, and property. Consequently, villagers seek blessings from the mountains during times of hardship and face consequences if they offend these deities. The sacred mountains are perceived as possessing human-like sentiments: they confer benefits when pleased and bring disasters when angered. As a shared traditional belief among Tibetans, the sacred mountain belief system has distinct local cultural characteristics and serves as a foundational framework for understanding the natural environment, including climate change (Fig. 1).

Fig. 1 Sacred Mountain Kawagebo (photoed by Yin Lun)

3.1.2 Traditional Perspectives on Climate

Traditional Tibetan society was profoundly engaged with the natural environment, with climate fluctuations significantly influencing their life patterns, strategies, and fortunes. The emphasis Tibetans place on weather influences is evident in various social institutions and practices, which are deeply rooted in a specific knowledge system and set of beliefs about nature. From a local perspective, climate is closely associated with reverence for the deities of the natural world (Huber & Pedersen, 1997).

Villagers hold a nuanced understanding of climate that is deeply connected to their traditional sacred mountain beliefs, which diverges notably from modern meteorological interpretations. They employ a variety of sensory inputs—such as sight, smell, hearing, and occasionally prophetic dreams—to observe and forecast climate changes. According to their beliefs, weather phenomena—including rain, hailstorms, snow, wind, clouds, fog, and lightning—are either controlled by sacred mountains or reflect their emotional states. Experienced monks and local experts interpret the spiritual significance of these mountains by analyzing cloud formations, wind directions, thunder sounds, fog patterns, and sunset colors. Additionally, villagers link climate changes to human actions. At the Buddhist Temple of Deqin, religious rituals are conducted to influence weather conditions. Monks perform ceremonies to invoke the sacred mountains' intervention in producing or controlling specific weather events, such as rain, frost, and snow.

These traditional perspectives on climate are rooted in a common spiritual foundation—worship and reverence for sacred mountains. In traditional Tibetan culture, climate is understood as a localized, qualitative system where human interactions with deities shape weather patterns. This contrasts with modern meteorology, which views climate as a global, quantitative system influenced by a variety of interacting factors (Huber & Pedersen, 1997). In Deqin, traditional climate perspectives are deeply intertwined with sacred mountain beliefs, where climate is seen as a manifestation of the mountains' spiritual power. Villagers use their senses—vision, smell, and hearing—to observe local climate conditions and natural phenomena. They conduct specific rituals to communicate with the sacred mountains and seek guidance or intervention regarding weather-related issues.

3.2 Cognition and Classification of Climate Change in Sacred Mountain Belief

Villagers' perspectives on climate are deeply influenced by their sacred mountain beliefs, which also shape their understanding of climate change. They interpret climate change as the consequence of interactions between human behavior and the spiritual power of sacred mountains. This perspective enhances our comprehension of the complex relationships among local society, the natural environment, and cultural beliefs.

Villagers view a reciprocal relationship among their community, climate change, and sacred mountains. Rising temperatures and extreme weather events are perceived as calamities inflicted by sacred mountains in response to human misconduct. Conversely, through rituals such as offerings, sutra chanting, and prayer, individuals seek the mountains' favor to modify the climate to better align with their needs and livelihoods. Consequently, villagers categorize climate change into two types: "punishing" and "beneficial."

3.2.1 The "Punishing" Type

In Deqin, the local belief system delineates various levels of sacred mountains, which influences how actions are perceived in terms of offending or honoring them. For example, climbing the sacred mountain Kawagebo is considered a grave desecration and is strictly prohibited by villagers. In contrast, for other sacred mountains, practices such as constructing incense tables and gathering on mountaintops during the Spring Festival are deemed appropriate forms of offering and sacrifice.

Despite the variations in practices associated with different sacred mountains, certain taboos are universally observed. These include prohibitions against hunting, cutting, digging, and polluting water sources such as lakes, springs, and rivers located on sacred mountains. Additionally, shooting or shouting in these areas is strictly forbidden. Villagers perceive these actions as transgressions against the sacred mountains, which, according to their beliefs, can incite the mountains' anger. This anger is thought to manifest as storms, hail, or abnormal droughts and snowfall—phenomena that align with what is recognized as extreme climate events in contemporary climate change theory.

Case one: avalanche in 1991

In Deqin, a frequently recounted story among villagers is the tragic avalanche that struck Japanese climbers in 1991, resulting in the deaths of 17 individuals. These climbers were attempting to reach the summit of Kawagebo, marking their third attempt. Their previous two attempts, as well as those made by American and European climbers throughout history, had all ended in failure. To this day, Kawagebo remains unconquered.

These events exemplify the intersection of culture and natural phenomena, illustrating the relationship between traditional beliefs and climate change. Such stories, like the avalanche that struck the Japanese climbers, are deeply ingrained in local folklore and have evolved into myths that underscore the power of sacred mountains in controlling climate and natural disasters. As Tibu Cili, an elderly villager, explained: "When the Japanese climbers began their ascent, Kawagebo was en route to Lhasa for the "Eight Sacred Mountains Meeting" in the Tibetan region. Upon returning from the meeting, Kawagebo discovered the climbers encroaching upon his domain and persisting in their attempt. Angered by their intrusion, he unleashed a powerful gust, resulting in a snowstorm and avalanche that drove the climbers away."

Dingzheng Duji, a local monk explains this:

Given that the Japanese mountaineers were also adherents of Buddhism, they visited a local monastery prior to their expedition to seek the Buddha's blessing for a successful ascent. Observing monks chanting scriptures and performing ceremonies, they mistakenly believed these acts were in their favor. In reality, the chanting and ceremonies were intended to inform Kawagebo of the climbers' intention to summit the mountain and to seek Kawagebo's forgiveness for their perceived transgressions. Simultaneously, prayers were offered to Kawagebo, acknowledging the mountain's status as both sacred and an incarnation of the God of War, requesting that it manifest its power by punishing the climbers for their ignorance and disrespect. The subsequent avalanche serves as a testament to Kawagebo's sanctity and resistance to human desecration. While we mourn the loss of the Japanese climbers, this event underscores Kawagebo's divine authority and acts as a broader warning about environ-

mental degradation, perceived as a reflection of the mountain's displeasure with humanity's environmental destruction.

Local villagers, monks, and living Buddhas have linked the avalanche to their traditional beliefs, attributing moral and ethical dimensions to climate phenomena. The avalanche has become a modern legend that exemplifies the belief that the climbers' desecration of Kawagebo provoked the mountain's wrath, resulting in the disaster as a form of divine retribution (Litzinger, 2004; Xu et al., 2005). Many interviewees regard this incident not only as a pivotal moment marking the onset of climate change but also as a clear illustration of the connection between the presence and actions of outsiders and subsequent climate conditions. The climbers' attempt to ascend the sacred mountain is perceived as the most extreme instance of ongoing disrespectful behavior by outsiders, whose presence is often viewed as a form of pollution (Huber & Pedersen, 1997).

Case two: avalanche in 2006

The Yubeng Sacred Waterfall, situated at the base of Kawagebo, holds profound spiritual significance in the traditional beliefs of the Tibetan people in Deqin. Cascading from the mountain's peak for hundreds of meters and blending with ice and snow, the waterfall is celebrated for its breathtaking grandeur. Each year, local Tibetans visit this sacred site to perform acts of worship. However, with the advent of local tourism, the waterfall has become a popular destination, attracting large numbers of visitors annually. In 2006, while a group of tourists was at the site, an unforeseen avalanche occurred, sending a cascade of ice and rocks down the mountain. This disaster resulted in the deaths of two tourists and injuries to seven others.

Regarding the avalanche, villagers provide explanations from both "material" and "spiritual" perspectives. From a "material" standpoint, they contend that the high summer temperatures and rainfall two days before the avalanche destabilized the glacier and snow at the top of the waterfall, increasing the likelihood of an avalanche. Although locals typically avoid visiting the waterfall under such conditions, tourists, who were unfamiliar with the local environment, did not recognize the associated risks. From a "spiritual" perspective, villagers attribute the avalanche to the tourists' improper behavior, which they believe provoked the sacred waterfall's retribution. As Liqing Peichu, a villager, explained:

The Yubeng waterfall holds profound sacred significance for the local community. During worship at the Yubeng sacred waterfall, practitioners maintain a quiet and respectful demeanor. They engage in reverent practices such as turning and reciting scriptures and burning incense. In contrast, the tourists exhibited a disregard for the sacred site and were unfamiliar with the local traditional beliefs. They shouted at the waterfall, disrobed, and took photographs—actions deemed disrespectful and offensive by the locals. According to local belief, such behavior angered the sacred waterfall, leading it to punish them with an avalanche.

Case three: avalanche in 2015

Not only do outsiders face consequences for disrespectful behavior, but local individuals are also subject to punishment from sacred mountains if they engage in what is considered "bad" behavior, such as hunting or cutting down trees. For example, in 2015, two local youths who went hunting near a sacred mountain called Garongyongba were killed in an avalanche. Local villagers believe that this tragic event occurred because their hunting activities angered Garongyongba, resulting in the sacred mountain exacting punishment through the avalanche. As Amaji, the father-in-law of one of the deceased, explained:

Hunting, particularly within sacred mountains, is prohibited both by government regulations and by our traditional beliefs. Despite this, these young individuals chose to disregard the rules and went hunting. The monks and living Buddhas at the monastery had previously cautioned them against hunting, warning that persistent disobedience could provoke retaliation from the sacred mountain. Nonetheless, they proceeded with their expedition in Garongyongba. When the snow melted two months later and their bodies were discovered, it became evident to me that this was a form of punishment and a stern warning from the sacred mountain. This incident serves as a lesson, with the hope that other young people will be deterred from hunting in future.

Case four: storm and hunting

Local villagers believe that offending sacred mountains can lead to not only avalanches but also other forms of climate change and natural disasters, including snowstorms, debris flows, and droughts. Cili Zhaba, a former hunter, recounted his experience:

> I was preparing to hunt near the Zhulaqueni sacred mountain when, suddenly, a dense fog enveloped the area, rendering visibility impossible and causing me to lose my way. During this ordeal, I injured my hand. Upon my return, a fierce storm erupted in my village, leading to a devastating debris flow. According to the elders, these calamities were a consequence of my hunting activities, which had provoked the sacred mountain. The storm persisted until I sought out the Living Buddha and vowed never to hunt again.

This story has become a legend, illustrating the villagers' belief that hunting on sacred mountains infringes upon the authority of Zhulaqueni. In their view, all animals and plants on sacred mountains are regarded as the mountain's private possessions and are thus inviolable. Activities such as personal hunting or deforestation are perceived as offenses that provoke the sacred mountain's wrath, resulting in natural retributions such as storms and debris flows.

In contrast to the localized and immediate weather effects caused by individual actions, broader and more sustained impacts—such as those from tourism development and subsequent pollution—can provoke prolonged and widespread anger from sacred mountains. This, in turn, may contribute to extended climate changes and rising temperatures over a more extensive area and period.

Shrinking glaciers and melting snow vividly illustrate the villagers' beliefs. A notable example is the Mingyong Glacier, located at the base of Kawagebo. Compared to historical photographs taken by American adventurer Joseph Rock a century ago, the glacier's ice tongue has significantly retreated and thinned, with glacier collapses occurring more frequently. Villagers attribute these changes to tourism development, viewing them as manifestations of the sacred mountains' dissatisfaction and warnings to the community. They believe that the influx of tourists—accompanied by numerous vehicles and substantial amounts of litter—has provoked this discontent. Concurrently, the perennial snow on the sacred mountains is melting, revealing dark rocks beneath. Snow is traditionally considered the "attire" of the sacred mountains, leading to the local saying: "It is getting hot, and the sacred mountains are wearing less".

The belief imposes strict limitations on villagers, who maintain that any offense against the sacred mountains will result in a "punishing" form of climate change, serving as a warning against improper behavior.

3.2.2 The "Beneficial" Type

In Deqin's belief system, villagers hold all levels of sacred mountains in equal reverence, although the specific rituals and sutras vary. For instance, the sutras dedicated to Kawagebo are uniquely composed to reflect its particular significance. Rituals and sutra chanting are adapted to address various weather conditions. These ceremonies are conducted by monks in the Buddhist monasteries of Deqin.

One type of ritual aims to ensure a generally favorable climate. Given the region's tridimensional montane climate, which results from varying elevations, weather conditions in Deqin are complex and unstable. Villagers seek a stable and beneficial climate, characterized by timely winds and rains. During prolonged droughts or snowstorms, monks perform sutra chanting and rituals to invoke the favor of the sacred mountains. Villagers also participate by reciting sutras at the white pagoda and incense table, praying for a swift improvement in adverse weather conditions. They direct their prayers to the sacred mountains of the highest or intermediate levels within their belief system, such as Kawagebo and Zhulaqueni.

Another type of ritual is livelihood-based, targeting specific weather conditions essential for traditional agro-pastoralism in Deqin, such as rainfall. Agro-pastoralism, which integrates agriculture and animal husbandry, constitutes the primary traditional livelihood in the region. Livestock provides essential fertilizer for crops, while straw serves as feed for animals. Additionally, harvested fields become valuable pastures during the winter. This livelihood is highly dependent on climate conditions, which are critical for both production and daily life. Villagers pray for specific weather conditions aligned with their agricultural and pastoral needs. During periods of drought or flooding, they typically request the Living Buddha and monks to perform rituals aimed at either invoking rain or mitigating its excess. Villagers also visit the incense table to recite sutras, seeking to correct abnormal weather patterns. These prayers are directed toward sacred mountains at the meso and micro levels of their belief system.

The classification of sacred mountains according to various climate change patterns is neither rigid nor absolute, as it can be determined by monks based on specific sutras and prevailing conditions. Consequently, different levels of sacred mountains may overlap or intersect. Essentially, due to the reciprocity inherent in sacred mountain beliefs, villagers hold that their reverence—expressed through sutra chanting and rituals—can positively influence the climate. This reverence is believed to bestow blessings upon their lives and livelihoods.

3.3 Analysis of the Impact of Climate Change on the Rituals of Sacred Mountain

The reverence and worship directed toward different levels of sacred mountains in local belief are fundamentally consistent among villagers. However, there are distinctions in the scale of ceremonies and the specific scriptures chanted. For example, Kawagebo has dedicated sacrificial scriptures specific to its veneration. Additionally, the choice of scriptures and rituals varies according to the specific weather phenomena being addressed. These rituals and the chanting of scriptures are conducted by monks.

The classification of sacred mountains and the associated prayers for various types of climate change is not rigidly defined. Instead, it is often determined by temple monks based on specific scriptures and contextual conditions, leading to considerable overlap between different levels. Villagers believe that through the chanting of scriptures and the performance of rituals, the sacred mountains will enact appropriate climate changes and safeguard their lives and livelihoods (Figs. 2 and 3).

There are four primary types of religious rituals directly associated with climate factors: requesting rainfall, halting rainstorms, preventing snowfall, and averting debris flows. Analysis of traditional belief rituals related to climate change, performed by villagers and monks over the past decade, reveals the significant impact of climate change on sacred mountain beliefs. This data not only reflects how climate change has influenced traditional practices but also corroborates observed trends in local climate change and extreme weather events from an alternative perspective (Table 1).

From the table above, it is evident that climate change has significantly influenced traditional belief rituals. The rainfall invocation ceremony has experienced the most rapid growth, with a notable surge in 2013, correlating with a severe local drought that lasted five years. Similarly, the ritual for halting rainstorms and snowfall has shown a steady annual increase, reflecting the rising frequency of unpredictable heavy rainfall in the region. Of particular note, the snow cessation ceremony saw significant growth in 2013, largely due to the severe snow disaster that occurred that year. In contrast, the debris flow prevention ceremony has remained relatively stable, showing no significant change over time.

4 Discussion

4.1 From Traditional Beliefs to Real Actions

Although villagers' perspectives on climate change are rooted in traditional sacred moun-

Fig. 2 Prayer flag dedicated to the sacred mountain (photoed by Yin Lun)

Fig. 3 Archery competition and religious sacrifices during the Sacred Mountain Festival (photoed by Yin Lun)

Table 1 Changes in the types and frequency of traditional religious rituals related to climate change. (The information is derived from field surveys conducted among Tibetan villagers and monks in monasteries.)

Ritual type	Disaster type	Number of rituals								
		2009	2010	2011	2012	2013	2014	2015	2016	2017
Asking for rainfall	Drought	53	43	51	57	73	79	81	84	87
Stopping rainstorm	Flood	37	34	45	43	55	57	61	52	59
Stopping snow	Snowstorm	28	23	23	34	56	42	44	47	43
Stopping debris flow	Landslide and debris flow	57	67	63	65	67	57	54	51	52

tain beliefs and differ from the Intergovernmental Panel on Climate Change (IPCC)'s modern scientific definition, notable similarities exist between the two views. Both recognize that climate change is driven by anthropogenic factors, suggesting that it can be mitigated through alterations in human behavior.

In the past decade, increasing interactions with the outside world have prompted the government and environmental organizations in Northwest Yunnan to launch a series of programs integrating poverty alleviation with environmental conservation. A notable example is the collaborative effort involving villagers, monasteries, local government, and environmental groups to address climate change impacts. These initiatives are deeply rooted in traditional cultures and beliefs but are also informed by modern environmental knowledge. Consequently, local villagers have started to incorporate contemporary ecological concepts into their traditional cultural frameworks, thereby reinterpreting modern theories through the lens of their traditions. This integration forms the basis for addressing climate change within the context of sacred mountain beliefs and is essential for understanding the evolving dynamics between local society, the natural environment, and traditional values.

In Deqin, the ecosystems of sacred mountains are notably better preserved and exhibit more abundant vegetation compared to ordinary mountains, largely due to the protective influence of traditional beliefs on the natural environment. A survey of vegetation on sacred mountains reveals that villagers and monks collect seeds and seedlings from 22 species growing at various elevations, including both sun-facing and shaded slopes. These seedlings are planted on barren mountains, accompanied by rituals designed to restore vegetation and close off the mountains for this purpose. White pagodas are constructed to protect the seedlings, and strict bans on tree cutting and harvesting are enforced, effectively transforming these barren mountains into sacred sites. Additionally, villagers and monks conduct comprehensive studies of Tibetan medicinal plants across different elevations and slopes throughout the year. These studies focus on valley areas below 2,500 m in the spring, forested regions between 2,500 and 3,500 m, and alpine meadows and scree areas ranging from 3,500 to 4,500 m. Throughout the year, 146 species of medicinal plants were identified, with 48 species found in valleys, 57 in forests, and 41 in alpine meadows and scree areas. This research has enabled the protective cultivation of 23 species of Tibetan medicinal plants. These collaborative efforts have not only restored vegetation on barren mountains near villages but have also enhanced the local microclimate and reduced the risk of secondary climate-related disasters, such as debris flows and landslides. Simultaneously, the cultivation of Tibetan medicinal plants has diversified local livelihoods and income streams, thereby mitigating the vulnerability of the traditional semi-agricultural and semi-pastoral lifestyle to climate change.

Through engagement with modern environmental protection knowledge, the traditional sacred mountain belief systems in Deqin have been enriched and expanded. What was once a form of pure nature worship has evolved into a comprehensive belief system that integrates traditional worldviews with contemporary ecological theories. This integration has led to practices that enable villagers to more effectively address and adapt to the impacts of local climate change.

4.2 From Tradition Beliefs to Modern Concepts

While some studies suggest that Tibetan traditional beliefs and ecological knowledge differ fundamentally from modern environmental protection concepts, arguing that traditional beliefs lack values equivalent to those of modern environmental conservation, these perspectives may overlook the nuanced interactions between the two systems. Modern knowledge typically conceptualizes the environment as an ecological space, focused on scientific parameters and quantitative assessments, whereas traditional Tibetan knowledge often perceives it as a moral space, emphasizing spiritual and ethical dimensions

(Huber & Pedersen, 1997). Moreover, other research suggests that traditional beliefs among local villagers may play a limited role in climate change adaptation. Despite their reliance on indigenous knowledge and the perceived reliability of native indicators, the scope of adaptive actions available to them is often constrained by various external factors beyond their control (Salick et al., 2012).

Nonetheless, we argue that through the active participation of local villagers, traditional beliefs can be effectively integrated with modern environmental concepts and knowledge, enabling them to play a significant role in climate change mitigation. Traditional belief systems are dynamic and evolving, much like how they incorporated the 1991 avalanche into their narratives, transforming it into a modern legend. These systems are similarly capable of integrating contemporary environmental concepts such as "climate change," "biodiversity," "Sustainable Development Goals (SDGs)" and "carbon neutrality" into their traditional frameworks. Guided by this integrated approach, we, in collaboration with local villagers, have initiated actions and practices designed to address the impacts of climate change.

In 2020, we supported local villagers in establishing the "Mentsunmo Climate Action SDG Laboratory," a project rooted in their traditional beliefs to combat climate change. Mentsunmo, meaning "Goddess Peak", is a revered sacred mountain in Deqin, integral to the traditional belief system. It is considered the consort of Kawagebo and stands over 6,000 m above sea level, remaining snow-covered year-round. As part of the Himalayas, its pristine snow is likened to the celestial attire of a goddess. Traditional beliefs hold that this snow-capped mountain has metamorphosed into a goddess who blesses local women and, more importantly, regulates the local climate. Villagers believe that favorable weather results when the goddess is pleased, while adverse weather and natural disasters arise if she is displeased. Consequently, the climate action laboratory is named Mentsunmo to honor the goddess. By doing so, villagers hope to receive her blessings, achieve favorable outcomes in their climate change mitigation efforts, and demonstrate that local women, empowered by their traditional knowledge and beliefs, are well-equipped to address climate change.

The Mentsunmo Climate Action SDG Laboratory undertook four major initiatives: the establishment of the "Mentsunmo Climate Action Carbon Neutral Forest", the implementation of integrated watershed management, the organization of "Climate Action Day" and a photography competition, and the creation of the "Mentsunmo Climate Action Micro Loan Fund". Following these initiatives, the laboratory assisted villagers in developing detailed implementation plans for each project and began to progressively advance these activities according to the plans. The first initiative to be executed was the construction of the "Mentsunmo Climate Action Carbon Neutral Forest".

Firstly, the Climate Action Laboratory concentrates on safeguarding existing forest ecosystems by integrating traditional ecological knowledge with sacred mountain beliefs. Drawing upon these beliefs and the sacred forest culture, villagers have established local regulations to designate hillside areas for afforestation while prohibiting activities such as tree cutting and forest destruction. Concurrently, the laboratory assists villagers in researching and documenting their traditional beliefs related to sacred mountains and forests. This research informs forest management decisions, including the selection of tree species, altitude, and hillside locations for afforestation efforts. Upon identifying specific sites, villagers initiated conservation activities, beginning with high-altitude areas. Traditional forest knowledge was applied in these efforts, such as the practice of removing side branches to ensure that tree trunks receive adequate water and nutrients, thereby promoting the overall growth of trees and the forest in high-altitude environments.

Secondly, on steep slopes and barren land, villagers select local cedar seedlings for afforestation. Cedars hold significant meaning in traditional beliefs; during rituals honoring the sacred mountain, villagers burn cedar branches and leaves as offerings to seek blessings. Afforesting with cedar trees helps mitigate the impacts of rising temperatures and drought, while also reducing the risk of landslides and debris flows caused by extreme weather events. The Climate Action Laboratory has established six "Mentsunmo Climate Action Carbon Neutral Forests" at elevations ranging from 2,200 to 2,500 m above sea level.

Southwest China, including Deqin, is one of the most recently recognized and significant carbon sink areas in the country. This region, renowned for its exceptional biodiversity, ranks among the most biologically diverse areas globally. The terrestrial biosphere here is acknowledged as the largest single carbon sink in China to date. Furthermore, Southwest China is marked by profound cultural diversity, with local ethnic groups maintaining traditional beliefs and practices related to forests. These cultural elements have significantly contributed to the preservation of native forests and provide the spiritual and cultural foundation for grassroots, spontaneous afforestation efforts. Such community-driven ecological restoration and emission reduction initiatives have substantially supported China's national emission reduction commitments (Lun 2020).

5 Conclusion

Generally speaking, the impacts of climate change on human society can be understood through two primary stages. The first stage involves physical manifestations, including alterations in temperature, precipitation, sea levels, and an increased frequency of extreme weather events. The second stage encompasses a broader spectrum of secondary consequences, such as ecological, social, and economic impacts. Additionally, we identify a third dimension in which climate change affects the spiritual realm, worldviews, religions, and belief systems. This dimension is particularly evident among ethnic minorities who uphold traditional beliefs and worldviews.

The case study of the relationship between sacred mountain beliefs and climate change carries significant anthropological implications, both theoretically and practically, as it underscores the interaction between traditional cultures and ecosystems. Some scholars view traditional cultures—particularly those of non-Western ethnic groups—as inherently aligned with nature, reflecting environmentally friendly worldviews and values. For example, the Han Chinese concept of the "oneness of heaven and humanity," which posits that humans are an integral part of nature, illustrates how traditional societies can support and protect the natural environment rather than oppose it. In the context of environmental and social crises driven by Western industrial civilization, it is crucial to reassess non-Western ethnic traditional cultures, explore their harmonious relationship with nature, and reconsider human development pathways from alternative perspectives (Dwivedi, 1990). However, some scholars argue that the living modes of so-called "non-Western groups," often depicted as being "in harmony with" nature and "protecting" the environment through traditional culture, may align more closely with Western concepts and logic. This perspective can reflect essentialism, where modern values are projected onto traditional practices. Traditional worldviews do not always directly correspond to Western concepts such as "nature,""environment," or "ecology".

Many traditional cultures view human society and the natural environment as interconnected rather than opposed; the idea of human society "conquering" or "utilizing" nature is often absent, and there is no explicit notion of traditional culture as "protecting" the environment. While traditional cultures have indeed contributed to ecological protection and resource management, this is not necessarily their core intent or primary objective (Huber & Pedersen, 1997). Traditional perspectives on climate, cognition, and the classification of climate change are embedded within villagers' sacred mountain beliefs, even though these perspectives do not directly align with modern definitions of weather or climate change. As interactions with the outside world have increased, villagers have developed new understandings of climate change and environmental protection based on modern scientific knowledge. They integrate these contemporary concepts into their traditional sacred mountain belief system, incorporating them into their worldview, and subsequently undertake practices to address climate change grounded in this integrated belief system.

The case study of sacred mountain beliefs and climate change leads to two critical considerations in examining the relationship between traditional cultures and the natural environment. First, traditional cultures should be understood as dynamic and evolving rather than static and closed systems. They integrate scientific knowledge and findings, which facilitates the emergence of new cultural paradigms. Second, Western and non-Western cultures can be seen as complementary rather than mutually exclusive. The relationship between them is not characterized by a binary of "either/or" or a scenario where one aspect diminishes as the other gains prominence. Instead, the future development of human society should embrace the fusion and innovation of multiple cultures to discover a sustainable way of living that is genuinely harmonious with nature.

References

Adger, W. N., Barnett, J., Brown, K., Marshall, N., & O'Brien, K. (2013). Cultural dimensions of climate change impactsand adaptation. *Nature Clinical Practice Endocrinology & Metabolism, 3*(2), 112–117.

Adger, W. N., & Kelly, P. M. (1999). Social vulnerability to climate change and the architecture of entitlements. *Mitigation and Adaptation Strategies for Global Change, 4*, 253–266. https://doi.org/10.1023/A:1009601904210

Armitage, D. R., Plummer, R., Berkes, F., Arthur, R. I., Charles, A. T., Davidson-Hunt, I. J., Diduck, A. P., Doubleday, N. C., Johnson, D. S., Marschke, M., McConney, P., Pinkerton, E. W., & Wollenberg, E. K. (2008). Adaptive co-management for social-ecological complexity. *Frontiers in Ecology and the Environment, 7*(2), 95–102.

Berkes, F., Colding, J., & Folke, C. (2000). Rediscovery of traditional ecological knowledge as adaptive management. *Ecological Applications, 10*(5), 1251–1262.

Berkes, F., & Folke, C. (2002). Back to the future: Ecosystem dynamics and local knowledge. In L. H. Gunderson & C. S. Holling (Eds.), *Panarchy: Understanding transformations in human and natural systems* (pp. 121–146). Island Press.

Berkes, F. (2012). *Sacred ecology: Traditional ecological knowledge and resource management*. Taylor & Francis.

Berkes, F. (2002). Epilogue: Making sense of arctic environmental change? In: Krupnik, I., & Jolly, D. (Eds.), *The Earth is faster now: Indigenous observations of arctic environmental change.* Arctic Research Consortium of the United States, Fairbanks, AK, pp. 334–349.

Brown, T., Budd, L., Bell, M., & Rendell, H. (2011). The local impact of global climate change: Reporting on landscape transformation and threatened identity in the English regional newspaper press. *Public Understanding of Science, 20*, 658–673.

Byg, A., & Salick, J. (2009). Local perspectives on a global phenomenon-climate change in Eastern Tibetan villages. *Global Environmental Change, 19*, 156–166. https://doi.org/10.1016/j.gloenvcha.2009.01.010

Deen (Samarrai), M. Y. I. (1990). Islamic environmental ethics: law and society. In J. R. Engel, & J. G. Engel (Eds.), Ethics of environment and development. London: Bellhaven Press.

Dwivedi, O. P. (1990). Satyagraha for conservation: awakening the spirit of Hinduism, In J. R. Engel, & J. G. Engel (Eds.), *Ethics of environment and development.* London: Bellhaven Press.

Furgal, C., & Seguin, J. (2006). Climate change, health and vulnerability in Canadian northern Aboriginal communities. *Environmental Health Perspectives, 114*, 1964–1970.

Gómez-Baggethun, E., Reyes-García, V., Olsson, P., Montes, C. (2012). Traditional ecological knowledge andcommunity resilience to environmental extremes: a case study in doñana, SW Spain. *Global Environmental Change, 22*(3), 640–650.

Hays, S. (1994). Structure and agency and the sticky problem of culture. *Sociol Theory, 12*, 57–72.

Holling, C. S. (2001). Understanding the complexity of economic, ecological, and social systems. *Ecosystems, 4*, 390–405. https://doi.org/10.1007/s10021-001-0101-5

Huber, T., & Pedersen, P. (1997). Meteorological knowledge and environmental ideas in traditional and modern societies: The case of Tibet. *The Journal of the Royal Anthropological Institute, 3*, 577–597. https://doi.org/10.2307/3034768

IPCC. (2001). Technical summary: climate change 2001: Impacts, adaptation, and vulnerability. A Report of Working Group II of the Intergovernmental Panel on Climate Change. Retrieved December 2008, from Source http://www.grida.no/clinmate/ipcc_tar/wg2/pdf/wg2TARtechsum.pdf.

Irish American Climate Project. (2008). Irish American climate project changing shades of green: The environmental and cultural impacts of climate change in Ireland.

Laidler, G. J. (2006). Inuit and scientific perspectives on the relationship between sea ice and climate change: The ideal complement? *Climatic Change, 78*, 407–444. https://doi.org/10.1007/s10584-006-9064-z

Litzinger, R. (2004). The mobilization of ""nature"": Perspectives from north-west Yunnan. *China Quarterly, 178*, 488–504. https://doi.org/10.1017/S030574100400027X

Luo, Y., Liu, J., & Zhang, D. (2009). Role of traditional beliefs of Baima Tibetans in 838 biodiversity conservation in China. *Forest Ecology and Management, 257*, 1995–2001.

Lun, Y. (2020). China's Afforestation Effort Helps Achieve Carbon Neutral. China Focus. http://www.cnfocus.com/china-s-afforestation-effort-helps-achieve-carbon-neutral

Momaday, S. (1976). A first American views his land. *National Geographic, 150*, 13–18.

Murphy, C., Tembo, M., Phiri, A., Yerokun, O., & Grummell, B. (2016). Adapting to climate change in shifting landscapes of belief. *Climatic Change, 134*, 101–114. https://doi.org/10.1007/s10584-015-1498-8

O'Brien, K. (2009). *Adapting to climate change: Thresholds, Values, governance* (pp. 164–180). Cambridge Univ. Press.

Orlove, B. (2009). Glacier retreat: Reviewing the limits of human adaptation to climate change. *Environment, 51*, 24–34.

Salick, J., Byg, A., & Bauer, K. (2012). Contemporary Tibetan cosmology of climate change. *Journal for the Study of Religion, Nature, and Culture, 6*(4), 447–476.

Samuel, G. (1993). *Civilized shamans: Buddhism in Tibetan societies*. Smithsonian Institution Press.

Schipper, E. L. F. (2010). Religion as an integral part of determining and reducing climate change and disaster risk: An agenda for research. In M. Voss (Ed.), *Der klimawandel: Sozialwissenschaftlicheperspektiven* (pp. 377–393). VS Verlag.

Swain, T. (1991). The Mother Earth conspiracy: An Australian episode. *Numen, 38*, 3–26. https://doi.org/10.1163/156852791X00024

Swidler, A. (1986). Culture in action: Symbols and strategies. *American Sociological Review, 51*, 273–286.

Wilbanks, T. J., & Kates, R. K. (1999). Global change in local places: How scale matters. *Climatic Change, 43*, 601–628. https://doi.org/10.1023/A:1005418924748

Xu, J., Ma, E. T., Tashi, D., Fu, Y., Lu, Z., & Melick, D. (2005). Integrating sacred knowledgefor conservation: Cultures and landscapes in southwest China. *Ecology and Society, 10*(2), 7.

Yin, S. T. (2001). *Man, and forest*. Yunnan University Press.

11 Study on Traditional Knowledge Issues Under the United Nations Framework Convention on Climate Change and China's Strategy

The report to the 20th National Congress of the Communist Party of China proposed that "China should actively participate in the global governance of climate change" (Xi, 2022). President Xi Jinping pointed out that it is necessary to implement a national strategy for actively responding to climate change, and promote and guide the establishment of a fair, reasonable, and win–win global climate governance system (CPC Publicity Department et al., 2022a). The *Paris Agreement*, adopted at the 21st Session of the Conference of the Parties (COP21) to the *United Nations Framework Convention on Climate Change* (UNFCCC), establishes the status of traditional knowledge in climate change adaptation, puts forward the topic of traditional knowledge, and thereby lays a foundation for China's traditional knowledge to participate in global governance for addressing climate change. President Xi Jinping pointed out that our Chinese civilization has been passed down for more than five thousand years and has accumulated rich ecological wisdom. These simple yet profound views of nature still provide people with profound warnings and inspirations (Party Literature Research Center et al., 2017). The ecological wisdom and ecological skills developed through intergenerational adaptation and accumulation are fully embodied in the traditional knowledge of various regions. Traditional knowledge, as the carrier of ecological wisdom, is undoubtedly capable of playing a unique role in ecological conservation (Yang, 2004). Particularly in practices related to ecological balance, biodiversity conservation, and climate adaptation, traditional knowledge possesses significant practical value (Luo and Mo, 2023). Therefore, it is necessary to promote the active participation of China's traditional knowledge in global governance for addressing climate change, vigorously facilitate the effective integration of traditional knowledge with climate change governance, enable traditional ecological knowledge to regain its vitality in contemporary climate change governance, and promote China's transition from a participant to a leader in the topic of traditional knowledge. While an international convention is undoubtedly a product of agreements reached by all countries, the specific implementation lies not only with each of the Parties but also, more importantly, with different regions and ethnic groups within the Parties. Traditional knowledge is a product of specific natural environments and unique historical processes, and there is no inherent inevitability regarding whether it can be compatible with the relevant provisions of the Convention. Therefore, the effective implementation of these provisions depends not only on understanding the full connotation of the Convention's provisions but also, more importantly, on grasping the specific essence of traditional knowledge among different regions and ethnic groups, identifying its scientific validity and rationality, and achieving effective alignment between traditional knowledge and global climate change governance. Only through these efforts can we truly recognize the value and role of tradi-

Y. Lun, *Climate Change and Traditional Knowledge*, Sustainable Development Goals Series,
https://doi.org/10.1007/978-3-032-12015-1_11

tional knowledge, enable the practices and actions of relevant ethnic groups to achieve compatible mutual benefit and coordinated coexistence with the specific provisions of the Convention, and prevent the participation of traditional knowledge in global climate change governance from remaining a mere formality.

However, given the complexity and breadth of the issue, it is obviously impossible to explore each case individually. Therefore, based on existing case studies of China, India, and other countries as well as their relevant ethnic groups, this paper discusses and analyzes the realization of effective alignment between traditional knowledge and global climate change governance. It aims to prove that to achieve global climate change governance, it is essential to fully recognize and utilize the traditional knowledge and ecological wisdom of different ethnic groups—only then can the ultimate goal of global climate change governance be truly achieved, and the tasks and objectives specified in specific provisions be fulfilled at the minimum cost. Although the scope covered in this paper is just a drop in the bucket, it intends to inspire broader insights and generalizations, so as to realize the common but differentiated responsibilities and efforts in global climate change governance. In this way, despite the grandeur of the governance goal, it becomes achievable and foreseeable.

1 Conceptual Analysis of Indigenous Peoples, Local Communities, and Ethnic Minorities

In recent decades, climate change has increasingly captured the attention of indigenous peoples, local communities, and ethnic minorities. This phenomenon is driven by the rising concentration of greenhouse gases in the atmosphere, which disrupts the climate system's energy balance and leads to various adverse natural phenomena. These include increased desertification, more severe droughts, floods, tropical cyclones, more frequent wildfires, rising sea levels, and melting glaciers (Gebremedhin and Theron, 2007). Such changes pose significant threats to indigenous peoples, local communities and ethnic minorities, who are disproportionately affected despite their minimal contribution to the current climate crisis.

Efforts have been made to assist indigenous peoples, local communities and ethnic minorities in adapting to climate change. Both governmental and non-governmental organizations (NGOs) have introduced various ideas, goals, and technologies aimed at supporting these communities in their adaptation efforts. However, many of these initiatives have proven to be unsustainable and unsuccessful. A primary reason for these failures is that these external interventions often neglect the traditional knowledge of indigenous peoples, local communities and ethnic minorities. They fail to recognize and leverage the value and potential role of traditional knowledge in the adaptation process to climate change.

Over millennia, indigenous peoples, local communities, and ethnic minorities have developed a rich body of traditional knowledge about their environments, essential for ensuring shelter, food, water, medicines, and other vital resources (Berkes et al., 2000; Lantz and Turner 2003; Jacob et al., 2010). This traditional knowledge has been employed by these communities to address climate change issues (Berkes and Jolly, 2002b; Turner & Clifton, 2009). Given their direct experience with the environmental impacts of climate change, their traditional knowledge is invaluable for understanding these impacts and identifying effective adaptation strategies (Reidlinger and Berkes, 2001). Traditional knowledge is particularly well-suited for climate change adaptation because it recognizes that environmental conditions are constantly evolving, necessitating responsive societal adjustments (Berkes et al., 2000).

Indigenous peoples, local communities, and ethnic minorities rely on traditional knowledge to sustain the ways of life that are vital to their communities. This knowledge is often rooted in their understanding of the environment, ecosystems, and biodiversity, and it forms the connections between humans and nature. The significance of traditional knowledge in addressing climate change stems from its dual-level structure in the context of climate change.

Firstly, through long-term practice, indigenous peoples, local communities and ethnic minorities have developed traditional knowledge that is closely tied to local weather and climate conditions. This knowledge, accumulated over generations, enables them to predict forthcoming weather patterns and climate changes, as well as to provide early warnings for extreme weather events and disaster risks. Consequently, this traditional knowledge can play a significant role in mitigating the risks and impacts associated with climate change.

Secondly, the livelihoods of indigenous peoples, local communities and ethnic minorities are predominantly dependent on their surrounding ecosystems and natural resources. Traditional knowledge, which is deeply rooted in local ecosystems and biodiversity, plays a crucial role in their sustainable use of these resources. This knowledge not only supports their immediate survival but also aids in achieving long-term sustainability. In the context of climate change, such traditional knowledge becomes even more vital, as it can significantly enhance the ability of indigenous peoples, local communities, and ethnic minorities to adapt to changing conditions and bolster their climate resilience.

Therefore, to understand traditional knowledge, especially those related to climate change, it is necessary to first clarify the concept of traditional knowledge owners, that is, what are indigenous peoples, local communities, and ethnic minorities? How do different countries define these populations, and how are they defined in international conventions?

1.1 Indigenous Peoples and Local Communities

The term "indigenous peoples and local communities" is mentioned in Article 8(j) of the Convention on Biological Diversity (CBD), as well as in numerous important provisions of the Nagoya Protocol regarding the access to genetic resources and the sharing of benefits, especially related to traditional knowledge (Xue, 2011). However, neither of these instruments provides a definition for this term, and there has yet to be an international consensus on its scope (Barten, 2015). As a result, parties to the agreements can only interpret and deduce its meaning based on the text of the conventions and the specific national contexts. The current general understanding includes both the "narrow" concept of indigenous peoples and local communities with colonial characteristics and the "broad" concept referring to those with indigenous traits. For China, it is of significant importance to clarify whether "indigenous peoples and local communities" corresponds to or differs entirely from the concept of Chinese ethnic minorities, especially in the context of fully implementing the Convention on Biological Diversity and the Nagoya Protocol (Li and Xue, 2021).

Due to differences in political systems and ethnic autonomy, the understanding and definition of "indigenous peoples and local communities" vary across countries, which poses a barrier to the full and accurate implementation of the Convention on Biological Diversity and the Nagoya Protocol (Eide, 2014). Currently, in South America, Africa, and most Asian countries, "indigenous and local communities" are understood as the native populations of the region, which essentially refers to the broad concept of "indigenous and local communities". This is in contrast to the "narrow" concept of indigenous peoples and local communities in countries like the United States, Canada, Australia, and New Zealand, which reflects colonial characteristics (Hannum, 1994; Van Cott, 2005). The definition of "indigenous" varies according to how different countries treat indigenous peoples. Given the complexity of human history and social organization, there is no single definition for indigenous peoples. Indigenous communities possess unique and vibrant cultures that are reflected through strong kinship and family networks, deeply rooted in traditional beliefs and customs. On the international stage, indigenous organizations have focused on addressing this issue together and have identified a more global and consistent approach (Dawson et al., 2021). A point of consensus within the international community is that it is up to indigenous peoples themselves to define who is indigenous, as self-determination is a fundamental characteristic of indigenous peoples (Corntassel, 2003; Sarivaara et al., 2013). In China, there is no issue of colonization, and it remains unclear whether the concept of ethnic minority communities is equivalent to or approximates the broad definition of "indigenous people and local communities".

1.1.1 Definition and Origin of Indigenous Peoples and Local Communities

The Chinese text of the Convention on Biological Diversity uses "Tuzhurenmin and Difangshequ" to correspond to "indigenous peoples and local communities". According to the 1983 edition of Ciyuan (Chinese Dictionary), "Tuzhu" originally referred to groups "permanently settled in one place", meaning that "indigenous" refers to people who live in one place for generations, as opposed to those who migrate. Thus, the term "Tuzhu" in Chinese primarily refers to people who have lived in a specific location for generations, in contrast to "migratory" groups (Li and Xue, 2021).

Both the Oxford Advanced Learner's Dictionary (2017 edition) and the Longman Dictionary (2016 edition) define "indigenous" as "native to a particular area", describing characteristics such as language and culture, without any discriminatory connotations. In English, the terms "aboriginal" and "native" are also used to translate "Tuzhu", "Aboriginal" in English is defined as "the 'original inhabitants' in relation to colonizers", usually referring to indigenous groups such as Native Americans, Māori in Oceania, and Australian Aboriginals. "Native" is defined in English as "those who arrived in a country before the white settlers", generally referring to the indigenous peoples of the Americas, with a colonial implication. Additionally, "tribal" also carries the meaning of "indigenous" or "native", often used to describe indigenous African tribes. The use of "indigenous people" in the Convention text avoids any political or discriminatory connotation, reflecting the precision and rigor of international conventions (Ramos-Castillo et al., 2017).

Indigenous peoples are native ethnic groups of a specific region, in stark contrast to groups that have recently settled, occupied, or colonized the area. Indigenous communities possess unique and vibrant cultures, reflected through strong kinship ties and family networks that embody deeply rooted traditional beliefs and customs (Berkes et al., 1994). Indigenous people make up 6.2% of the global population, with over 476 million indigenous residents across 90 countries. Among them, there are more than 5,000 indigenous groups (UN, 2025). However, these numbers mask significant differences in legal recognition, social, and political realities. The definition of "indigenous" varies according to how different countries treat indigenous peoples. Given the complexity of human history and social organization, there is no single definition of indigenous peoples (Borrini et al., 2004).

"Local community" is translated as "Difangshequ", referring to a collective formed by social groups living in a specific geographic area with shared life connections. According to international legal conventions, a "community" cannot act as a legal entity with corresponding rights; only "indigenous peoples" can hold legal rights as such (Hill et al., 2020). Whether "local communities" have certain privileges depends on national legislation.

1.1.2 Discussion of the Term Indigenous Peoples and Local Communities in the Convention

Article 8(j) of the Convention and related provisions have been the subject of multiple discussions by the working group established by the member countries. These discussions have led to some phase-based conclusions, mainly including the expert group report on the scope of "indigenous peoples and local communities" under Article 8(j), and the analysis report on the impact of using the term "indigenous peoples and local communities" on the Convention on Biological Diversity and the Nagoya Protocol, presented at the second Conference of the Parties (SCBD, 2017). These reports reached a preliminary consensus: Indigenous peoples are groups living in diverse cultural and political contexts who rely on oral traditions and intergenerational transmission of traditional knowledge, beliefs, and practices, are dependent on nature, and play a key role in the management of biological resources and the conservation of biodiversity. And "local communities" are small geographical areas closely related to the survival of "indigenous peoples" and reflect their unique political, economic, and cultural factors. Discussions on the scope of application of the term "indigenous peoples and local communities" continue. The Conference of the Parties also called on other international organizations, such as the International Labor Organization and the World Intellectual Property Organization, to promote research and discussion on issues related to "indigenous peoples and local communities" (Pruim, 2014). Therefore, under the Convention on Biological Diversity and the Nagoya Protocol, "indigenous peoples and local communities" is a composite legal term that can be understood as a collection of special groups of people living within a specific geographical area that can be identified through technical indicators.

1.1.3 Understanding of Indigenous Peoples and Local Communities by Other International Organizations

International organizations and forums have been discussing the issue of "indigenous peoples and local communities" for decades, but no definitive solution has been reached. The discussion around indigenous peoples remains controversial, and there is still no unified standard. Countries that are parties to international agreements can make reasonable inferences and interpretations regarding the definition of indigenous peoples. The international community increasingly uses the term "indigenous peoples" to describe beneficiaries, although some international agreements use other terms such as "aboriginals" or "tribal peoples". For single-ethnic indigenous groups, terms like "Native Americans" in the U.S. and "First Nations" in Canada are used. Currently, a widely accepted viewpoint in the international community is that it is up to indigenous peoples themselves to define who is considered indigenous, based on self-identification (Canessa, 2007; Bertrand and Xu, 2023).

The World Council of Indigenous Peoples (WCIP) was established in 1975 and holds consultative status with the United Nations. Its working definition provides another perspective on the contemporary political/legal status of indigenous peoples: Indigenous peoples are those who live in countries composed of different ethnic or racial groups, where these groups are descendants of the earliest surviving populations of the region and, as a group, do not control the government of the country in which they reside (Crossen, 2014). The WCIP's working definition closely aligns with that of the United Nations, reflecting its close relationship with the UN as a pioneer of the global indigenous rights movement in the 1970s. Like the UN, the WCIP emphasizes self-identification as its official policy, although its working definition rarely discusses indigenous peoples' culture, land claims, or identity.

In 1982, the World Bank issued a policy statement on Tribal Peoples in Bank-Funded Projects, which identified the following characteristics to recognize indigenous peoples within specific regions: 1. A close connection to ancestral lands and the natural resources in these areas; 2. Self-identification and recognition by others as members of a distinct cultural group; 3. Indigenous languages, usually different from the national language; 4. The existence of traditional social and governance systems; 5. A production method primarily focused on subsistence. The policy emphasized the uniqueness, cultural identity, dependency on land resources, and vulnerability of indigenous groups, adding the linguistic dimension, excluding colonial backgrounds, and placing self-cultural identification as a core element of recognition (MacKay, 2001; Kenrick and Lewis, 2004). This approach reflects respect for their subjectivity and self-determination based on their place of origin.

Before the declaration of the "International Decade of Indigenous Peoples" (1995–2004), the United Nations appointed Martinez Cobo to submit a report to the UN Committee on the Elimination of Racial Discrimination (Canessa, 2007). In his report, he defined indigenous peoples as follows: Indigenous communities, peoples, and nations are those groups who have a historical continuity with pre-invasion and pre-colonial societies that developed on their territories, and who consider themselves distinct from other social groups currently or historically occupying these territories. This definition became an important reference for other international organizations, countries, and anthropologists (Saugestad, 2001; Kenrick and Lewis, 2004) in defining indigeneity. In further elaborating on this definition in his work Study of the Indigenous Peoples (1987), Martinez Cobo emphasized that indigenous peoples should have the right to self-identify and be accepted as members by their communities. He argued that indigenous groups are those who maintain historical continuity with pre-invasion societies in their territories, who view themselves as distinct from the dominant social groups in the current society, and who are committed to maintaining their ancestral territories and cultural characteristics according to their own cultural systems, laws, and governance. Indigenous identity is based on self-identification and community acceptance, with the right to determine membership.

In 1989, the International Labour Organization (ILO) adopted the Indigenous and Tribal Peoples Convention (No. 169), which emphasizes the right of indigenous and tribal peoples to self-identification and equal rights. This convention provides a unique, inclusive approach to addressing diversity issues, with the recognition of self-distinction as a basic standard for indigenous or tribal peoples (ILO, 1989). Unlike the World Bank's definition, ILO Convention No. 169 highlights the right to self-identification as the "fundamental standard" for indigenous identity. The convention also offers an initial definition of indigenous peoples: 1. Tribal peoples in independent countries whose social, cultural, and economic conditions distinguish them from other parts of the national society, and whose status is partly or entirely governed by their own customs, traditions, or specific laws or regulations; 2. Peoples in independent countries who are considered indigenous because they are descendants of the populations that lived in those countries or geographical regions before colonization, conquest, or the establishment of current national borders, and who, regardless of their legal status, have maintained some or all of their own social, economic, cultural, and political systems (Swepston, 1990).

Understanding the Key Features of "Indigeneity": Martinez Cobo proposed that historical continuity can be manifested through the following characteristics: 1. Occupation of ancestral lands or part of them; 2. Shared ancestry with the original inhabitants; 3. Cultural expressions (such as religion, tribal life, clothing, production methods, etc.); 4. Use of language; 5. Residence in a specific country or region; 6. Other related factors.

Although the United Nations has not provided a unified definition of "indigenous", its understanding of the characteristics of indigenous peoples includes: 1. Self-identification as indigenous, and acceptance as a member by the community; 2. Historical continuity with pre-colonial or pre-settler societies; 3. Close connection to territories and natural resources; 4. Distinct social, economic, or political systems; 5. Use of a different language; 6. As a distinct people and community, committed to maintaining and developing the environment and systems passed down by their ancestors (Assembly UNG, 2007).

Regarding "Local communities": The International Union for Conservation of Nature (IUCN) defines "community" as "a social collective composed of members living within a specific geographical area, sharing natural resources, production knowledge, and technical practices, among other life elements. The members of the community generally have a stable population size, social organization, and cultural characteristics (including language, behavioral norms, and values). Communities often operate as small political entities with specific capabilities (Andrade and Rhodes, 2012). Local or traditional communities are not necessarily indigenous but are usually linked to specific lands, lifestyles, and cultural traditions. The Convention on Biological Diversity uses the term "Indigenous Peoples and Local Communities" (ILCs), acknowledging the longstanding connection of these groups with their traditional lands and waters.

Like the definitions provided by other academic and intergovernmental organizations discussed in this paper (including those from WCIP, the World Bank, UN Cobo, ILO, etc.), any definition of indigenous peoples carries the risk of being incomplete in historical, cultural, political, and economic dimensions. Such definitions can also risk concretizing indigenous peoples, continuing to place "difference under the lens of identity".

1.1.4 Definitions of Indigenous Peoples and Local Communities by Scholars

In her acclaimed work The Indigenous Voice in World Politics, Franke Wilmer was among the first social scientists to systematically examine the global historical process of moral exclusion perpetrated by Western powers against indigenous peoples. She defines indigenous peoples broadly as: 1. Peoples with tradition-based cultures; 2. Those who were politically autonomous before colonization; 3. Those who, in the aftermath of colonization and/or decolonization, continue to struggle for the preservation of their cultural integrity, economic self-reliance, and political independence by resisting the assimilationist policies of nation-states. Listing the core elements of indigenous identity (such as the three above) provides maximum flexibility in identifying approximately 5,000 indigenous groups worldwide (Wilmer, 1993). However, the definition is so general that it is difficult to determine whether indigenous peoples differ in terms of their cultural worldviews and goals from other minority groups across the world. In a more recent article co-authored with Alfred, Wilmer revised her original three-part definition to correct some of the previous ambiguities: 1. They are descended from the original inhabitants of the geographic areas they continue to occupy, hence, they are aboriginal; 2. They wish to live in conformity with their continuously evolving cultural traditions; 3. They do not now control their political destiny, and consequently, are frequently subjected to policies arising from the cultural hegemony originally imposed by an 'outside' force (Alfred and Wilmer, 1997).

This three-part definition emphasizes the importance of geographic homelands and evolving cultural traditions for indigenous peoples. However, Part 3 narrowly focuses on their political destiny and implies that a lack of "control" is part of being indigenous. Additionally, there is no mention of group or collective rights emphasized by indigenous peoples, nor specific cultural traditions such as language and ceremonial cycles, which distinguish them from other populations. On the other hand, this definition is broad enough to encompass the approximately 350 million indigenous peoples worldwide. Indigenous scholars, such as Alfred and S. James Anaya, tend to advocate for broad and inclusive definitions of indigenous groups in order to avoid downplaying the variation between and within groups. For example, Anaya refers to indigenous peoples as:

The living descendants of pre-invasion inhabitants of lands now dominated by others. They are indigenous because their ancestral roots are imbedded in the lands in which they live, or would like to live, much more deeply than the roots of more powerful sectors of society living on the same lands or in close proximity. Furthermore, they are indigenous to the extent they comprise distinct communities with a continuity of existence and identity that links them to communities, tribes, or nations of their ancestral past (Anaya, 2004).

Anaya's definition of indigenous peoples highlights the continued colonial domination of indigenous homelands and the ancestral roots of these "pre-invasion inhabitants". He also acknowledges indigenous peoples as distinct communities with extensive kinship networks, clearly distinguishing them from minority groups. However, unlike the Alfred and Wilmer definition, Anaya's version includes little discussion of culture or distinct worldviews. Overall, both Alfred/Wilmer and Anaya's definitions, while ultimately incomplete, are inclusive and demonstrate strategic flexibility in defining the world's indigenous peoples.

1.1.5 Identifications of Indigenous Communities in Different Countries

(1) Narrow definition of indigenous communities

Indigenous peoples are also regarded as the descendants of those who lived in a country or geographic region before different cultural or ethnic groups arrived. These newcomers later established dominance through conquest, settlement, or other means. For example, in countries with colonial histories, such as the Americas, indigenous groups like Native Americans, Inuit, First Nations in Canada, Aboriginal and Torres Strait Islander peoples in Australia, and Māori in New Zealand (Formaini, 2009; Reid et al., 2019). These indigenous communities, like most other indigenous communities, retain distinctive characteristics that are clearly different from the rest of the national population. Due to the colonial process, which involved the usurpation of native lands and the disruption of political and social lives, these communities form special groups in these countries, making them part of the narrower definition of indigenous communities.

(2) Broader definition of indigenous communities

In Africa, Asia, and Europe, indigenous communities are sometimes defined in the context of conquest or internal marginalization, and are often societies of different ethnic groups united within one country, or in other cases, formed by loosely connected tribes or isolated communities. Similar definitions of "indigenous" and their scope have been discussed in African countries (Twikirize and Spitzer, 2019), Latin American countries (Canessa, 2007), and South American nations (Rappaport, 2004).

It is estimated that two-thirds of the world's indigenous peoples live in Asia, meaning approximately 260 million people representing 2,000 different civilizations and languages. The use of the term "indigenous" and the official recognition of their status varies across Asian countries. In regions like Asia, where there is no colonial history or long-lasting colonial experience, the term "indigenous" needs to be interpreted more broadly to describe the characteristics of groups who do not fit the colonial narrative (Castellino, 2014).

1.2 Definition and Origin of Ethnic Minorities

In 1995, the Chinese government officially stated that the term "indigenous" has a colonial connotation, and its use in China is subject to specific limitations (Sturgeon et al., 2017). All 56 ethnic groups in China are considered to be long-dwelling groups, and the Chinese Communist Party provides a more flexible definition of ethnic minorities: "An ethnic group is a stable community of people formed at a certain historical stage of development. Ethnic groups generally share a common historical origin, mode of production, language, culture, customs, and a sense of unity. For some ethnic minorities, religion has also played an important role in their formation and development".

There is no international consensus or agreement on the definition of "ethnic minorities", and the distinction between minorities and indigenous peoples may lie in equal social rights and the political will for self-determination. Although "minorities" and "indigenous peoples" are two distinct concepts, there are commonalities and connections between them. When discussing ethnic minorities in China, government representatives use the term "long-dwelling groups". Although indigenous peoples and ethnic minorities are not equivalent in legal terms, the protective measures for both can be mutually informative. In China's National Report on the Implementation of the Convention on Biological Diversity, when addressing the achievements of indigenous peoples and local communities, the report uses ethnic minorities and local communities to address indigenous and local community issues. Therefore, the Chinese government's understanding of the relationship between "indigenous" and "ethnic minorities" is also a gradual and dynamic process.

China's ethnic development differs from most countries with indigenous groups. While there are certain similarities with indigenous approaches, ethnic unity has always been the core of policy, and ethnic minorities in China enjoy autonomous rights in autonomous regions. There are four main international approaches to managing ethnic issues: minority rights law, ethnic self-determination law, human rights law, and indigenous law. It is argued that the identification and definition of ethnic minorities in China should be based on cultural rights measures, ensuring that all ethnic minorities are equally protected by law (Wang, 2014).

In the early 1950s, during China's ethnic identification process, factors such as language and culture were considered, but the willingness of ethnic minority groups was also fully respected, with self-identification being an important principle. From the perspective of traditional knowledge, "ethnic minorities and minority regions, while utilizing and managing biodiversity, have maintained their traditional culture, production methods, and lifestyles, with socio-economic activities largely governed by traditional customary law, although lacking defined characteristics and processes, they possess a sense of self-identification (Miranda et al., 2002; Huang et al., 2024).

The distribution of ethnic minority populations in China is closely related to natural resource management. Much of the land inhabited by ethnic minorities contains a significant portion of China's natural resources. This geographical feature not only reflects the political and economic importance of ethnic minorities in China but also forms the basis for the formulation of ethnic minority policies. Traditional knowledge is the wisdom accumulated by local ethnic minorities over long periods of production and life, playing a crucial role in the management of local biological resources, the protection of biodiversity, and the rational use of land resources. Therefore, the traditional culture of ethnic minorities holds particular value in the conservation of biodiversity and the sustainable use of biological resources (Long and Zhou, 2001; Shao et al., 2021; Yin, 2022).

1.3 Understandings of Ethnic Minorities, Indigenous Peoples, and Local Communities in Different Countries

1.3.1 General Understanding of "Indigenous Peoples" and "Ethnic Minorities" in China

China is a unified multi-ethnic country and does not have the colonial connotation of the narrow definition of "indigenous" or the indigenous issues seen in other regions like the Americas and Australia. Due to challenges in language translation and other factors, the Chinese government and scholars are sensitive to the term "indigenous", generally believing that China does not have the same colonial history or indigenous peoples issues as regions such as the Americas and Australia. Domestic scholars often approach the study of "indigenous" and "ethnic minorities" from anthropological and ethnological perspectives. They commonly argue that China's ethnic minorities still retain relatively primitive and traditional production and lifestyle practices that are closely linked to their living environment. These characteristics bear many similarities to indigenous peoples in the international community, and there should be mutual learning in terms of rights protection (Heath and Yu, 2005).

1.3.2 Views of Other Countries on Indigenous Peoples and Local Communities

Southeast Asia is a diverse and heterogeneous region, marked by substantial ethnic, cultural, and religious variety. Broadly, the region hosts three main types of minorities: ethnic and linguistic minorities, religious minorities, and indigenous peoples. In this context, ethnic, linguistic, and religious minorities are grouped under the term "ethnic minorities", which is distinct from indigenous peoples. Generally, the difference between ethnic minorities and indigenous peoples is clear (Clarke, 2001). Indigenous peoples are those who are native to a region, or are descendants of the earliest known inhabitants of a territory.

In contrast, ethnic minorities are usually settler populations with historical ties to the land (sometimes extending back for centuries) and share a common identity with other groups in at least one neighboring country. For example, in the Philippines, the Moro Muslims in Mindanao are considered ethnic minorities due to their strong cultural and religious connections with Malaysia and Indonesia, while groups such as the Igorots of Northern Luzon and the Lumad in Mindanao are regarded as indigenous peoples. Similarly, in Malaysia, distinctions are made between the Orang Asli, the original inhabitants of the Malay Peninsula, and the Dayaks from the eastern states of Sabah and Sarawak, who share ethnic ties with groups in Kalimantan, Indonesia (Harrisson, 2017).

However, in Southeast Asia, the distinction between these categories can be complex and blurred, reflecting long-standing and unresolved political tensions. For instance, in Vietnam, the government does not use the term "indigenous peoples", instead classifying all minority groups as ethnic minorities, distinct from the majority Kinh ethnic group. In several parts of Vietnam, however, what are considered ethnic minorities are more indigenous than the Kinh. For example, the Khmers have inhabited the lower Mekong Delta for thousands of years, in contrast to the Kinh, who settled in the region primarily from the nineteenth century (Sinh et al., 2009). This raises a broader issue. In mainland Southeast Asia, many ethnic minorities are culturally or linguistically linked with groups in neighboring countries (both minority and majority groups), leading them to be perceived as "foreigners" by governments and dominant populations, especially when these groups appeal to external support to maintain or protect their identity. Examples of such groups include the Hmong in southern China, Vietnam, Laos, and Thailand, the Chams in Vietnam and Cambodia, and the Nagas in Bangladesh and Myanmar. These communities are seen as ethnic minorities in their respective countries, distinct from the majority ethnic groups and smaller groups officially recognized as indigenous, despite their long-standing presence in specific parts of the national territories (Bentley, 1986; Chua et al., 2019).

In Bolivia, although the Constitution mentions "indigenous peoples" and "local communities", the term "indigenous" is not specifically defined. However, the Law of People's Participation (1994) provides an explanation in its first article, stating that indigenous communities are: "Groups of people who are descendants of populations that settled before conquest and colonization, residing within the current national borders; they share a history, organization, language or dialect, and other cultural traits, through which they identify as members, recognize themselves as part of the same socio-cultural unit, and maintain territorial ties through the management of their settlement area and their social, economic, political, and cultural institutions "(Canessa, 2007).

2 The Formulation of the Topic of Traditional Knowledge in Global Climate Change Governance

As an international issue, traditional knowledge is also referred to as traditional ecological knowledge, local knowledge, indigenous ecological knowledge, indigenous knowledge, and so on, and there is still no unified definition for it so far. In Western academic circles, traditional knowledge is defined as a cumulative body of knowledge, practices, and beliefs concerning the relationships between living organisms (including humans) with one another and with the environment, which evolves through adaptive processes and is passed down from generation to generation through cultural inheritance (Berkes, 2009). Traditional knowledge is not only a way of knowing but also a way of acquiring information. It is dynamic, grounded in experience, and adapts to ongoing changes (Berkes, 2017). In China, traditional knowledge is regarded as a knowledge system developed for the natural and social contexts of specific regions by specific ethnic groups, constructed through intergenerational accumulation. It also serves as a knowledge repository adapted to the natural and ecological systems inhabited by specific ethnic groups or local communities (Yang and Sun, 2016). Traditional knowledge in China refers to the knowledge, technological innovations, and traditional practices that have been developed in the long-term production and daily life practices of various ethnic groups, suited to the local culture and environment, and passed down from generation to generation through oral transmission or written records (Xue, 2019).

In 1992, the United Nations Conference on Environment and Development (UNCED), held in Rio de Janeiro, Brazil, issued the *Rio Declaration on Environment and Development* (the Rio Declaration), which emphasized the importance of traditional knowledge and called on all countries to attach importance to it. The conference adopted three pivotal conventions, namely the United Nations Framework Convention on Climate Change (UNFCCC), the Convention on Biological Diversity (CBD), and the *United Nations Convention to Combat Desertification* (UNCCD), with the issue of traditional knowledge running through all three conventions (Cheng, et al., 2012). This also marked the first time that the international community, with the United Nations as its representative, linked traditional knowledge to climate change. Moreover, the issue of traditional knowledge began to be incorporated into the UNFCCC, which recognized that traditional knowledge can play a role in addressing current global climate change and other environmental issues, and possesses scientific value in assisting human society in tackling climate change. Given the value and role of traditional knowledge in addressing climate change, the *Intergovernmental Panel on Climate Change* (IPCC) recognizes that traditional knowledge serves as an extremely critical and valuable foundation for developing response strategies and conducting natural resource management to adapt to environmental and climate changes (IPCC, 2007a). It explicitly states that traditional knowledge from around the world should be drawn on and respected, covering three key aspects: traditional agricultural practices can resist climatic and non-climatic stresses; local knowledge systems and practices enable local people to adapt to various climate changes; and initiatives based on such traditional knowledge have a foundation of legitimacy within local communities (IPCC, 2022a, 2022b).

In the global governance of climate change, the emergence of the issue of traditional knowledge can be attributed to two levels of reasons. First, the value of traditional knowledge in global environmental governance and sustainable development has gained attention and recognition. Particularly in traditional ethnic communities around the world, traditional knowledge has always been an important means for these communities to address ongoing environmental changes and achieve sustainable livelihoods (Berkes & Jolly, 2002a). Second, in the process of humanity's comprehensive response to climate change, the role of traditional knowledge has been gradually discovered and utilized. An increasing number of scientists have found that traditional knowledge can serve as a supplement to global climate change scientific research: it constitutes information and data on climate change at the local or regional scale, and on this foundation, traditional knowledge can provide support for the formulation of relevant climate change policies and adaptation strategies (Yin, 2024). Particularly in ancient civilizations like India and China—ones with an extremely profound historical heritage—traditional knowledge has already played an important role in addressing climate change.

For instance, in the East Godavari district of Andhra Pradesh, India, the local Konda Reddis utilize traditional agricultural techniques such as shifting cultivation, intercropping, and relay cropping of different crop varieties to minimize the risk of crop failure caused by varying weather conditions and climate change. When extreme weather events like hurricanes and droughts lead to poor harvests and food shortages, they promptly adjust their livelihood strategies: while practicing shifting cultivation, they also engage in forest foraging—for example, extracting starch from the pith of local palm species belonging to the *Caryota* and *Arenga* genera to produce food. This enables them to effectively address the food shortage resulting from extreme weather events and prevent the local people from going hungry. When they identify that relying on traditional single palm species poses higher risks under the current impacts of climate change, they proactively experiment with and adopt new palm varieties, such as date palms and sugar palms, to expand the range of alternative food sources. At the same time, they are no longer content with collecting wild palm trees; instead, they have begun large-scale artificial cultivation of palm trees, and in terms of varieties, they have chosen heat-tolerant and drought-resistant palm species for planting. Because these palm trees have objectively increased forest cover, enabling greater absorption of carbon dioxide, these actions of the Konda Reddis—based on traditional knowledge—align with the Government of India's National Action Plan on Climate Change. India plans to increase its forest cover from 24 to 33% by 2030 to absorb 250 million to 3 billion metric tons of carbon dioxide, and its confidence stems from the fact that the local climatic and ecological environment is highly suitable for the growth of palm and banana family plants. By incorporating such crops into the framework of agricultural production, economic activities can proceed normally while also ensuring the steady expansion of forest cover. This outcome, in turn, achieves the seamless integration of traditional knowledge with climate change governance. Therefore, in the aforementioned process of addressing climate change, the traditional knowledge of the Konda Reddis has gained social recognition and been more effectively passed down and expanded. These traditional knowledge systems have helped the Konda Reddis, enabling them to exert their own initiative in addressing climate change without having to rely on the market or external assistance—and have also become a new highlight in India's National Action Plan on Climate Change (Kodirekkala, 2018). In rural areas of Rajasthan, India, the traditional knowledge of local communities serves as an important tool for environmental protection and natural disaster management. This traditional knowledge includes insights gained through local communities' long-term observations of cloud formation, lightning, wind direction, and rainfall occurrence. Based on this experience, local people can predict quite accurately the rainfall patterns throughout the season and their impacts on agriculture. Villagers in Rajasthan face numerous climate disasters, with droughts and floods being the most prominent. These disasters lead to famine, food shortages, and poverty. Therefore, villagers draw on traditional wisdom accumulated over generations and adopt a range of measures to combat climate disasters. Examples include growing drought-tolerant and early-maturing indigenous crop varieties, avoiding flood- or landslide-prone areas when building houses, and refraining from herding livestock into valleys during spring floods. To control erosion and flooding during monsoon seasons, villagers convert hillsides into flat terrace fields and construct drainage outlets to regulate the slow flow of water from one terrace field to the next, thereby reducing soil erosion caused by running water. They build check dams and pond networks to slow down rainwater runoff and store water for the dry season. Additionally, they plant trees to stabilize slopes and prevent gully erosion. Additionally, villagers engage in diversifying livestock farming, improving building techniques, enhancing water quality monitoring, harvesting rainwater for supplementary irrigation, developing traditional agricultural techniques for watershed conservation, adjusting the timing and scope of hunting and foraging, diversifying crops and livelihoods, and implementing community-based climate forecasting and disaster risk reduction measures. These approaches to addressing extreme weather and climate change—grounded in traditional knowledge—are effective and relatively low-cost, enabling local villagers to proactively tackle the risks and challenges posed by climate change with little or no external support (Pareek & Trivedi, 2011). The aforementioned traditional knowledge not only serves the people of all ethnic groups in local areas comprehensively but also constitutes an integral component of climate change governance in the entire nation of India and even on a global scale. If countries and ethnic groups in other regions can also do the same, not only will traditional knowledge gain global significance, but the costs that relevant countries need to incur to respond to the various requirements of the conventions will also be significantly reduced—and fulfilling the conventions will no longer become a burden or liability for any country. This kind of contemporary value needs to be further widely promoted, in order to eliminate the ideological burdens and public opinion pressure at the national level of state parties.

Similarly, China is also rich in traditional knowledge. Various ethnic groups in the country have accumulated a wealth of traditional knowledge and ecological wisdom related to weather and climate change through long-term production practices. For instance, the Jinuo people, who reside in Jinuo Mountain in Xishuangbanna, Yunnan Province, have developed unique adaptation strategies after living with mountainous climates over an extended period. Based on altitude and climatic conditions, they carefully divide their land into three categories, namely "*Zhexiao*", "*Zhejiao*", and "*Dieta*", and adopt distinct farming methods tailored to each type of land. For instance, on fertile first-category land, they adopt a long-term crop rotation system, using oxen for plowing. They regularly rotate the crop species planted on an annual basis, which ensures the balanced utilization of soil fertility. This allows crops to maintain stable and high yields without additional artificial fertilization, and more importantly, guarantees year-round full coverage of green vegetation on the farmland. Thereby, without additional input, the ineffective evaporation and waste of soil water resources are significantly reduced, ensuring the long-term stability of temperature and relative humidity in the local climate. While the traditional knowledge in these regions may seem insignificant, if mountain-dwelling ethnic groups across the globe could apply such traditional knowledge, the normal water cycle in the atmosphere would be able to maintain a long-term and stable continuation. In this way, it also makes a positive contribution to global climate change governance.

On the other hand, on the infertile third-category land, they practice non-rotational slash-and-burn agriculture: they cultivate one plot each year and follow a 13-year fallow cycle, allowing sufficient time for vegetation regeneration. The scientific rationale behind this practice lies in the fact that although trees are cut down during the slash-and-burn process, the tips of the tree stumps are shaped into a conical form. Moreover, the tips are even coated with a dilute starch solution, which forms a hard film after dehydration—preventing the stumps from being infected by fungi and bacteria. As a result, new shoots sprout from the stumps within a month after the trees are cut down, and the stumps do not die. Furthermore, due to the difference in growth rate between these new shoots and the planted crops, the shoots do not interfere with the crops' growth. After crop harvesting, the farmland surface is also not exposed to direct sunlight. Its scientific effect is similarly to reduce the ineffective evaporation of water resources in the soil as much as possible. The traditional knowledge and ecological wisdom embedded in slash-and-burn agriculture have made a positive contribution to the implementation of relevant provisions in the conventions on global climate change governance. Failing to recognize this fact and instead blindly criticizing slash-and-burn agriculture as backward and primitive is obviously unfair and unjust.

The Jinuo people also carefully select and arrange crop varieties based on climatic seasons and land types. They grow a variety of crops, including upland rice, corn, and cotton, and by means of intercropping and relay cropping, they improve land use efficiency and ensure a stable food supply. Meanwhile, they make every effort to ensure that the vegetation formed by the crops maintains full coverage of the land surface at all times—whose functional principle ultimately achieves the same effect as the technical operation of slash-and-burn agriculture.

In addition, the Jinuo people rely on their rich phenological knowledge to guide agricultural production—for example, by observing natural phenomena such as bird calls and plant flowering to determine key farming timings, including land preparation and sowing. Local Jinuo experts who possess traditional knowledge play a crucial role in adapting to climate change: relying on their extensive production experience and religious authority, they organize farming rituals and guide villagers in arranging production activities rationally. This farming culture, which integrates natural observation and religious beliefs, not only reflects the Jinuo people's profound understanding of the environment but also demonstrates their wisdom and resilience in the face of climate change. Through these measures, the Jinuo people have successfully survived and thrived in the mountainous climate, achieving harmonious coexistence between humans and nature (Huang and Yin, 2024). Such effective cases are applicable not only in China but also can be promoted for use by all mountain-dwelling ethnic groups in the tropical zones (zones near the Tropics). Therefore, the relevant provisions of the UNFCCC can achieve effective implementation without substantial investment.

In summary, the value of traditional knowledge has gradually gained attention and recognition in the global governance of addressing climate change. Particularly in traditional societies around the world, traditional knowledge has always been an important means for them to cope with the ongoing climate change and achieve sustainable livelihoods. Moreover, in the process of humanity's comprehensive response to global climate change, the role of traditional knowledge may be further gradually discovered and utilized in more aspects. In this regard, an increasing number of scientists have found that traditional knowledge can serve as an effective means for biodiversity conservation and addressing climate change. As a Party to the UNFCCC, China plays an increasingly important role in the global governance of addressing climate change and exerts a pivotal influence on the formulation and adoption of relevant issues and decisions. Traditional knowledge, as a key issue under the UNFCCC, means that related decisions and policies will also impact China's active participation in the global governance of climate change. Given this, it is necessary to carefully study and clarify the issue of traditional knowledge to determine China's specific approaches and convention implementation actions on this issue—ensuring that all ethnic groups in China can each play their respective roles and derive corresponding benefits from it. In this way, the investment required for China to fulfill its convention obligations can be significantly reduced, while the achievements attained will instead gain recognition and praise from global public opinion. This is precisely the essence of China's innovative approach to convention implementation.

3 The Stance of UN-Related Environmental Conventions Toward Traditional Knowledge

Among UN-related environmental conventions represented by the UNFCCC, the CBD, and the UNCCD, the inclusion of the traditional knowledge issue stems from two key reasons. First, the value of traditional knowledge has gained attention and recognition (Berkes & Jolly, 2002b), particularly in traditional societies around the world, traditional knowledge has always been an important means for these societies to cope with ongoing environmental changes and achieve sustainable livelihoods (Turner & Clifton, 2009). Second, in the process of humanity's comprehensive response to global climate change, the role of traditional knowledge has been gradually discovered and utilized. An increasing number of scientists have found that traditional knowledge can serve as an effective means for biodiversity conservation and addressing climate change (Riedlinger & Berkes, 2001).

3.1 The Positioning of Traditional Knowledge in UN-Related Environmental Conventions

First, the Rio Declaration on Environment and Development was the first to explicitly affirm the importance of traditional knowledge in UN-related environmental conventions. Released at the 1992 UNCED in Rio de Janeiro, Brazil, Principle 22 of this Declaration emphasizes the significance of traditional knowledge, stating: "Indigenous peoples and their communities, as well as other local communities, play a vital role in environmental management and development due to their knowledge and traditional practices. States shall recognize and appropriately support their identities, cultures and interests, and enable them to effectively participate in achieving sustainable development." This Conference also adopted three important conventions, namely the UNFCCC, the CBD, and the UNCCD, and Principle 22 of the Declaration has been consistently embedded in these three aforementioned conventions (Cheng et al., 2012). This marked the first time that the international community represented by the United Nations turned its attention to traditional knowledge and began to introduce the issue of traditional knowledge into UN-related environmental conventions (Yin, 2021). Currently, UN-related environmental conventions and strategies—such as the CBD, the UNFCCC, the UNCCD, the Convention on Wetlands of International Importance Especially as Waterfowl Habitats (Ramsar Convention), the United Nations International Strategy for Disaster Reduction (UNISDR), and the Sendai Framework for Disaster Risk Reduction 2015–2030—all attach a certain degree of importance to traditional knowledge and contain relevant textual content.

The CBD has provided a clear definition and positioning of traditional knowledge, and incorporated it into its operational provisions. The CBD defines traditional knowledge as knowledge, innovations, and practices that have been developed on the basis of long-term experience and are adapted to local cultures and environments, and at the same time positions traditional knowledge as practical science. The CBD has established an intersessional open-ended ad hoc working group on traditional knowledge. This working group is tasked with implementing the provisions on traditional knowledge set out in Article 8(j), which stipulates that each Contracting Party shall in accordance with national legislation, respect, preserve and maintain knowledge, innovations and practices of indigenous and local communities embodying traditional lifestyles relevant for the conservation and sustainable use of biological diversity and promote their wider application, with the approval and involvement of the holders of such knowledge, innovations and practices and encourage the equitable sharing of the benefits arising from the utilization of such knowledge, innovations and practices. Beyond this, the CBD also incorporates the issue of traditional knowledge in Article 10 (c), Article 17 (2), and Article 18 (4).

Subsequently, the IPCC — building on the CBD — extended traditional knowledge from the domain of biodiversity to that of climate change, while recognizing and acknowledging the importance of traditional knowledge. In its assessment report *Climate Change: Impacts, Adaptation and Vulnerability*, the IPCC states that traditional knowledge constitutes an extremely critical and valuable foundation for developing response strategies and managing natural resources to adapt to environmental and climate change (IPCC, 2007b). Working Group II of the Sixth Assessment Report (AR6) of the IPCC explicitly states that traditional knowledge from around the world should be drawn on and respected, identifying three key areas: first, traditional agricultural practices can build resilience against climatic and non-climatic stresses; second, local knowledge systems and practices enable local communities to adapt to a range of climate changes; and third, initiatives based on traditional knowledge have a basis of legitimacy within local communities (IPCC, 2022a, 2022b).

Ultimately, through a series of programs and agreements, the UNFCCC has further established traditional knowledge under the theme of "climate change adaptation". Traditional knowledge and its engagement are recognized as a key component of achieving successful adaptation to climate change (Erbaugh et al., 2020). To date, over 60 decisions adopted by the Conference of the Parties (COP) to the UNFCCC, as well as reports issued by its subsidiary bodies, have explicitly referenced traditional knowledge. Of course, the Convention's process of understanding traditional knowledge was not accomplished overnight; instead, it went through three phases: recognition, attention, and gradual integration.

First, starting with the Marrakech Accords, traditional knowledge began to be recognized. In 2001, the 7th Session of the Conference of the Parties (COP7) to the UNFCCC adopted the Marrakech Accords. The text of the Accords referred to traditional knowledge related to adapting to the impacts of climate change and acknowledged the need for traditional knowledge to be involved in climate change adaptation measures. Specifically, Item 6 of the Marrakech Accords—entitled "Structure of National Adaptation Programme of Action (NAPA) Documents"—stipulates in Section B ("Framework for Adaptation Programmes") that historical information and traditional knowledge hold certain value in addressing climate change (UNFCCC, 2002). At COP11 in 2005, "climate change adaptation" was designated as a high priority for all countries, and the importance of local and traditional knowledge for adaptation was highlighted (UNFCCC, 2005).

Second, traditional knowledge began to receive focused attention with the launch of the Nairobi Work Programme. At the 12th Session of the Conference of the Parties (COP12) to the UNFCCC in 2006, traditional knowledge was explicitly highlighted. The Conference adopted the Nairobi Work Programme on Impacts, Vulnerability and Adaptation to Climate Change (NWP), which systematically addressed traditional knowledge in its analysis of climate change impacts and adaptation strategies. Specifically, the NWP integrated traditional knowledge into adaptation planning and practices, emphasizing its relevance to climate change resilience. In developing adaptation plans and practical measures, the Programme recommended collecting, analyzing, and disseminating information—including historical and current adaptation actions and practices—with a particular focus on local and traditional knowledge (UNFCCC, 2006). At the 16th Session of the Conference of the Parties (COP16) to the UNFCCC held in 2007, discussions on "Reducing Emissions from Deforestation and Forest Degradation (REDD)" emphasized traditional knowledge related to forest management and biodiversity conservation in the context of climate change mitigation (Mollicone et al., 2007). At the 15th Session of the Conference of the Parties (COP15) to the UNFCCC in 2009, which is crucial for formulating climate policies, the importance of protecting ecosystems and integrating traditional knowledge into climate solutions was emphasized (UNFCCC, 2010).

Finally, the Cancún Agreements gradually incorporated traditional knowledge into climate change adaptation actions. At the 16th Session of the Conference of the Parties (COP16) to the UNFCCC held in Cancún, Mexico, in 2010, as well as in its related decisions, traditional knowledge and practices were mentioned on multiple occasions. Item 2 ("Enhancing Adaptation Actions") of the Cancún Agreements—adopted at the Conference—states that enhancing climate change adaptation actions shall be based on and guided by the best available science, and take into account traditional knowledge as appropriate. The Cancún Agreements recognized the value that traditional knowledge holds in the adaptation and mitigation of climate change, and proposed integrating traditional knowledge into national climate strategies. Article 20 of the Agreements decided to establish an Adaptation Committee to promote and strengthen the implementation of adaptation actions—including enhancing and consolidating the exchange of relevant information, knowledge, experience, and good practices at the local, national, regional, and international levels—and, in doing so, take into account traditional knowledge and practices as appropriate (Liu, 2011). The topic of traditional knowledge in the Cancún Agreements marks a shift in the knowledge systems and information sources deemed to guide the UNFCCC in taking adaptation actions. Specifically, approaches to climate change adaptation no longer rely solely on scientific knowledge, but also attach importance to traditional knowledge and its relevant practices (Ford et al., 2016). In discussions related to mitigation and adaptation, the decisions and mandates unanimously adopted by all Parties provide guidance for traditional knowledge to participate in the design and implementation of climate policies at the local, national, regional, and international levels.

3.2 Traditional Knowledge in China's Policies and Regulations

As a contracting party to relevant United Nations conventions, China has always attached great importance to the value and role of traditional knowledge. In September 2010, the 126th Standing Committee Meeting of the CPC Central Committee and the State Council approved and released China's National Strategy and Action Plan for Biodiversity Conservation (2011–2030). Under its "Strategic Tasks" section, the document explicitly states that China will "draw on advanced international experience, carry out pilot demonstrations, strengthen research on the value assessment and management system of biological genetic resources, conduct emergency protection and inheritance of relevant traditional knowledge, and improve the traditional knowledge protection system." In "Action 8: Conducting Surveys and Cataloging of Biological Genetic Resources and Relevant Traditional Knowledge", it proposes to "conduct surveys on traditional knowledge, innovations, and practices related to biological genetic resources in ethnic minority areas, and establish a database". In 2014, China's Ministry of Ecology and Environment (MEE) released the Technical Specifications for the Classification, Survey, and Cataloging of Biodiversity-Related Traditional Knowledge (Trial Implementation). This document not only provides a clear definition of China's traditional knowledge but also establishes a classification system for it. Based on the attributes and uses of traditional knowledge, the document classifies biodiversity-related traditional knowledge into 5 categories and 30 items.

In January 2024, the Ministry of Ecology and Environment (MEE) of China released the China National Biodiversity Conservation Strategy and Action Plan (2023–2030) (hereinafter referred to as the Action Plan), which references traditional knowledge 30 times throughout the text. Examples include: "conduct research and draft regulations on the protection of biodiversity-related traditional knowledge"; "for eco-products with distinctive features and significant advantages in typical regions, deeply explore biodiversity-related traditional knowledge and cultivate and develop regional public brands"; and "conduct surveys and assessments of urban ecosystems, species, genetic resource diversity, and related traditional knowledge". Furthermore, the Action Plan explicitly proposes to "adopt measures such as carrying out surveys and cataloging of biodiversity-related traditional knowledge and establishing protection and registration systems for traditional knowledge, so as to enhance the public's cultural confidence in biodiversity-related traditional knowledge. By 2030, initially form a policy system for the protection of biodiversity-related traditional knowledge, and significantly strengthen the capabilities of protecting, inheriting and innovating traditional knowledge". The Action Plan has established the "Priority Projects for the Protection and Inheritance of Traditional Knowledge," proposing: "Emergency protection of biodiversity-related traditional knowledge, including the establishment of a traditional knowledge database to provide data and information support for the protection, inheritance and application of traditional knowledge"; "Demonstration of the protection and inheritance of biodiversity-related traditional knowledge: select representative ethnic groups, regions or typical species to conduct research on biodiversity-related traditional knowledge, explore the value of traditional knowledge in biodiversity conservation, climate change adaptation and rural revitalization. and carry out demonstrations of ecosystem adaptive management based on the integration of traditional knowledge and modern technical methods".

3.3 Analysis of China's Traditional Knowledge Participation in UN-Related Environmental Conventions

As a contracting party to UN-related environmental conventions, China plays an increasingly important role and exerts a pivotal influence on the formulation and adoption of relevant topics and decisions. Meanwhile, China has formulated relevant policies and regulations to respond to the international conventions it has signed. As a key topic in relevant United Nations environmental conventions, traditional knowledge means that relevant decisions and policies under these conventions also exert an impact on China. Therefore, China has developed corresponding domestic policies and regulations, in which it highlights the topic of traditional knowledge and defines its strategies and compliance actions regarding this topic. However, there are still certain limitations in China's participation in relevant United Nations environmental conventions through traditional knowledge at present. Based on this, this paper puts forward the following suggestions.

First, it is necessary to establish China's independent traditional knowledge system and enhance its status and role in international conventions. At present, China's discourse system of traditional knowledge is basically established within the context of UN-related environmental conventions and follows the paradigm of these international conventions. For instance, the definition of China's traditional knowledge is largely consistent with that of traditional knowledge in the CBD, sharing the same attributes and positioning. Admittedly, as a contracting party, China should comply with the provisions of relevant international conventions; however, it is undeniable that due to the long-standing Western-centrism stance of Western countries, these United Nations environmental conventions were significantly influenced by Western environmentalism and ecological thought during their formulation process, while lacking sufficient attention to the ecological thoughts and cultures of other regions, including China. Therefore, in response to the current limitations in the concepts and definitions of traditional knowledge in international conventions, it is essential to propose China's independent traditional knowledge system, so as to strengthen China's in-depth participation in the topic of traditional knowledge.

Second, it is necessary to strengthen the participation of domestic philosophy and social sciences in research. According to the current definition of China's traditional knowledge, the owners of traditional knowledge are people of all ethnic groups and local communities, which includes relevant traditional cultures, religious beliefs, and ecological ethics. Based on this definition, in the task division of the China National Biodiversity Conservation Strategy and Action Plan (2011–2030), there are two priority actions related to traditional knowledge. However, at present, the departments responsible for these priority actions are mainly environmental and resource departments, with no participation from cultural and ethnic affairs departments. In particular, the research institutions and disciplinary guidance for these priority actions are dominated by natural sciences, lacking the participation of philosophy and social sciences. This has led to limitations in the current application and implementation of traditional knowledge in China under the framework of relevant policies and regulations. The insufficient participation of philosophy and social sciences has restricted the further development of work related to China's traditional knowledge. Therefore, it is essential to further strengthen the participation of domestic philosophy and social sciences in relevant research.

It is worth noting that in recent years, China places significant emphasis on climate change. The central government has established the National Coordinating Agency for Climate Change. Aligned with the national sustainable development strategy, the state has implemented a series of policies and measures related to climate change, which positively contribute to mitigating and adapting to its impacts. China actively participates in the international negotiation process of the Global Climate Convention (GCC) and in global climate governance. To fulfill its commitments to the GCC, the central government has developed China's National Climate Change Programme (CNCCP). In accordance with GCC requirements, China has submitted the "National Evaluation Report of Climate Change" on schedule four times. The most recent report was published in 2018. From 2009 to 2019, China consistently published the "China Climate Change Policy and Annual Action Report" over this decade to demonstrate the progress of China's adaptation efforts on the global stage.

However, in China, traditional knowledge has still not been incorporated into current climate change policy and lacks legislative protection. Fortunately, existing policy on biodiversity conservation provides a useful precedent for enhancing future climate change policies. For instance, in September 2010, the State Council approved the "China Biodiversity Conservation Strategy and Action Plan (2011–2030)", which emphasizes the importance of strengthening research on the preservation of traditional knowledge related to biodiversity and its adaptation to climate change (Yin and Xue, 2013). In 2014, the Ministry of Environmental Protection (now the Ministry of Ecology and Environment) issued the "Regulations on the Classification, Investigation, and Cataloging of Traditional Knowledge on Biodiversity", which established the first national standard for the traditional knowledge system. At the local level, regional governments are also focusing on traditional knowledge and legislation for biodiversity conservation. For example, in 2018, the Yunnan Provincial Government enacted and published China's first local law on biodiversity conservation, titled "Yunnan Provincial Biodiversity Conservation Laws and Regulations." This not only addresses biodiversity conservation but also establishes specific protective measures for traditional knowledge related to biodiversity.

Accordingly, it is necessary to clarify the relationship between traditional knowledge and climate change, with regard to climate change policy and adaptation action plans in China. For instance, traditional knowledge should be integrated into the National Evaluation Report on Climate Change and the Action Plan for Adapting to Climate Change. At the same time, formal institutions need to be established to incorporate traditional knowledge into policy-making.

For future global climate negotiations, China should establish an appropriate mechanism to incorporate and integrate traditional knowledge into the Paris Agreement and global climate governance, building on the Agreement, its resolutions, and the Local Communities and Indigenous Peoples Platform (LCIPP).

Firstly, the value and impact of traditional knowledge in adapting to climate change should be acknowledged and emphasized. Leveraging the Local Communities and Indigenous Peoples Platform (LCIPP)'s role in recording and sharing knowledge and experiences, we can systematically collect, organize, and establish a classification system and database for traditional knowledge related to climate change. Traditional knowledge in this context can be categorized into two main types: traditional climate knowledge and traditional phenological knowledge.

Traditional climate knowledge encompasses the insights and understandings developed by ethnic minorities through their long-term experiences and practices. This knowledge is based on observations of weather patterns, natural phenomena, and environmental changes. It can be analyzed from three perspectives: the traditional definitions of climate change, the perceptions of climate change, and the experiences and technologies related to climate change.

Traditional phenological knowledge, on the other hand, is derived from observations in traditional livelihoods. Local communities track changes in crops, livestock, and environmental conditions in farmlands and pastures, identifying interactions with climatic variations. This knowledge helps reveal the natural laws governing seasonal changes. It can be examined from three aspects: the traditional definitions of phenological changes, the understanding of these changes, and the associated experiences and technologies.

Secondly, in accordance with the "Paris Agreement" and its resolutions, it is imperative to solidify the role of traditional knowledge in climate governance. Leveraging the functions of the Local Communities and Indigenous Peoples Platform (LCIPP), which include integrating knowledge systems and promoting the incorporation of traditional knowledge into national climate decision-making, future Action Plans for Climate Change should be guided not only by scientific advancements but also by the value and contributions of traditional knowledge. By integrating traditional and scientific knowledge, traditional knowledge can be effectively incorporated into social, economic, and environmental policies and action plans, thereby facilitating China's transition to a green and low-carbon economy.

Thirdly, it is essential to encourage ethnic minorities to actively participate in the negotiation process of the Global Climate Convention (GCC). Leveraging the Local Communities and Indigenous Peoples Platform (LCIPP), capacity-building initiatives should be implemented to

Fig. 1 The traditional knowledge and climate change project led by Prof. Yin Lun has been selected for the second Paris Peace Forum (photoed by Zhang Xiaohan)

empower ethnic minorities, enabling them to participate as representatives in the GCC negotiation process. In the context of global climate governance, it is crucial to highlight the value and role of traditional knowledge in climate adaptation, while also sharing China's traditional experiences and successful case studies (Fig. 1).

4 Traditional Knowledge Issues in the United Nations Framework Convention on Climate Change (UNFCCC)

The Paris Agreement formally acknowledged the status and the value of traditional knowledge within the UNFCCC framework and its resolutions. This agreement guaranteed the traditional knowledge to participate in global climate governance and recognized the significance of traditional knowledge in action plans. However, the "Paris Agreement" represents merely the initial step for traditional knowledge.

4.1 The Status and Evolution of Traditional Knowledge in the Paris Agreement

At the 21st Conference of the Parties (COP 21) to the UNFCCC, held from November 30 to December 11, 2015, nearly 200 Parties unanimously agreed to adopt the Paris Agreement. The Agreement aims to establish a global framework for climate action post-2020. The Paris Agreement marks a significant milestone in the international governance of climate change, symbolizing a crucial step forward in the global collective response to climate change (Frédéric, 2018).

For traditional knowledge, the Paris Agreement also holds great significance as a milestone. The Paris Agreement explicitly and formally positions traditional knowledge within the framework of the UNFCCC. This achievement did not come overnight, but rather developed gradually over time.

The first significant development originated from the UNFCCC itself. During the 20th Conference of the Parties (COP 20) held in Lima, Peru, in December 2014, discussions focused on the draft of a new climate agreement. Among these discussions was a proposal specifically addressing traditional knowledge. In February 2015, the UN climate negotiations held in Geneva, Switzerland, aimed to prepare for the new agreement to be finalized later that year at the Paris climate conference. The Geneva Text, the official negotiation document adopted at the meeting, also referenced traditional knowledge. A proposal concerning implementation provisions indicated that traditional knowledge was one of the elements constituting the foundation of climate change adaptation, placing it alongside other measures for adapting to climate change. The Geneva Text also highlighted that traditional knowledge is foundational for technological development, particularly in the sections addressing capacity building and technology.

In summary, the Geneva Text acknowledged the role of traditional knowledge in adapting to climate change, referencing participatory approaches to adaptation. Although it did not provide specific guidelines for how traditional knowledge should be used or protected, it formally recognized the role of traditional knowledge within the climate change regime. This laid the groundwork for its formal positioning within the Paris Agreement and helped encourage Parties to give greater consideration to traditional knowledge.

Moreover, other UN conventions have also contributed to the positioning and development of traditional knowledge within the framework of the UNFCCC. The Convention on Biological Diversity (CBD) places the greatest emphasis on traditional knowledge. The Convention requires its Parties to respect, protect, and maintain traditional knowledge related to the conservation and sustainable use of biodiversity. It further mandates that its application be promoted with the approval and participation of the knowledge holders, while encouraging the fair and equitable sharing of benefits arising from its use.

The Nagoya Protocol on Access and Benefit-Sharing stipulates that Parties must ensure the use of traditional knowledge related to genetic resources for research and development purposes, while obtaining the prior informed consent of indigenous peoples and local communities. These communities must also be involved in decision-making processes, and the benefits derived from the use of this knowledge should be shared with them.

Similarly, the International Treaty on Plant Genetic Resources for Food and Agriculture by the Food and Agriculture Organization (FAO) recognizes that farmers have the right to participate equitably in the benefits derived from the use of plant genetic resources. The treaty proposes the establishment of a benefit-sharing fund, with a focus on supporting traditional knowledge related to the management and use of plant genetic resources in the context of climate change adaptation.

The World Intellectual Property Organization (WIPO)'s Intergovernmental Committee on Intellectual Property and Genetic Resources, Traditional Knowledge, and Folklore is also working on drafting international laws for the protection of traditional knowledge from an intellectual property perspective.

The United Nations Convention to Combat Desertification (UNCCD) requires its Parties to exchange information on local and traditional knowledge in line with their national legislation and policies, protect traditional knowledge, and ensure the fair and equitable sharing of benefits. Furthermore, the Convention calls for Parties to support research on the protection, integration, enhancement, and verification of traditional knowledge, as well as the protection, promotion, and utilization of traditional knowledge, practices, and experiences.

The recognition of traditional knowledge by these UN conventions and international organiza-

Fig. 2 The gender and climate change project led by Prof. Yin Lun has been selected for the 3rd Paris Peace Forum (photoed by Zhang Xiaohan)

tions has played a pivotal role in encouraging the UNFCCC to give greater attention to traditional knowledge, thereby laying a foundation for its formal recognition in the Paris Agreement (Fig. 2).

4.2 Traditional Knowledge in the Paris Agreement

The Paris Agreement emphasizes the value and role of traditional knowledge in climate change adaptation actions, establishing objectives for its inclusion. In the section on climate change adaptation, the Agreement states, "The Parties recognize that adaptation action should follow a country-driven, gender-sensitive, participatory, and fully transparent approach, taking into account the needs of vulnerable groups, communities, and ecosystems, and should be based on and guided by the best available science, as well as relevant traditional knowledge, indigenous knowledge, and local knowledge systems, in order to integrate adaptation as appropriate into relevant social, economic, and environmental policies and actions." To clarify what is meant by "as appropriate" and how traditional knowledge should be integrated, the Decision adopting the Paris Agreement at the 21st Conference of the Parties (COP 21) provided further guidance.

The Paris Agreement formally recognizes the value of traditional knowledge in the international climate governance process and establishes mechanisms for its participation. Paragraph 135 of the Agreement refers to traditional knowledge, calling for the UNFCCC Parties to recognize the need to strengthen the knowledge, technologies, practices, and efforts of local communities and indigenous peoples in addressing and responding to climate change (Matemilola, et al., 2020). This can be understood as the Paris Agreement setting objectives for what needs to be achieved concerning traditional knowledge within the UNFCCC. However, this section does not specify how these objectives should be realized or how traditional knowledge issues should be strengthened within the framework of the Convention.

Additionally, the Paris Agreement calls for the establishment of a platform to exchange and share experiences and best practices related to both mitigation and adaptation in a comprehensive and integrated manner. This proposal marks the formal introduction of a platform for traditional knowledge within the Agreement. The Paris Agreement, along with its related decisions, thus establishes a new mechanism for the participation of traditional knowledge in international climate governance by formally creating the "Local Communities and Indigenous Peoples Platform" within the UNFCCC framework. This platform aims to promote the exchange and sharing of experiences and practices related to traditional knowledge and commits to strengthening efforts in traditional knowledge, technologies, and practices (Belfer et al., 2019).

Although the Paris Agreement calls for the establishment of this platform, it does not provide specific details on how the exchange and sharing of traditional knowledge should occur or on the goals for achieving such actions. Furthermore, it does not explicitly define the platform's role in strengthening the knowledge, technologies, practices, and efforts related to traditional knowledge.

The Paris Agreement acknowledges that, in addition to scientific and technological knowledge, diverse human knowledge systems—including traditional knowledge—can play a significant role in addressing climate change. This recognition of diverse knowledge systems within the Paris Agreement paves the way for the inclusion of Chinese traditional knowledge in global climate change governance based on its alignment with the Agreement's goals.

4.3 Analysis of the Traditional Knowledge Issue in the Paris Agreement

Currently, in discussions within the UNFCCC on how different knowledge systems can contribute to climate change adaptation under the Paris Agreement, three distinct tendencies have emerged among the Parties.

First, Western Developed Countries (e.g., Europe and North America). Countries such as those in Europe and North America emphasize the informational value of traditional knowledge, indigenous knowledge, and local knowledge systems. They argue that these forms of knowledge should remain supplementary to existing scientific and technological frameworks. While local climate information may complement global climate governance with additional data and insights, these knowledge systems are considered as supporting, rather than replacing, mainstream scientific knowledge. This reflects the Paris Agreement's provision of "as appropriate" integration of different knowledge systems, where traditional knowledge is viewed as a complementary tool to established scientific approaches.

Second, Indigenous Communities in the Americas, Africa, and Oceania. In contrast, indigenous communities in the Americas, Africa, and Oceania emphasize the rights-based attributes inherent in indigenous knowledge, particularly related to land, resources, and autonomy. When discussing the integration of different knowledge systems in the Paris Agreement, they prefer to focus on indigenous knowledge and indigenous peoples. Many of these communities, having been colonized by Western countries, often refer to contemporary scientific knowledge as Western Knowledge or Colonial Knowledge, positioning it in opposition to Indigenous Knowledge. For these communities, indigenous knowledge is not merely data or information belonging to a knowledge system, but a cultural and value-driven perspective that reflects the deep relationship between indigenous peoples and the land, forests, mountains, lakes, and other natural elements. This approach highlights the worldview that is tied to their cultural identity and stewardship of the environment.

Third, Emerging Economies (e.g., China and India). Finally, emerging economies, particularly China and India, adopt a more inclusive approach. In discussions on knowledge systems in the Paris Agreement and other related UN conventions, these countries use terms like "traditional knowledge" and "local communities." They assert that all human cultures and knowledge systems should be respected and treated equally in climate change adaptation and biodiversity governance. These nations emphasize the ecological attributes of traditional knowledge, which is formed based on natural environments and ecosystems. Both countries have undertaken significant efforts to investigate, protect, and utilize traditional knowledge related to biodiversity. For example, China is building a traditional knowledge database, while India has already established a digital library for traditional knowledge. This inclusive stance underlines the importance of recognizing diverse knowledge systems from a cultural and ecological perspective, ensuring that the role of traditional knowledge is not only acknowledged but actively integrated into the framework for climate change governance under the Paris Agreement.

4.4 Deepening Traditional Knowledge Under the UNFCCC Post-Paris Agreement

Since 2017, the UN Framework Convention on Climate Change (UNFCCC) has further deepened and institutionalized the participation mechanisms for traditional knowledge in climate governance. Building upon existing platforms and decision-making bodies, it has promoted a systematic framework for integrating traditional knowledge. The institutional developments during this period can be divided into the following key stages.

At COP23 in Bonn, Germany (2017), Decision 2/CP.23 defined the three core functions of the Local Communities and Indigenous Peoples Platform (LCIPP): (1) Facilitate the exchange of experiences and best practices; (2) Build capacity and provide technical support; (3) Enhance the engagement of indigenous communities in climate governance policies. This decision laid a crucial foundation for the institutionalization of the LCIPP (Hermwille et al., 2017).

COP24 (2018) formally established the Local Communities and Indigenous Peoples Platform's implementing body, the Facilitative Working Group (FWG). During this phase, the FWG focused on integrating traditional knowledge with climate policy technical frameworks, emphasizing respect for the principle of Free, Prior and Informed Consent (FPIC) to prevent the marginalization of indigenous communities and the dispossession of their land rights.

COP25 in Madrid, Spain in 2019 approved the FWG's first two-year work plan (UNFCCC, 2020). Centered on the three pillars of "Equitable Participation," "Capacity Building," and "Knowledge Sharing," the plan explicitly supported Indigenous Peoples in strengthening their own capacity building, ensuring the effective inclusion of traditional knowledge in climate adaptation and mitigation policies, and promoting their equitable participation in UNFCCC processes.

COP26 in Glasgow, UK marked the official recognition of Indigenous Peoples as "Knowledge Holders". The large-scale invitation of 28 indigenous representatives into formal negotiations significantly boosted their influence in global climate governance. 14 country and philanthropic donors pledged at least $1.7 billion from 2021 to 2025 to advance Indigenous Peoples' and local communities' forest tenure rights and support their role as guardians of forests and nature (UNFCCC, 2023). The LCIPP platform's functions were further strengthened with the release of its 2022–2024 three-year work plan, which added measures for technical support and fostering collaboration among trans-regional indigenous organizations, thereby enhancing indigenous leadershipx. This COP also featured the first Pavilion led by indigenous organizations, providing a platform for indigenous communities to exchange ideas and conduct public education, amplifying their collective voice and increasing international recognition of their ecological wisdom and community resilience. Multiple joint statements called for integrating FPIC and traditional knowledge protection into Nationally Determined Contributions (NDCs) and National Adaptation Plans (NAPs) to safeguard indigenous land rights and self-governance.

COP29 in Baku, Azerbaijan adopted the "Baku Workplan," which continues and deepens the LCIPP and FWG's three core tasks: knowledge exchange and dissemination, capacity building and support mechanisms, and integrating diverse value systems. The work plan specifically emphasizes support for linguistic diversity, enhancing cultural inclusivity, and ensuring the equal participation of indigenous groups from different language backgrounds. To strengthen institutional continuity and representativeness, the membership tenure mechanism was optimized to preserve institutional knowledge while ensuring the participation of emerging indigenous organizations. Concurrently, efforts were made to direct technical and financial support toward indigenous communities, collaborating with mechanisms like the Green Climate Fund (GCF) to prioritize climate adaptation capacity-building projects tailored to indigenous needs.

Furthermore, the UNFCCC enhanced coordination with other environmental multilateral mechanisms, such as the Convention on Biological Diversity (CBD), promoting the cross-sectoral integration of traditional knowledge in climate and biodiversity conservation policies to achieve synergies.

Overall, between 2017 and 2024, the institutionalization of traditional knowledge within the UNFCCC system underwent significant maturation and strengthening. The LCIPP platform and the FWG implementing body have been continuously improved, with increasingly scientific, systematic, and multi-functional work plans, demonstrating the following key characteristics:

- Gradual Institutional Refinement: Ensuring traditional knowledge participate in negotiations and policy making through balanced representation mechanisms.
- Deep Integration of Traditional Knowledge Systems & Climate Policy: Official recognition of the critical role of traditional knowledge in adaptation and mitigation strategies.
- Strengthened Protection of Cultural & Linguistic Diversity: Promoting inclusive and equal participation.

5 Integrating Traditional Knowledge into Global Climate Governance

Currently and in future, it is essential to further mainstream the traditional knowledge in the context of climate change. This integration will encourage humanity to reflect on, adjust, and improve the processes of social development, ultimately striving for sustainable development (Fig. 3).

Integrating traditional knowledge into global climate governance consists of three aspects: (1) the integrating of traditional knowledge and scientific knowledge; (2) the integrating of traditional knowledge and mitigation and adaptation strategies; (3) the integrating of traditional knowledge and climate policy.

5.1 Linking Traditional Knowledge with Scientific Knowledge

In the process of coping with climate change, traditional knowledge and scientific knowledge are facing challenges, and both knowledge systems have their own weaknesses and shortcomings. For example, for traditional knowledge, due to the variability of climate and the intensity and frequency of current climate change, traditional knowledge based on observation, subjective awareness and qualitative description has become increasingly inaccurate in forecasting weather change. For scientific knowledge, however, scientific models and data analysis also have shortcomings: a reductionist approach focusing on

Fig. 3 The project team led by Prof. Yin Lun participated in the Paris Peace Forum (photoed by Vincent Podvin)

a limited set of variables; extrapolations from narrow data sets; a restricted set of metrics; focus on mean values when the extremes may be the critical factor. In other words, the analysis based on models and data can provide a macro picture of climate change, but it is difficult to reduce the data scale to meet the needs of local regions. The effective integrating of the two knowledge systems can make up for their shortcomings, so that the decision-making and action of climate change can be based on the "best available knowledge". However, the prerequisite for such integrating is "respect",that is to understand the knowledge of indigenous peoples and local communities on climate change impacts and adaptation in a comprehensive and mutually respectful way. Considerable efforts and originality should be made, interdisciplinary collaboration should be carried out, and new measurement standards should be developed. At the same time, the research of traditional knowledge should be further improved and promoted. For example, based on the close relationship with biodiversity, traditional knowledge should be classified, the index system of traditional knowledge should be found, and the qualitative parameters of traditional knowledge should be transformed into quantitative research and analysis indicators, so as to find the "objectivity" of traditional knowledge.To bridge the gap between traditional knowledge and scientific knowledge and realize the integrating of the two knowledge systems to help human beings cope with climate change effectively requires mutual understanding between traditional knowledge and scientific knowledge, as well as the joint efforts of natural science and social science.

5.2 Integrating Traditional Knowledge into Mitigation and Adaptation Strategies

Many strategies have been made to help indigenous peoples and local communities cope with the climate changes. Both governmental and nongovernmental organizations (NGOs) have brought projects, technique and infrastructure to these regions in order to assist them in the mitigation and adaptation process. Unfortunately, these strategies have been frequently unsustainable and have been abandoned. There may be many reasons for the failure of these strategies, but the most fundamental reason is that in the process of mitigating and adapting to climate change, indigenous peoples and local communities are regarded as a weak, passive group waiting for external help, rather than a capable, active and spontaneous group. These strategies ignore the resources available in the local community, especially the resources of the ecosystem and social system. These strategies are mainly external and depend on the so-called "advanced" external concepts, technologies and objectives. These strategies are still based on the development concept of modern society, without considering the local cultural background and socio-economic conditions. The goals of these strategies do not reflect the wishes and needs of the local people. In other words, these strategies do not recognize the value of traditional knowledge of indigenous peoples and local communities. Traditional knowledge proves useful in recognizing potentials of some mitigation and adaptation strategies as well as the efficiency of cost, participation and sustainable development of the community in the context of climate change impacts. Traditional knowledge can improve understanding of the impact of climate change, and provide ideas to develop the equitable and effective mitigation and adaptation strategies for climate change. For the indigenous peoples and local communities, compare with the externally driven mitigation and adaptation options, they are more willing to take the community-based mitigation and adaptation strategies, because such strategies will enable them to apply their traditional knowledge according to the geographical location and environment of their areas, so that their actions to cope with climate change will more effective. It should be pointed out that traditional knowledge can play an important role in community-based mitigation and adaptation strategies, but it does not mean the exclusion or substitution of external modern scientific knowledge and technology. In order to produce "best practices" for mitigation and adaptation, internal leadership and external support must complement each other and learn from each other. Therefore, the basis of "best practice" is to realize the combination of integrating traditional knowledge with modern science and technology on the basis of community leading.

5.3 Involving Traditional Knowledge in the Climate Policy

Although indigenous peoples and local communities have suffered more severe impacts of climate change, until recently, they were still marginalized in climate policy-making. For a long time, there has been little mention of indigenous peoples and local communities in climate policy at the international and national levels. Moreover, in the relevant climate policies that refer to indigenous peoples and local communities, they are also regarded as hopeless victims in the process of climate change, they are considered unable to cope with the impact, and their traditional knowledge is rarely mentioned. Fortunately, in recent years, at the international level, relevant climate policies have been paying more and more attention to this topic, and indigenous peoples and traditional knowledge have begun to integrate into relevant international climate policies, such as the "Paris Agreement". The Paris agreement formally recognizes the value and role of traditional knowledge in adaptation, that is," by the best available science and, as appropriate, traditional knowledge, knowledge of indigenous peoples and local knowledge systems, with a view to integrating adaptation into relevant socioeconomic and environmental policies and actions, where appropriate". But this is only a beginning, especially for traditional knowledge, which is far from enough. Traditional knowledge is not only a subordinate position, but also an equal position with scientific knowledge. To integrate traditional knowledge into relevant climate policies at the international and national levels, we should first recognize the value of traditional knowledge, the role of traditional knowledge in mitigating and adapting to climate change, and the ability of traditional knowledge to reduce vulnerability in the face of extreme climate and other pressures, so as to enhance the resilience of indigenous peoples and local communities. Second, in the process of climate policy-making, we should try to take a bottom-up approach to encourage the highest level of participation of traditional knowledge. Thirdly, in climate policy, indigenous peoples and local communities should be regarded as equal stakeholders, and traditional knowledge should be further given correct and equal status.

In the process of traditional knowledge mainstreaming, we should also pay attention to two issues: (1) the protection of traditional knowledge of indigenous peoples and local communities; (2) the rights of indigenous peoples and local communities based on traditional knowledge.

5.4 The Protection of Traditional Knowledge

Indigenous peoples and local communities are holders of traditional knowledge, which means that traditional knowledge can be regarded as part of the property of indigenous peoples and local communities. At present, in the process of mainstreaming traditional knowledge, that is, integrating scientific knowledge, mitigation and adaptation strategies, and climate policies, indigenous peoples and local communities are basically open-minded and willing to share their traditional knowledge. Climate scientists and policy makers are increasingly using the traditional knowledge, either to supplement scientific research or to enhance the effectiveness of mitigation and adaptation strategies. Climate scientists, scholars and policy makers regard traditional knowledge more as a valuable data and its source channel. In other words, the reason why traditional knowledge is conducive to understanding and adapting to climate change is that they provide baseline data and strategic information, which climate scientists usually cannot access in their databases (Fig. 4).

However, those who view the value of traditional knowledge from the perspective of data may not understand the significance of traditional knowledge to indigenous peoples and local communities, especially the value of traditional knowledge to local people in terms of culture and world outlook. Therefore, improper sharing of traditional knowledge related to climate change may have adverse effects on the cultural and spiritual aspects of local society. At the same time, the traditional knowledge of indigenous peoples and local communities is often closely related to the environment, natural resources and ecosystems in which they live, especially the local biodiversity and genetic resources. Therefore, when the traditional knowledge is shared openly, it often brings potential harm and risk to indigenous peoples and local communities. It is in this context that traditional knowledge needs to be protected, that is to say, corresponding principles need to be formulated to regulate the access, sharing and use of traditional knowledge. These principles can draw lessons from the principles of access and benefit sharing of genetic resources and related traditional knowledge in the Convention on biological diversity, these principles should include:

(1) The principle of "prior informed consent". To formulate the prior informed consent standard for acquiring and agreeing to use traditional knowledge associated with climate change, especially traditional knowledge associated biodiversity, that is, when using traditional

Fig. 4 Prof. Yin Lun conducted field work on traditional knowledge and climate change (photoed by Zhang Xiaohan)

knowledge, it is necessary to formally inform the owner of traditional knowledge through channels in advance and obtain their consent.

(2) The principle of "mutually agreed conditions". Develop mutually agreed conditions for access to and consent to use of traditional knowledge associated with climate change. When using traditional knowledge, the local community should be informed formally through channels, and relevant conditions should be agreed with the local community.

(3) The principle of "no harm condition". Establish non harmful conditions for access to and consent to the use of traditional knowledge associated with climate change. That is, the use of traditional knowledge should not cause spiritual, cultural, property and rights harm to the local community.

(4) The principle of "benefit sharing". Develop benefit sharing approaches to access and agree to use traditional knowledge associated with climate change. Through compulsory means such as direct payment to local communities and establishment of community development funds, relevant users should share the benefits of traditional knowledge, especially traditional knowledge associated with biodiversity, in the process of use and after.

6 Reflections on the Participation of Chinese Traditional Knowledge in Global Climate Change Governance

General Secretary Xi Jinping has emphasized that we must deeply participate in global ecological and environmental governance, enhance China's discourse power and influence in the global environmental governance system, and actively guide the direction of international order reform (CPC

Fig. 5 Prof. Yin Lun has become the Winner of the 2024 CAREC Gender Climate Award (photoed by Zhang Xiaohan)

Publicity Department et al., 2022b.). To encourage greater participation from diverse stakeholders in the global governance process for addressing climate change, the Paris Agreement recognizes the value and role of diverse human knowledge systems in adaptation to climate change and establishes a mechanism for their involvement, including traditional knowledge. Against this backdrop, Chinese traditional knowledge can also be integrated into global climate change governance, thereby enhancing the influence of Chinese traditional knowledge in the global discourse on climate change response.

On September 3, 2016, the Standing Committee of the National People's Congress of China issued a decision to approve the Paris Agreement, marking China's formal accession to the agreement. The Paris Agreement created a mechanism for the participation of traditional knowledge, encouraging stakeholders to engage in global climate governance based on their traditional knowledge. This presents an opportunity for China to actively participate in the UNFCCC's discussions on traditional knowledge issues and contribute its traditional knowledge to global climate change governance. By incorporating traditional knowledge into international climate governance processes, China not only contributes to the diversity of perspectives in addressing climate change but also strengthens its role and influence in the global governance system, aligning with its broader goals in international environmental diplomacy (Fig. 5).

6.1 Recognition and Utilization of Chinese Traditional Knowledge in Global Climate Change Governance

The Paris Agreement recognizes that, beyond modern science and technology, human knowledge systems- including traditional knowledge—can play an important role in addressing climate change. Traditional knowledge is derived from experience and trial-and-error, rather than from logical deduction or formal scientific methodologies. As a result, its formation process and mechanisms differ significantly from those of contemporary mainstream scientific systems. However, traditional knowledge, like modern science, is grounded in an understanding and utilization of the truth of natural phenomena. Therefore, it is not devoid of scientific validity or reason. In the context of global climate change governance, traditional knowledge contributes to both mitigation and adaptation—the two primary human strategies for addressing climate change. By analyzing concrete cases and examples of traditional knowledge applications, it becomes evident that these knowledge systems hold contemporary value for global climate change governance. This aligns with the objectives set forth in the Paris Agreement. Through fieldwork conducted in various ethnic communities and villages, the author has found that traditional knowledge plays a significant role in both mitigating and adapting to climate change, offering unique value and insights. This knowledge is embedded in local practices, environmental stewardship, and community resilience strategies, which have been honed over generations. By effectively utilizing traditional knowledge, China can contribute to global climate change governance while also demonstrating how indigenous practices can complement modern scientific approaches. In this context, Chinese traditional knowledge has great potential to be promoted globally as a valuable asset in climate change response, consistent with the goals of the Paris Agreement to integrate diverse knowledge systems. It also highlights the growing importance of inclusive, culturally sensitive approaches to environmental management and climate change adaptation.

In the context of mitigating climate change, traditional knowledge plays a crucial role in reducing greenhouse gas emissions, increasing carbon sequestration, and enhancing carbon alternatives. Ecosystems such as forests, widely recognized as significant carbon sinks, capture and store large amounts of carbon dioxide from the atmosphere. Therefore, ecosystems like forests can play a vital role in reducing greenhouse gas emissions and alleviating climate change risks. Chinese traditional knowledge contains abundant wisdom and practices related to natural resource conservation and ecosystem management. For example, the Dai people's traditional knowledge, centered around the belief in the Dragon Mountain, encompasses rich practices for maintaining forest ecosystems. This unique knowledge system has, in practice, contributed to the protection of the tropical rainforests of Xishuangbanna, enhancing the forest's carbon sink capacity. This provides valuable insights for modern forest conservation practices. Similarly, the Dai people's traditional forestry knowledge reflects their understanding of mitigating climate change. Through rational logging and afforestation practices, they have maintained ecological balance and sustained the carbon sink function of the forests. The Tibetan people's belief in sacred mountains has, in an objective sense, formed a unique ecosystem protection system. Their belief system treats humans and nature as one integrated entity and imposes strict taboos, such as prohibiting hunting or tree cutting on sacred mountains. This system has, in effect, protected the sacred mountains and their surrounding ecosystems, enhancing the carbon sequestration capacity of the environment, reducing the unnecessary evaporation of surface water resources, and stabilizing atmospheric temperatures. These practices contribute to mitigating the adverse effects of global warming, and the traditional knowledge embedded in these beliefs offers valuable alternatives to modern scientific and technological interventions. which often require significant human and material resources and yield limited and unstable results. The Jino people's traditional agricultural practices also exemplify climate change mitigation strategies. Through crop rotation, intercropping, and other farming methods, they not only improve land use efficiency but also help maintain soil's carbon sequestration potential. Additionally, the Blang people utilize traditional knowledge in breeding agricultural genetic resources to develop livelihood plans that adapt to climate change. By selecting crop varieties that are resilient to climate change, they have reduced the use of chemical fertilizers and pesticides, consequently lowering the greenhouse gas emissions associated with agricultural activities.

These examples demonstrate how traditional knowledge plays an important role in mitigating climate change by enhancing the carbon sequestration capacity of ecosystems and improving land management practices. In contrast to modern scientific approaches, which often require high investment and produce limited and uncertain results, traditional knowledge offers a sustainable, culturally rooted solution that is more adaptive to local contexts and ecosystems. Integrating such traditional wisdom into modern climate governance can enhance the global response to climate change, offering valuable contributions from indigenous knowledge systems.

In the context of adapting to climate change, traditional knowledge holds particular value in reducing climate vulnerability and enhancing climate resilience. Many indigenous groups have developed knowledge systems that enable them to anticipate, respond to, and mitigate the effects of climate variability and extreme weather events. For example, the Kazakh people possess traditional knowledge that allows them to predict and interpret weather and climatic changes on the steppe. This knowledge enables them to determine the timing for grazing and relocating livestock, adjusting these activities according to climatic conditions. They can also foresee extreme weather events such as droughts or snowstorms, assess their potential severity and impacts, and take preventive measures to reduce the risks these events pose to livestock farming. The Wa people, through their traditional knowledge of agricultural biodiversity, select different crop genetic resources to adapt to environmental changes induced by climate change, such as variations in soil quality, temperature, slope, orientation, and sunlight. This traditional knowledge related to plant genetic resources allows for agricultural diversification, reducing the impacts of climate change on farming. By incorporating these practices, the Wa people are better equipped to adapt to climate variability and ensure food security despite changing environmental conditions. Faced with the challenges posed by climate change and water scarcity, the Uyghur people in the arid regions of Xinjiang have developed traditional knowledge for effective water resource management. One example is their use of Karez irrigation systems—an ancient underground water channel system that minimizes evaporation and makes efficient use of limited freshwater resources. This system has allowed Uyghur communities to thrive in a desert environment, supporting sustainable livelihoods and development. The Karez system provides valuable lessons for water management in other arid regions globally, especially when addressing the challenges of drought and water scarcity, offering concrete strategies for adaptation to extreme climate phenomena. The Yi people have developed a comprehensive set of practices to mitigate climate disaster risks. After heavy rainfall, they observe changes in river color, water flow speed, and listen to the sounds of flowing water and mountain movements to predict the likelihood of floods or landslides. Based on these observations, they implement disaster prevention measures, enhancing their ability to respond to extreme weather events. This traditional knowledge has significantly improved disaster resilience in local communities, helping reduce the risks posed by climate-related disasters.

These examples illustrate how traditional knowledge systems are directly linked to local ecosystems and communities, enabling better adaptation to climate change through practices that are often rooted in centuries of observation and experience. Such knowledge is not only scientifically grounded but also deeply connected to cultural and environmental contexts, offering valuable contributions to modern climate adaptation efforts. By recognizing and integrating traditional knowledge into global climate governance frameworks like the Paris Agreement, countries and communities can strengthen their collective resilience to climate change, fostering sustainable development and reducing vulnerabilities on a global scale.

6.2 Prospects for the Participation of Chinese Traditional Knowledge in Global Climate Governance

As an integral component of China's active engagement in global climate governance, the issue of traditional knowledge deserves significant attention. In China's future climate change policies and action plans, the relationship between traditional knowledge and climate change should be explicitly recognized, and relevant policies and plans should be improved accordingly. For example, a dedicated section on traditional knowledge could be included in the National Assessment Report on Climate Change, and traditional knowledge could be integrated into national climate action plans. At the same time, formal mechanisms and channels should be established to ensure the participation of traditional knowledge in the policy-making process, thereby echoing the emphasis placed on this issue in the Paris Agreement. Consequently, this study proposes several considerations for enhancing the participation of Chinese traditional knowledge in global climate governance.

6.2.1 Participation in Discussions on Traditional Knowledge Issues

In line with the Paris Agreement and subsequent Conference of the Parties (COP) decisions, it is necessary to analyze mechanisms for the participation of traditional knowledge and actively engage in related discussions. Based on the three principal functions of the Local Communities and Indigenous Peoples Platform (LCIPP), which was established under the Paris Agreement to facilitate the integration of traditional knowledge, a mechanism should be developed for the participation of Chinese traditional knowledge in both the Paris Agreement and the broader global climate governance process. The value and role of Chinese traditional knowledge in addressing climate change can be explored following the COP decisions. Furthermore, given that the LCIPP functions as a platform for recording and sharing knowledge and experiences, China could draw on its classification systems related to biodiversity-associated traditional knowledge to systematically collect, organize, and establish a classification system and database of traditional knowledge related to climate change.

6.2.2 Strengthening the Integration of Traditional Knowledge with Climate Policy

By integrating traditional knowledge with scientific knowledge, incorporating it into mitigation and adaptation strategies, and recognizing its value and role in climate policymaking, traditional knowledge can serve as one of the knowledge systems for addressing climate change. Climate policies should not rely solely on scientific, technological, and economic measures; traditional knowledge should also be considered, and its value in climate response should be fully harnessed and incorporated into modern policy frameworks. This approach is essential for sustainable development, particularly in ethnic minority regions (Zhu and Wei, 2024). Moreover, it can provide new perspectives for developing China's mitigation and adaptation strategies. The status of Chinese traditional knowledge in addressing climate change should be clarified, and knowledge system integration should be promoted through participatory mechanisms to ensure traditional knowledge is incorporated into national decision-making processes. In China's future climate change actions and plans, it will be crucial to adhere to the requirements of the Paris Agreement, which emphasizes not only scientific and technological guidance but also the recognition of traditional knowledge. The integration of both systems should be pursued, and traditional knowledge should be incorporated into China's socioeconomic and environmental policies and actions aimed at transitioning toward a green and low-carbon economy.

6.2.3 Promoting the Participation of Traditional Knowledge in Global Climate Governance

Global climate governance encompasses not only political and economic dimensions but also ecological and cultural dimensions. In the context of climate change, exploring the relationship between culture and nature is of particular importance, especially in ecologically fragile regions. By combining traditional ecological wisdom with modern technology, it is possible to identify a balance between ecological protection and socioeconomic development (Meng and Yang, 2023). Following the Paris Agreement and subsequent COP decisions, pathways should be proposed and explored for involving Chinese traditional knowledge in global climate governance. Such efforts should encourage the participation of Chinese traditional knowledge in global climate processes, and fully draw upon traditional practices in environmental protection, climate adaptation, natural resource management, legal norms, production technologies, and ecological philosophies (Yin, 2022). In global climate governance, the value and potential contributions of Chinese traditional knowledge, techniques, and practices should be highlighted. Furthermore, China should actively share its models, experiences, and successful practices related to traditional knowledge, thereby contributing to the eventual establishment of a Chinese traditional knowledge system for participating in global climate governance.

References

Alfred, G. R., & Wilmer, F. (1997). Indigenous peoples, states, and conflict[J]. *Wars in the Midst of Peace, 1997*, 26–44.

Anaya, S. J. (2004). *Indigenous peoples in international law*. USA: Oxford University Press.

Andrade, G. S. M., & Rhodes, J. R. (2012). Protected areas and local communities: an inevitable partnership toward successful conservation strategies? *Ecology and society, 17*(4).

Assembly UNG (2007). United Nations declaration on the rights of indigenous peoples. *Un Wash,* 1–18.

Barten, U. (2015). What's in a Name: Peoples, Minorities, Indigenous Peoples, Tribal Groups and Nations. *JEMIE, 14*, 1.

Belfer, E., Ford, J. D., Maillet, M., et al. (2019). Pursuing an indigenous platform: Exploring opportunities and constraints for indigenous participation in the UNFCCC. *Global Environmental Politics, 19*(1), 34–46.

Bentley, G. C. (1986). Indigenous states of southeast Asia. *Annual Review of Anthropology, 15*, 275–305.

Berkes, F. (2009). Indigenous ways of knowing and the study of environmental change. *Journal of the Royal Society of New Zealand, 39*(4), 23–45.

Berkes, F. (2017). *Sacred ecology* (4th ed., pp. 1251–1271). New York: Routledge.

Berkes, F., & Jolly, D. (2002a). Adapting to climate change: Social-ecological resilience in a Canadian western Arctic community. *Conservation Ecology, 5*(2), 45–48.

Berkes, F., & Jolly, D. (2002b). Adapting to climate change: Social-ecological resilience in a Canadian western Arctic community. *Conservation Ecology, 5*(2), 87–90.

Berkes, F., Colding, J., & Folke, C. (2000). Rediscovery of traditional ecological knowledge as adaptive management. *Ecological applications, 10*(5), 1251–1262.

Berkes, F., Folke, C., & Gadgil, M. (1994). *Traditional ecological knowledge, biodiversity, resilience and sustainability//Biodiversity conservation: problems and policies*. Papers from the biodiversity programme beijer international Institute of ecological economics royal Swedish academy of sciences. Dordrecht: Springer Netherlands, 269–287.

Bertrand, J., & Xu, C. (2023). Indigenous groups and Ethnic minorities//Routledge handbook of civil and uncivil society in Southeast Asia. *Routledge,* 242–259.

Borrini, G., Kothari, A., & Oviedo, G. (2004). *Indigenous and local communities and protected areas: Towards equity and enhanced conservation: Guidance on policy and practice for co-managed protected areas and community conserved areas*. IUCN.

Canessa, A. (2007). Who is indigenous? Self-identification, indigeneity, and claims to justice in contemporary Bolivia. *Urban Anthropology and Studies of Cultural Systems and World Economic Development,* 195–237.

Castellino, J. (2014). *International law and self-determination: peoples, indigenous peoples, and minorities*.

Cheng, G., Wang, C., & Xue, D. (2012). Attitude to the traditional knowledge of international governmental organizations and suggested strategies of China. *Biodiversity Science, 20*(4), 505–511. (In Chinese)

Chua, R. Y., Kadirvelu, A., & Yasin, S., et al. (2019). The cultural, family and community factors for resilience in Southeast Asian indigenous communities: A systematic review. *Journal of Community Psychology, 47*(7), 1750–1771.

Clarke, G. (2001). From ethnocide to ethnodevelopment? Ethnic minorities and indigenous peoples in Southeast Asia. *Third World Quarterly, 22*(3), 413–436.

Corntassel, J. (2003). Who is indigenous? 'Peoplehood' and ethnonationalist approaches to rearticulating indigenous identity. *Nationalism and ethnic politics, 9*(1), 75–100.

Van Cott, D. L. (2005). Building inclusive democracies: Indigenous peoples and ethnic minorities in Latin America. *Democratisation, 12*(5), 820–837.

Crossen, J. (2014). *Decolonization, indigenous internationalism, and the World Council of Indigenous Peoples*. University of Waterloo.

CPC Publicity Department, IPHLCPCCC, CICG (2022a). *Xi JINPING: The governance of China* (3rd ed.). Beijing: Foreign Languages Press, 372 (In Chinese).

CPC Publicity Department, IPHLCPCCC, CICG (2022b). *Xi JINPING: The governance of China* (3rd ed.). Beijing: Foreign Languages Press, 364 (In Chinese).

Dawson, N. M., Coolsaet, B., Sterling, E. J., et al. (2021). The role of Indigenous peoples and local communities in effective and equitable conservation. *Ecology and Society, 26*(3), 19.

Dayuan, X. (2019). Conservation and outlook of traditional knowledge associated with biodiversity. *Biodiversity Science, 27*(7), 705–707. (In Chinese).

Eide, A. (2014). United Nations standard-setting regarding rights of minorities and indigenous peoples. *Europa Ethnica, 71*(3–4), 51–61.

Erbaugh, J. T., Pradhan, N., Adams, J., et al. (2020). Global forest restoration and the importance of prioritizing local communities. *Nature Ecology & Evolution, 4*(11), 90–101.

Ford, J., Maillet, M., Pouliot, V., et al. (2016). Adaptation and indigenous peoples in the United Nations framework convention on climate change. *Climatic Change, 139*, 87–98.

Formaini, H. (2009). Contesting the'we'of'we': The rights of Indigenous peoples in Australia. *Cosmopolitan Civil Societies: An Interdisciplinary Journal, 1*(2), 96–110.

Frédéric Gilles Sourgens (2018). Paris agreement regained or lost? *Initial Thoughts*, 97–105

Gebremedhin, S. H., & Theron, F. (2007). Locating community participation in a water supply project—the Galanefhi Water Project (Eritrea). *Anthropology Southern Africa, 30*(1–2), 20–28.

Gong, C., Cheng, W., & Dayuan, X. (2012). Attitude to the traditional knowledge of international governmental organizations and suggested strategies of China. *Biodiversity Science, 20*(4), 505–511. (In Chinese).

Hannum, H. (1994). Minorities, indigenous peoples, and self-determination. *Stud. Transnat'l Legal Pol'y, 26*, 1.

Harrisson, T. (2017). Tribes, Minorities and the Central Government in Sarawak Malaysia. *Southeast Asian Tribes, Min orities and Nations,* 2.

Heath, A. F., & Yu, S. (2005). Explaining Ethnic Minority. *Understanding social change,* 187.

Hermwille, L. et al. (2017). UNFCCC before and after Paris–what's necessary for an effective climate regime?. *Climate policy 17*(2), 150–170.

Hewuliji, M., & Tingshuo, Y. (2023). The shared perception of water resources of nomadic pastoralists in the inland arid regions from the perspective of the term "Tala" – validation from wetland ecosystem variation. *Journal of Ethnic Culture, 15*(1), 16–26. (In Chinese).

Hill, R., Adem, Ç., Alangui, W. V., et al. (2020). Working with indigenous, local and scientific knowledge in assessments of nature and nature's linkages with people. *Current Opinion in Environmental Sustainability,* 8–20.

Huang, Q., & Yin, L. (2024). Achievements and prospects of biodiversity conservation in Xishuangbanna Prefecture, Yunnan Province. *Journal of West China Forestry Science, 53*(4), 7–12. (In Chinese).

Huang, Q., Kuang, Y., Zhou, H., et al. (2024). Biodiversity conservation in Xishuangbanna, China: diversity analysis of traditional knowledge related to biodiversity and conservation progress and achievement evaluation. *Diversity, 16*(5), 260.

International Labour Organization (ILO). Indigenous and Tribal Peoples Convention, 1989 (No. 169). https://www.ilo.org/sites/default/files/wcmsp5/groups/public/@ed_norm/@normes/documents/publication/wcms_205225.pdf. (accessed on 2025-08-05).

IPCC. (2007a). *Climate Change 2007: Impacts, Adaptation and Vulnerability* (pp. 23–45). Cambridge: Cambridge University Press.

IPCC. (2007b). *Climate Change 2007: Impacts, Adaptation and Vulnerability* (pp. 98–100). Cambridge: Cambridge University Press.

IPCC (2022a). *Climate change 2022: Impacts, adaptation and vulnerability, working group II contribution* (pp 87–90). Online.

IPCC (2022b). *Climate change 2022: Impacts, adaptation and vulnerability, working group II contribution* (pp 67–69). Online.

Jacob, M., et al. (2010). Leaf litter decomposition in temperate deciduous forest stands with a decreasing fraction of beech (Fagus sylvatica). *Oecologia, 164*(4), 1083–1094.

Jinping, X. (2022). Hold high the great banner of socialism with chinese characteristics and strive in unity to build a modern socialist country in all respects—Report to the 20th national congress of the communist party of China. People's Daily (In Chinese).

Kangzhi, L., & Mingxian, Mo. (2023). On the dialectical and unified relationship between the transformation of China's agricultural modernization and compliance with nature. *Journal of Ethnic Culture, 15*(5), 11–19. (In Chinese).

Kenrick, J., & Lewis, J. (2004). Indigenous peoples' rights and the politics of the term 'indigenous'. *Anthropology today, 20*(2), 4–9.

Kodirekkala, K. R. (2018). Cultural adaptation to climate change among indigenous people of South India. *Climatic Change, 147*(1), 299–312.

Li, B. P., & Xue, D. Y. (2021). The Applicability and Evaluation Index System of the Term "Indigenous and Local Communities" in the Convention on Biological Diversity in China . *Biodiversity, 29*(02), 193–199.

Lantz, T. C., & Turner, N. J. (2003). Traditional phenological knowledge of Aboriginal peoples in British Columbia. *Journal of Ethnobiology, 23*(2), 263–286.

Liu, J. (2011). The Cancun agreements. *Environmental Law Review, 13*(1), 67–89.

Long, C. L., & Zhou, Y. (2001). Indigenous community forest management of Jinuo people's swidden agroecosystems in southwest China. *Biodiversity & Conservation, 10*(5), 753–767.

Lun, Y. (2021). Traditional knowledge as a response to climate change: Connotation, value and approach. *Journal of Eco-Civilization Studies, 2*, 82–98. (In Chinese).

Lun, Y. (2022). Ethnic ecological customary law and biodiversity conservation: Theory, value and approach. *Journal of Ethnic Culture, 14*(5), 15–28. (In Chinese).

Lun, Y. (2024). Theories and methods of climate anthropology. *Journal of Yunnan Normal University (Humanities and Social Sciences Edition), 56*(2), 115–125. (In Chinese).

MacKay, F. (2001). Universal Rights or a Universe unto Itself? Indigenous Peoples' Human Rights and the World Bank's Draft Operational Policy 4.10 on Indigenous Peoples. *Am. U. Int'l L. Rev.*, 527.

Matemilola, S., Fadeyi, O. & Sijuade, T. (2020). Paris agreement. *Encyclopedia of Sustainable Management* 98–105

Miranda, J., Lawson, W., & Escobar, J. (2002). Ethnic minorities. *Mental Health Services Research, 4*(4), 231–237.

Mollicone, D., Freibauer, A., Schulze, E. D., et al. (2007). Elements for the expected mechanisms on 'reduced emissions from deforestation and degradation REDD' under UNFCCC. *Environmental Research Letters, 2*(4), 34–56.

Pareek, A., & Trivedi, P. C. (2011). Cultural values and indigenous knowledge of climate change and disaster prediction in Rajasthan, India. *Indian Journal of Traditional Knowledge, 10*(1), 183–189.

Party Literature Research Center, CPC Central Committee. (2017). *Xi JINPING: Selected excerpts on socialist ecological civilization construction*. Beijing: Central Party Literature Press, 6. (In Chinese).

Pruim, S. (2014). Ethnocide and indigenous peoples: Article 8 of the declaration on the rights of indigenous peoples. *Adel. L. Rev.*, 269.

Ramos-Castillo, A., Castellanos, E. J., & Galloway McLean, K. (2017). Indigenous peoples, local communities and climate change mitigation. *Climatic Change, 140*(1), 1–4.

Rappaport, J. (2004). Between Sovereignty and Culture: Who is an indigenous intellectual in Colombia? *International Review of Social History, 49*(S12), 111–132.

Reid, P., Cormack, D., Paine, S. J. (2019). Colonial histories, racism and health—The experience of Māori and Indigenous peoples. *Public Health*, 119–124.

Riedlinger, D., & Berkes, F. (2001). Contributions of traditional knowledge to understanding climate change in the Canadian Arctic. *Polar Record, 37*(203), 315–328.

Sarivaara, E., Maatta, K., & Uusiautti, S. (2013). Who is indigenous? Definitions of indigeneity. *European Scientific Journal*.

Saugestad, S. (2001). *The inconvenient indigenous: remote area development in Botswana, donor assistance and the first people of the Kalahari*. Nordic Africa Institute.

Secretariat of the Convention on Biological Diversity (SCBD). Glossary of relevant key terms and concepts within the context of article 8(j) and related provisions, 2017. https://www.cbd.int/doc/c/1d3f/4110/c92254 9b825d2fd4e589cf79/wg8j-10-03-en.pdf. (accessed on 2025-08-10).

Shao, H., Hill, R., Xue, D., et al. (2021). In situ conservation of traditional vegetable diversity in Wa homegardens in southwestern Yunnan, China. *Journal of Ethnobiology and Ethnomedicine, 17*(1), 54.

Sinh, B. T., Lebel, L., Tung, N. T. (2009). *Indigenous knowledge and decision making in Vietnam: Living with floods in An Giang province, Mekong Delta, Vietnam.* Indigenous knowledge and disaster risk reduction: From practice to policy.

Sturgeon, J. D., Folsom, A. R., Longstreth, Jr. W. T., et al. (2017). Risk factors for intracerebral hemorrhage in a pooled prospective study. *Stroke, 38*(10), 2718–2725.

Swepston, L. (1990). A new step in the international law on indigenous and tribal peoples: ILO convention no. 169 of 1989[J]. *Okla. City UL Rev, 15*, 677.

Tingshuo, Y. (2004). On the ecologic value of regional knowledge. *Journal of Jishou University (Social Sciences), 3*, 23–19. (In Chinese).

Tingshuo, Y., & Qingzhong, S. (2016). Ecological anthropology and the study of indigenous ecological knowledge: An Interview with Professor Yang Tingshuo. *Journal of China Agricultural University (Social Sciences), 33*(1), 5–23. (In Chinese).

Turner, N. J., & Clifton, H. (2009). It's so different today: Climate change and indigenous lifeways in British Columbia, Canada. *Global Environmental Change, 19*(2), 67–89.

Twikirize, J. M., & Spitzer, H. (2019). Indigenous and innovative social work practice: Evidence from East Africa. *J. Twikirize, & H. Spiitzer, Social Work Practice in Africa Indigenous and Innovative Approaches*, 1–20.

UNFCCC (2002). *The Marrakesh ministerial declaration* (pp 87–95)

UNFCCC (2005). *Tiohtiá: ke declaration* (pp 87–98)

UNFCCC (2006). *The Nairobi work programme: UNFCCC knowledge-to-action hub on adaptation and resilience* (pp 34–37)

UNFCCC (2010). *Report of the conference of the parties on its fifteenth session* (pp 90–101)

UNFCCC (2020). *United Nations climate change annual report 2019*. Bonn: UNFCCC Secretariat.

UNFCCC (2023). *United Nations climate change annual report 2022*. Bonn: UNFCCC Secretariat.

United Nations. Indigenous Peoples. https://www.un.org/en/fight-racism/vulnerable-groups/indigenous-peoples (accessed on 2025-08-10).

Wang, L. Z. (2014). The identification of minorities in China. *APLPJ, 16*, 1.

Wilmer, F. (1993). *The indigenous voice in World Politics Newbury Park[J]*. Cal.: Sage Publications.

Xue, D. Y. (2011). Analysis for the main elements and potential impacts of Nagoya Protocol. *Biodiversity Science, 19*, 113–119.

Xue, D. (2019). Conservation and outlook of traditional knowledge associated with biodiversity. *Biodiversity Science, 27*(7), 705–707. (In Chinese).

Yang, T., & Sun, Q. (2016). Ecological Anthropology and the Study of Indigenous Ecological Knowledge: An Interview with Professor Yang Tingshuo. *Journal of China Agricultural University (Social Sciences), 33*(1), 5–23. (In Chinese).

Yin, L. (2022). Ethnic Ecological Customary Law and Biodiversity Conservation: Theory, Value and Approach. *Journal of Ethnic Culture, 14*(5), 15–28. (In Chinese)

Yin, L., & Xue, D. (2013). National ecological conservation policies and the revitalization of local traditional knowledge: A case study of the grazing withdrawal and grassland restoration program in Hongpo Village, Diqing Prefecture, Yunnan Province. *Journal of Southwest Minzu University (Humanities and Social Sciences), 34*(3), 43–46.

Yingzhan, Z., & Tingting, W. (2024). Multidimensional practices and contemporary values of traditional ecological views of southwest mountain ethnic groups—Taking the Jinuo Ethnic group as an example. *Journal of Ethnic Culture, 16*(5), 105–117. (In Chinese).

12 Reflections on the Participation of Chinese Traditional Knowledge in Global Climate Change Governance Within the Framework of Ecological Civilization

The report of the 20th National Congress of the Communist Party of China proposes to "actively participate in global governance on climate change". General Secretary Xi Jinping emphasized, "We must implement a proactive national strategy to address climate change, promote and guide the establishment of a fair, reasonable, and mutually beneficial global climate governance system" (Xi, 2020a, 2020b). The Paris Agreement, adopted at the 21st Conference of the Parties to the United Nations Framework Convention on Climate Change, established the role of traditional knowledge in climate change adaptation and raised the issue of traditional knowledge, laying the foundation for China's traditional knowledge to participate in global climate change governance. General Secretary Xi Jinping pointed out, "We must follow the goals and principles of the United Nations Framework Convention on Climate Change and its Paris Agreement, and strive to implement the 2030 Agenda for Sustainable Development" (Xi, 2022). As a carrier of ecological wisdom, Chinese traditional knowledge has unique value and significance in biodiversity conservation and climate change adaptation, and has potential value in global climate change governance.

The ecological civilization knowledge system, guided by Xi Jinping's ecological civilization thought, aims to construct an autonomous Chinese knowledge system for building a modern China in which humans and nature live in harmony (An & Lin, 2023). As China's international discourse power and influence in ecological civilization construction continue to grow, "building an ecological civilization knowledge system with Chinese characteristics, style, and temperament requires efforts from multiple dimensions, such as theoretical paradigms, path selection, and indicator systems. This system should not only systematically explain the experiences and laws of China's ecological civilization construction but also strive to extract original results that are globally relevant and universal. It should tell the story of China's ecological civilization construction well and contribute more to the creation of a beautiful China where humans and nature coexist harmoniously and to the joint effort of building a clean and beautiful world" (Zhang, 2022). In the new era, to achieve the innovation and preservation of Chinese traditional knowledge within the ecological civilization knowledge system, we must actively promote the participation of Chinese traditional knowledge in global climate change governance and facilitate China's transition from a participant to a leader in traditional knowledge issues.

1 Traditional Knowledge and Chinese Culture in the Context of Climate Change

Climate scientists, academics, policymakers, and politicians may tend to view traditional

Y. Lun, *Climate Change and Traditional Knowledge*, Sustainable Development Goals Series,
https://doi.org/10.1007/978-3-032-12015-1_12

knowledge primarily as data sources, useful mainly for providing baseline information and strategic insights that are difficult to obtain through scientific methods alone. This perspective reduces traditional knowledge to a mere tool or data set. However, for Chinese, especially the ethnic minorities in China, traditional knowledge encompasses much more than this. It is deeply embedded in their culture and way of life. This knowledge emerges from the interplay between local conditions, natural environments, and social contexts, and is often honed through long-term experience. It is integrated into a broader cultural framework, encompassing information about the location and timing of culturally significant species, hunting and gathering sites, and sacred places. Traditional knowledge frequently holds profound spiritual or cultural significance for its bearers, making it not just culture-specific but also an integral aspect of their identity.

Based on traditional knowledge, Chinese peoples have developed what can be termed as “climate culture”. Climate culture encompasses a profound understanding of the relationship between climate and human society, including worldviews, values, and beliefs held by Chinese peoples. In the context of global climate change, climate culture holds significant value not only for Chinese peoples, but also for the broader global society. It encourages a philosophical reflection on climate change, fosters a deeper understanding of the human-nature relationship, and guides efforts toward sustainable development. Ultimately, climate culture can contribute to the creation of a low-carbon, environmentally friendly society.

In the context of climate change, this chapter also attempts to discuss climate-related culture based on traditional knowledge, and the value of climate culture in the future human response to climate change.

This part explores the intersection of culture and climate change through the lens of traditional knowledge. It reviews literature on cultural aspects of climate research, with a particular focus on traditional Chinese philosophy and the cultural understandings of climate among various Chinese peoples. Building on this review, the chapter analyzes climate-related cultural aspects and introduces the concept of climate culture. It further conducts qualitative research to examine the relationship between climate culture and climate change.

Traditional knowledge offers invaluable insights into how climate influences the environment and affects cultures reliant on local ecological systems. Many Chinese peoples maintain traditional lifestyles that are closely aligned with seasonal and inter-annual patterns of preparation, harvesting, and utilization, as reflected in their practices, narratives, and languages. Thus, climate is not merely a natural phenomenon; it is also a social and cultural construct. This section reviews and discusses climate culture from various perspectives, including definitions, theoretical frameworks, research emphases, and methodologies. Understanding climate culture is fundamental to comprehending the relationship between humans and climate, the impacts of climate change on human societies, and effective adaptation strategies. Currently, the goal of research in climate culture is to establish a robust epistemology and methodology that facilitate a deeper understanding of, and responses to, climate environments, as well as strategies for coping with climate change and associated disasters.

The phenomenon of climate change is increasingly severe, with frequent climate-related disasters posing significant challenges to the sustainable development of human societies. While we have implemented various protective policies and management projects to mitigate climate change, it is crucial to pay close attention to climate culture, which can play an essential ideological role in helping people adapt to these changes. In recent years, researchers across diverse fields, particularly in social sciences and ecological anthropology, have begun to focus on the intersection of climate change and cultural studies, leading to the emergence of a specialized branch known as climate anthropology. Given the nature of this research, studies on climate change must be grounded in an understanding of climate culture, which is a core component of climate anthropology.

The study of climate culture can be approached from both theoretical and methodological perspectives. This section will review historical traditional knowledge regarding climate change and examine relevant research conducted by national and international scholars on the relationship between climate change and culture. Based on this review, the discussion will further elucidate the concept of climate culture through its definitions, theoretical frameworks, research emphases, and methodologies. The aim is to highlight the significance of climate culture in enabling human societies to adapt to climate change and to enhance the academic discourse surrounding climate culture from both theoretical and methodological standpoints.

1.1 Traditional Knowledge, Culture, and Climate

In China, ethnic minorities possess a rich and profound traditional culture that includes significant insights related to climate change. The livelihoods and production practices of these communities are directly influenced by climatic conditions, leading to the development of distinct climate cultures as they adapt to environmental changes. Given the diversity of locations and ecosystems, different ethnic minorities offer varied interpretations and expressions of climate culture, reflecting their unique relationships with the environment.

The Han people possess a rich and distinctive climate culture that shapes their farming practices and livelihoods, enabling them to adapt to varying climatic changes and disasters. This traditional climate culture embodies the ideology of the climate and reflects the Han perspective on the relationship between humanity and nature. Central to this worldview is the concept of "nature" (Tian), which emphasizes that humans are an integral part of the natural world. In Han traditional culture, "nature" (Tian) encompasses two significant meanings: it refers to both "natural phenomena" (Tian Xiang) and "natural law" (Tian Dao).

Early Han scholars and thinkers understood "natural phenomena" as the natural manifestations and objective realities of the world. In Xunzi's "Nature Theory", the concept of "natural phenomena" encompasses the regular cycles and changes observed in the sun, moon, stars, seasons, day and night, as well as wind and rain—elements that collectively reflect climate change. This theory posits that these natural phenomena exist independently of human influence and cannot be controlled. Building on this notion, Han scholars developed the concept of "natural law" (Tian Dao). In Han language, "Dao" signifies a path and is conceptualized as methods, rules, laws, and virtues. Thus, "natural law" embodies people's subjective understanding and cultural interpretation of nature.

In "Mister Lv's Spring and Autumn Annals", specifically in the chapter titled "Shunmin", it is recorded that although Leader Tang defeated the ruler of the Xia Dynasty, Heaven was not pleased and punished the people with severe drought. To alleviate the drought, Tang sacrificed himself to pray for rain from Heaven. This narrative underscores the belief that nature possesses power and will govern climate change. Additionally, the sections on Chunqiu Theory assert that the transitions between the four seasons are the results of changes in natural law. Thus, climate change is intricately linked to this natural law (Niu et al. 2011). The interpretation of nature (Tian) reflects the Han people's understanding of climate change. The concepts of natural phenomena and natural law effectively illustrate the progression of Han people's awareness of climate, evolving from a purely natural understanding to a more cultural interpretation.

Except for the Han people, ethnic minorities in China also have developed rich and diverse interpretations, traditional knowledge, culture and philosophy of climate, with their production and livelihoods closely linked to climate change. Such changes directly influence livelihood strategies and have far-reaching impacts on politics, economy, society, and culture. For the Tibetan people, climate change is perceived not only as a natural phenomenon but also as having significant social implications. Their understanding of climate is rooted in a unique knowledge and religious system that emphasizes reverence for deities

(Toni and Poul, 1997). For instance, Tibetans residing in the northwest of Yunnan Province base their indigenous climate knowledge on the worship of God Mountain, viewing the climate as a manifestation of the spirit and power of this sacred site. Local people meticulously observe climatic conditions through visual, olfactory, and auditory cues, engaging in ritual ceremonies to communicate with the God of the mountain (Yin and Xue, 2013). Similarly, the Kazak people derive their climate knowledge from their ancestral practice of Shamanism. Their forebears could predict climate changes by closely observing natural phenomena, and the twenty-four solar terms in Kazak culture provide a rich record of climatic variations.

The narrative of the twenty-four solar terms illustrates how climate disasters can arise from humanity's underestimation of climate change (Fang, 2006). For the Bai people, renowned for their agricultural practices, rainfall and the twenty-four solar terms are critical elements for their production and livelihood. Their religious beliefs and indigenous knowledge vividly reflect their understanding of climate, particularly evident in their veneration of the Dragon King. The Bai believe that the Dragon King governs sunshine, wind, and rain, and thus, to comprehend climatic phenomena, people must seek his protection through prayer (Yang, 2009). Similarly, for the Mongolian people, the religion of Mongke-Tangri embodies their climate perceptions. They attribute control over various natural elements—such as the sun, moon, stars, wind, thunder, fog, lightning, and clouds—to different Mongke-Tangri gods. For instance, lightning is viewed as a manifestation of the MongkeTangri god's anger, while thunder during storms signifies conflict between HuodaiMelguet (another MongkeTangri deity) and adversaries. GulbanBorron, yet another MongkeTangrigod, is believed to govern cloudy weather, snowfall, and winds (Nan, 2004).

1.2 Culture and Climate Change

Historically, scholars have examined the relationship between climate and culture, with early investigations by figures such as Hippocrates in ancient Greece exploring how climate change influenced society and human activities (Harris, 1968). However, it was not until the early 1990s that researchers began to analyze climate from an anthropological perspective. In the context of carbon dioxide emission reduction and global warming, Western scholars from both social and natural sciences have engaged in climate research since the 1980s. In the paper "Social Science Research and Climate Change", (Richard and William, 1983) posited that culture could serve as a framework to guide societal adaptation to climate change. At the beginning of this century, interest in climate culture surged among Western scholars, reflecting a growing recognition of its importance in addressing climate change.

In the book Weather, Climate, Culture, (Sarah and Benjamin, 2003) thoroughly examine the interrelationships among weather, climate, and climate change from an anthropological perspective, drawing on 13 field-based case studies from around the world. The authors underscore the significance of this research in understanding these dynamics. Over the past decade, Western anthropologists and ethnic ecologists have conducted various studies based on specific case contexts. For instance, Neeraj Vedwan's research in the northwest Himalayas of India analyzes the connections among culture, weather, and environment, emphasizing indigenous knowledge of climate change (Vedwan, 2006). Similarly, Jennifer Couzin investigates indigenous climate knowledge in the Arctic region, exploring how local communities adapt to climate change (Couzin, 2007). (Anja Byg and Jan Salick, 2009) also focus on Tibetan indigenous knowledge of climate in the Himalayas, specifically in the northwest region of Yunnan Province, China.

In 2009, Susan A. Crate and Mark Nuttall published Anthropology and Climate Change: From Encounters to Actions. The first chapter, titled "Climate and Culture", features four articles that examine the relationship between culture and climate change through the lenses of archaeology, cultural anthropology, and disaster analysis. This chapter also discusses how to conduct climate change research from a theoretical perspective.

The second and third chapters utilize 20 case studies to analyze the interplay between climate change and culture across various ethnic regions (Crate and Nuttall, 2009). Additionally, in the third edition of Sacred Ecology published in 2012, (Fikret Berkes, 2012) explores the connection between climate change and indigenous knowledge using a community-based methodology.

In China, research on climate and culture is conducted concurrently by scholars in both natural and social sciences. From the perspective of natural science, meteorology examines the climate and its relationship to traditional culture through philosophical and cultural lenses. For instance, Luo Shengzhou explored the definition of climate culture and utilized a case study in Qinghai to illustrate the significance and purpose of constructing meteorological culture (Luo, 2006). In Introduction to the Philosophy of Meteorology, Niu Zhoubao and colleagues discussed ancient Chinese philosophers' understandings of meteorology (Niu and Liu, 2011). Additionally, in a case study from Shanxi, Wang Wanruianalyzed traditional meteorological culture and its origins (Wang, 2012).

Conversely, within the realm of social sciences, anthropologists investigate the interplay between climate and culture through the lenses of "indigenous knowledge" and "climate anthropology". In the book Culture and Environment in China, edited by Yin Shaoting, the authors highlight the relationship between indigenous knowledge and climate in the chapter titled "The Climate and Indigenous Knowledge". They utilize twelve case studies involving ethnic groups such as the Tujia, Dong, Zhuang, Bai, and Dai to illustrate how these communities employ indigenous knowledge to adapt to climate change (Yin, 2010). From a climate-anthropological perspective, Yin Lun analyzed Tibetan people's understanding of climate and their application of traditional culture for adaptation, focusing on the case study of Tibetan God Mountain Religion in Deqin County of Shangri-La, Yunnan (Yin and Xue, 2013). Additionally, (Liu, et al., 2013) explored the effects of climate change on indigenous knowledge and local adaptation strategies.

In summary, an increasing number of national and international scholars from both natural and social sciences are focusing on climate culture. However, it is noteworthy that a clear definition of climate culture remains elusive, and existing research tends to be confined to single-subject studies rather than embracing interdisciplinary approaches. Therefore, there is a pressing need for a systematic analysis of the definitions, theories, and methodologies pertaining to climate culture.

1.3 Climate Culture

Climate is a key element influencing human evolution and social development. Drastic climate changes can significantly impact the rise, development, decline, and even destruction of human civilizations. As climate change intensifies, it inevitably affects global environments, international politics, world economies, and international trade. The issue of climate change transcends mere environmental concern; it represents a critical challenge for sustainable development.

Relying solely on single-subject research, technological advancements, policy-making, or treaty agreements is insufficient. Therefore, it is essential to understand climate from a cultural perspective and construct a relevant climate culture. This concept encompasses the ways in which societies perceive, utilize, and govern climate throughout history. Climate culture includes not only people's cognitive and religious interpretations of climate but also reflects traditional knowledge and techniques for climate governance. It spans multiple domains, including religion, literature, arts, institutions, law, and behavior.

Climate culture represents the interactive relationship between humans and the climate, manifesting across various natural environments, historical contexts, social formations, countries, regions, and ethnic groups. Understanding climate culture can be approached from two levels.

Firstly, the climate influences social development as well as people's production and livelihoods. Observations and analyses of these impacts shape our cognition, beliefs, and religious perspectives regarding climate.

Secondly, through the processes of utilizing and governing the climate, communities develop indigenous knowledge, scientific technologies, and institutional frameworks. Climate culture, therefore, embodies humanity's understanding and reflections on the climate and its relationship with human existence. It constitutes a vital component of human culture and serves as a fundamental basis for adapting to climate change.

1.3.1 Levels of Climate Culture

Analysis of Climate Culture from Four Levels.

I. Ideological Level

Human interaction with the climate fosters cognitive development, leading to the formation of values and ways of thinking. This climate culture is reflected in various domains, including religion, literature, and the arts, and it serves as a guiding framework for practical actions in daily life.

II. Behavioral Level

While climate conditions undoubtedly shape human living environments, individuals do not merely adapt passively. Instead, they actively create indigenous knowledge and technological innovations to manage and enhance their climatic circumstances, demonstrating agency in response to environmental challenges.

III. Institutional Level

In the process of climate governance, societies develop customs, norms, regulations, and laws. Establishing effective institutions is essential for standardizing and regulating behaviors related to climate interactions, thereby fostering sustainable environmental practices.

IV. Material Level

Through the integration of ideological guidance, practical implementation, and institutional oversight, communities achieve tangible outcomes that contribute to their cultural heritage related to the climate. This material aspect encompasses the physical manifestations of climate culture, including sustainable practices, technologies, and artifacts.

1.3.2 Characteristics of Climate Culture

The climate culture encompasses human generality, national character, and various natural, cultural, global, regional, historical, and epochal characteristics.

I. Human Generality and Ethnic Character

The climate culture is a construct of humanity, characterized by its inherent human generality. Different traditional cultures worldwide share similar traits in climate culture, particularly in terms of common symbols and interpretations of the climate. Simultaneously, as different ethnic groups inhabit varying climatic conditions globally, their observations, cognitions, and interpretations of the climate exhibit significant diversity. Thus, the climate culture demonstrates a pronounced national character.

II. Natural and Social Character

The climate culture reflects the interactive relationship between climate (a natural phenomenon) and humanity (a social construct). Unlike traditional social sciences, the study of climate culture is inherently interdisciplinary, bridging natural and social sciences. Diverse natural environments provide the foundational basis for the climate culture, while various societies cultivate distinct cultures and belief systems that naturally influence their climate culture.

III. Global and Regional Character

Climate is a global phenomenon that profoundly affects human life. Given that different geographical environments lead to varying climatic conditions, climate culture exhibits diverse regional characteristics, originating from different settings such as plains, plateaus, islands, oases, and deserts. Furthermore, with rapid social development, urban climates differ markedly from rural climates, resulting in distinct climate cultures between the two.

IV. Epochal and Continuity Character

Throughout history, human cognition of climate has evolved alongside natural, social, economic, and cultural changes. Consequently, the climate culture possesses both historic and epochal characteristics. While the climate culture may differ across various historical periods, it also maintains a degree of cultural continuity. Presently, modern climate culture represents a comprehensive integration of historical climate cultures.

In summary, the climate culture possesses a distinct theoretical structure, which can be elaborated upon through ideological, behavioral, institutional, and material dimensions. It embodies human generality and national character while also reflecting natural, cultural, global, regional, historical, and epochal characteristics. Research on climate culture should embrace diversity and transcend disciplinary boundaries.

1.3.3 Research Fields of Climate Culture

Based on its theoretical structure, research of the climate culture covers five fields as follows:

I. Immaterial and material climate culture

Referring to ancient civilizations worldwide, there exists a rich tapestry of traditional cultures related to the climate. This climate culture exhibits both immaterial and material characteristics. The immaterial aspects are well-reflected in philosophy, indigenous knowledge, religion, literature, and art across various historical periods. Conversely, the material dimension is represented through engineering constructions, instruments, facilities, and equipment from ancient times. Researching the climate from both material and immaterial perspectives can enhance our comprehensive understanding of climatic phenomena, thereby enabling us to address climate-related challenges and manage climatic influences effectively in the future.

II. Cognition of the climate from cultural level

People's cognition of the climate varies based on their cultural contexts. For instance, in ancient societies, individuals worshiped climatic forces, establishing their relationship with the climate through religious practices and ritual ceremonies. In contrast, contemporary society seeks to adapt to and govern the climate through the lens of modern industrial culture. By integrating ancient and contemporary climate cultures, we can foster a more harmonious relationship between humanity and the climate.

III. Climate's effects on culture

The climate significantly influences cultural types and their developmental trajectories. Human beings create various types of cultures in response to differing climatic environments. We observe rich and diverse cultures originating from distinct climate zones, such as the frigid polar regions, plateaus, dry temperate zones, moist tropical zones, and lake regions. Simultaneously, climate change, exacerbated by hostile climatic conditions, adversely affects cultural development. For instance, the disappearance of the ancient Mayan civilization in Central America was directly linked to prolonged drought (Larry and Gerald, 2005). The instability of the climatic environment also played a crucial role in the decline of the ancient Andes culture in South America (Michael et al., 1997). Furthermore, the decline and eventual collapse of the ancient Loulan culture in Xinjiang, China, were closely associated with climatic disasters (Gao et al. 2004). Inevitably, human society and its development in ancient China were profoundly influenced by climatic factors (Ge et al. 2010). Contemporary research on these effects can inform our efforts to mitigate and adapt to climate-related disasters, thereby promoting sustainable development in our society moving forward.

IV. Traditional knowledge of the climate

Traditional climate culture is rooted in people's observations of local climatic and phenological phenomena. This culture aids individuals in predicting and adapting to climate changes. The traditional culture surrounding climatic phenomena encompasses human observations of weather, nature, and living environments, as well as people's understanding of the relationship between climate and their production and livelihoods. Conversely, the traditional culture related to phenological phenomena pertains to indigenous knowledge derived from local observations of crop growth, livestock management, and environmental changes in farmland and grassland. Through daily practices, individuals discern natural laws governing crop cultivation, livestock breeding, and their interactive relationships with climate change. This culture of phenological phenomena reflects a long accumulative process of indigenous knowledge. Research into this indigenous climate knowledge holds significant importance for ethnic communities, as it helps them comprehend the effects and challenges posed by climate change and identify appropriate adaptation strategies.

V. Climate culture and climate change

Currently, climate change significantly impacts both ecological environments and human societies, emerging as a pressing international issue that encompasses scientific, economic, political, cultural, and other fields. It is not merely a meteorological problem; rather, it is a critical topic concerning the sustainable development of human society. Research into climate culture effectively illuminates the intricate relationship between culture and ecology. Presently, humanity faces challenges posed by climate change and the social risks engendered by industrial civilization. Therefore, it is imperative to focus on climate culture, reflect on the relationship between humanity and the natural environment, and explore appropriate developmental pathways from a cultural perspective.

There are two significant methodologies for conducting research on climate culture: positivist methodology and humanistic methodology. Research in this area is inherently interdisciplinary. On one hand, it must be grounded in observation and objective facts to facilitate quantitative analysis. On the other hand, it involves the cultural interpretation of climate, necessitating empirical analysis. Both positivist and humanistic methodologies underscore the importance of comprehensive research and binary thinking.

Specific research methods encompass fieldwork, experimental studies, literature reviews, and action research. Fieldwork involves researchers visiting villages or local communities to conduct surveys through questionnaires, structured interviews, semi-structured interviews, and participatory observation, thereby collecting and analyzing data on climate culture. Experimental studies entail analyzing climate culture by modeling data from both climatic and cultural perspectives to reflect the interactive relationships among correlated variables. Literature reviews aim to analyze and synthesize ancient texts, documents, and datasets. Action research refers to practices aimed at recovering, inheriting, and developing traditional climate culture at the local level based on community-based surveys. Climate culture can be preserved not only through academic papers, documentary records, research institutes, or museums but also through active inheritance and development via action research.

1.3.4 Research Methodology of Climate Culture

Two significant methodologies underpin the research on climate culture: positivism and humanism. This research is inherently interdisciplinary. On one hand, it must be grounded in observation and objective facts for quantitative analysis. On the other hand, it involves cultural interpretations of climate, which require empirical analysis. Both methodologies emphasize comprehensive research and binary thinking.

Specific research methods encompass fieldwork, experimental studies, literature reviews, and action research. Fieldwork involves researchers visiting villages or local communities to conduct surveys through questionnaires, structured interviews, semi-structured interviews, and participatory observation, as well as to collect and analyze data on climate culture. Experimental studies entail modeling data from both climate and culture to elucidate the interactive relationships among correlated variables. Literature reviews focus on analyzing and summarizing ancient texts, documents, and data. Action research pertains to the proactive recovery, inheritance, and development of traditional climate culture at the local level, informed by community-based surveys. Climate culture can be preserved not only through academic papers, documentary records, research institutes, or museums, but also through active inheritance and development facilitated by action research.

From the above, it is evident that climate culture reflects the interactive relationship between human beings and the climate, existing across different historical periods and human societies. Research on climate culture serves not only as a fundamental basis for understanding and protecting the climatic environment but also as a core element in managing climate change. Climate culture can be interpreted as a tradition to be inherited and developed by individuals, a discipline to be constructed by researchers, or a framework to govern climate change. Emphasizing the research value of climate culture and its practical applications is both valuable and meaningful.

Currently, research on climate culture remains in its early stages. However, its potential value and significance are already apparent. It is crucial to advance theoretical and methodological research on climate culture while conducting fieldwork and action research to strengthen its academic foundation. In the future, climate culture is expected to evolve into a comprehensive cultural system encompassing ecological, social, political, and economic dimensions. This development will enhance interdisciplinary communication and cooperation.

Climate culture embodies multiple values, rich knowledge systems, constructive behavioral patterns, and robust institutions. It can aid in addressing climate change and disasters, fostering harmonious relationships with the environment, thereby underscoring the notion that humanity is an integral part of nature.

In the face of climate change and extreme climate disasters, the most direct, effective, and cost-efficient adaptation methods often draw from the traditional knowledge and culture of Chinese peoples. Similar to their ecosystems and natural resources, these cultural practices can enhance resilience to climate-related challenges.

2 The Innovation and Preservation of Traditional Knowledge Under the Ecological Civilization Knowledge System

General Secretary Xi Jinping emphasized, "Accelerating the construction of a philosophy and social science with Chinese characteristics ultimately involves building an autonomous Chinese knowledge system". To construct an autonomous Chinese knowledge system and build a modern society where humans and nature coexist harmoniously, an important task is to adhere to Xi Jinping's ecological civilization thought as guidance and to construct an autonomous ecological civilization knowledge system in China (Zhang and Wang 2024a). The ecological civilization knowledge system, centered on Xi Jinping's ecological civilization thought, is based on a framework that couples the holistic concept of the community of life between humans and nature with the three dimensions of organization, space, and time (Yang, 2023). Through long-term production and life practices, and based on the diverse natural environments and ecosystems, the Chinese nation has nurtured a rich and colorful array of excellent traditional knowledge. Traditional knowledge refers to the knowledge, innovations, and practices created, inherited, and developed by various ethnic groups and local communities over long periods of traditional production and life practices, which are beneficial for biodiversity conservation and sustainable use.[1]

[1] Technical Provisions on Classification, Investigation and Cataloguing of Traditional Knowledge Related to Biodiversity (Trial Implementation), Ministry of Ecology and Environment of the People's Republic of China https://www.mee.gov.cn/ywgz/fgbz/bz/bzwb/stzl/201406/t20140606_276594.shtml.

Chinese traditional ecological wisdom is the ethnic soil and cultural gene of Xi Jinping's ecological civilization thought (Huang, 2018). Traditional knowledge is the core part of Chinese traditional ecological wisdom, as it is closely related to the ecological environment and biodiversity, based on unique terrains, diverse ecosystems, and abundant natural resources. Therefore, it is a natural cornerstone for ecological civilization construction and an important source of the ecological civilization knowledge system. Conversely, the ecological civilization knowledge system also provides direction for the preservation and innovation of traditional knowledge. Guided by Xi Jinping's ecological civilization thought, we must drive the creative transformation and innovative development of traditional knowledge in the new era, explore its role and value in the modernization of harmonious coexistence between humans and nature, and ensure that it plays a part in both ecological civilization construction and China's participation in global climate change governance, ultimately becoming an important and beneficial component of the ecological civilization knowledge system.

2.1 Traditional Knowledge as an Important Source of the Ecological Civilization Knowledge System

The contemporary environmental protection movement originated in the West, with various environmentalist ideologies emerging in European and American countries, which also influenced China's environmental protection practices. While these movements have had a positive impact, there has also been a phenomenon of always referring to Western environmentalism, as if in this field, we can only follow the various environmental ideologies and theories from the West. General Secretary Xi Jinping pointed out, "The primary task of strengthening cultural confidence is to base ourselves on the great historical and contemporary practices of the Chinese nation, summarize China's experiences with Chinese principles, and elevate these experiences into Chinese theories, without blindly following various dogmas or copying foreign theories" (Xi, 2024).[2] Traditional knowledge is the result of long-term practices by the Chinese nation, developed based on local ecological environments. It contains rich experience in natural resource management, ecosystem maintenance, biodiversity conservation, and sustainable use. This knowledge is a major source of the ecological civilization knowledge system, and it is precisely the core value of traditional knowledge.

Constructing an autonomous ecological civilization knowledge system in China must adhere to a global perspective, drawing from the achievements of all human civilizations with an open and inclusive mindset (Yang, 2023). Traditional knowledge can contribute to the construction of the ecological civilization knowledge system in three aspects: inclusiveness, unity, and innovation. From the perspective of inclusiveness, traditional knowledge originated from the ongoing natural process in which various ethnic groups adapted to and managed the ecological environment, as well as from the cultural exchanges, interactions, and integration between these groups. In both of these processes, ethnic groups constantly learned from, absorbed, and drew on each other's traditional knowledge, eventually forming a rich body of traditional knowledge. Therefore, the inclusiveness of traditional knowledge not only facilitates harmonious coexistence between various ethnic groups and nature, but also fosters harmonious relations among the ethnic groups themselves. From the perspective of unity, although traditional knowledge appears diverse on the surface, it essentially has a strong sense of unity. This unity has a natural attribute, as different ethnic groups live in similar or identical ecosystems and use similar or identical natural resources in their daily production and life. Unity also has a social attribute, because ethnic groups living in the same or similar ecosystems have developed common customs, beliefs, and worldviews. From the perspective of innovation, traditional knowledge is the accumulated cultural experi-

[2] Xi Jinping's Important Speeches (2023 Edition), Beijing: People's Publishing House, 2024.

ence of various ethnic groups over generations regarding the ecological environment and natural resources. It is a dynamic process of historical progress and innovation. Innovation provides ecological vitality and cultural dynamism to traditional knowledge. Over the long course of history, the Chinese nation has always adapted to the ever-changing ecological and social environments with continuously innovating traditional knowledge. The innovation of traditional knowledge is the foundation for the sustainable development of the Chinese nation. In the new era, this innovative capability will be further enhanced and developed. Therefore, outstanding traditional knowledge is an important source of the ecological civilization knowledge system and serves as the solid foundation for the construction of ecological civilization.

2.2 The Ecological Civilization Knowledge System Leading the Innovation and Preservation of Traditional Knowledge

General Secretary Xi Jinping emphasized, "Only by adhering to preservation can we avoid losing ourselves and our direction, and only through innovation can we grasp and lead the times". This assertion of "persisting in both preservation and innovation" is the fundamental principle by which the ecological civilization knowledge system guides the development of traditional knowledge.

First, the ecological civilization knowledge system defines the basic direction for the innovation and preservation of traditional knowledge. On one hand, the system must adhere to the fundamental role of Marxism in guiding ideology. First and foremost, there is the principle of preservation, where General Secretary Xi Jinping emphasized, "Preservation means not deviating from Marxism and socialism" (Xi, 2020a, 2020b). Therefore, traditional knowledge must stay true to the "correct" principles of Marxism. Second, there is the principle of innovation. General Secretary Xi Jinping stated, "Ideology determines the direction and path of cultural advancement" (Xi, 2017a, 2017b). In the process of innovating traditional knowledge, it must always be guided by Marxism, especially by the Marxist view of ecology. Within this framework, the innovation and creativity of traditional knowledge can be stimulated and elevated. On the other hand, we must apply the dialectical method of Marxism and its principle of sublation. Regarding historical and cultural matters, General Secretary Xi Jinping emphasized, "We must uphold the principle of using the old for the new, push forward with new ideas, and approach them with discernment and selectivity". Traditional knowledge is not a static body of knowledge, but rather a knowledge system that has evolved and been passed down over generations. As times change, outdated and historically limited aspects of this system are gradually eliminated, while new knowledge that meets the demands of the era continues to emerge. Therefore, the innovation and preservation of traditional knowledge requires the application of Marxist dialectics: extracting its essence and discarding its dross. This dynamic approach ensures that traditional knowledge remains relevant, uniting history with the present, as well as ecology with culture.

Second, the ecological civilization knowledge system points out the fundamental path for the innovation and preservation of traditional knowledge. General Secretary Xi Jinping emphasized, "Only by remembering history can we open up the future, and only by being good at inheritance can we be good at innovation" (Xi, 2017a, 2017b). To inherit and promote traditional knowledge, it is essential to adhere to creative transformation and innovative development, elevating traditional knowledge to new heights within the ecological civilization paradigm (Zhuang, 2024). Creative transformation involves endowing traditional knowledge with new contemporary connotations and modern forms of expression in line with the requirements of the ecological civilization knowledge system. For example, the protective value of traditional knowledge such as the Hani sacred mountain forests, Yi dense forests, Tibetan sacred mountains, and Dai dragon mountains should be fully explored. Simultaneously, the meanings and forms of expression can be aligned with the concept of "traditional knowledge" in international agreements such as the Convention on Biological Diversity (CBD) and the United Nations Framework Convention on Climate Change (UNFCCC).

Innovative development refers to supplementing, expanding, and refining the connotations of traditional knowledge in the new era, thereby enhancing its influence. For instance, the Hani rice terraces' four-element system of rivers–terraces–villages–forests, the Naxi Dongba view that "humans and nature are brothers", the Tibetan reverence for sacred mountains, and the Dai belief that plants and animals are companions. When incorporated into the ecological civilization knowledge system, the connotations of these traditional knowledge systems will be expanded, enriched, and elevated, gaining new modern meanings of harmonious coexistence between humans and nature. Thus, the ecological civilization knowledge system provides the correct path for continuing and advancing the preservation and innovation of traditional knowledge.

2.3 Traditional Knowledge Enriches the Ecological Civilization Knowledge System

Traditional knowledge is the crystallization of the wisdom of all ethnic groups, nurturing a wealth of ecological wisdom and cultural traditions related to the relationship between humans and nature. Over the course of long-term natural and social co-evolution, it has shaped worldviews and value systems regarding nature, society, and the community of life, providing the Chinese nation with a powerful cultural source and spiritual support for its continuous existence and sustainable development. Traditional knowledge contains important insights into addressing contemporary ecological crises, such as biodiversity conservation and climate change, offering valuable guidance for global climate governance. In the new era, the integration of the basic principles of Marxism with the practice of China's ecological civilization construction and with the nation's excellent traditional ecological culture further enriches Xi Jinping's thought on ecological civilization (Huang, 2023). Similarly, for traditional knowledge to serve as a foundation or support for ecological civilization, it must undergo creative transformation and innovative development, which requires a deep "chemical reaction" with the Marxist ecological outlook. On the one hand, it should elevate the core contents of traditional knowledge through Marxist ecological perspectives in philosophy and thought, guiding its dynamic cultural process in the right direction. On the other hand, the ecological wisdom and cultural spirit inherent in traditional knowledge should provide the ecological-cultural soil for the practical application of the Marxist ecological outlook in China, thereby enriching the "cultural life of Marxism". On this basis of mutual integration, traditional knowledge can become a vital part of the ecological civilization knowledge system and further enrich it.

Traditional knowledge is a living body of knowledge, playing an important role in the spiritual and cultural inheritance of various ethnic groups, the management of natural resources, and ecological environmental protection. Therefore, it is crucial to clarify the historical origins, development context, and future directions of traditional knowledge, uncover its unique values and characteristics, and enrich the ecological civilization knowledge system. At the national level, traditional knowledge, as part of the ecological civilization knowledge system, should actively serve the construction of ecological civilization, reflecting its unique value and role. For example, elements such as techniques, systems, customs, and practices in traditional knowledge can be transformed and developed to support ecological environmental protection, ensure ecological and biological security, and promote ecological civilization education. At the international level, biodiversity conservation and climate change adaptation are major global issues related to the future of humanity. The creative transformation and innovative development of traditional knowledge can play a significant role in addressing these issues. For instance, the fair and equitable sharing of genetic resources and related traditional knowledge has become one of the key concerns under the Convention on Biological Diversity. Traditional knowledge is regarded as a practical science for biodiversity conservation, and both the United Nations Framework Convention on Climate Change (UNFCCC) and the Paris Agreement explicitly recognize traditional knowledge as part of the knowledge system for human adaptation to climate change, alongside scientific and technological knowledge. In summary, traditional knowledge enriches the ecological civilization knowledge system, and by promoting it, China can actively engage with traditional knowledge issues in the Convention on Biological Diversity and the UNFCCC, offering concepts, propositions, and solutions grounded in Chinese wisdom. Therefore, traditional knowledge serves as both a cultural gene and a precious resource for constructing the ecological civilization knowledge system.

3 Constructing a Discourse System for Chinese Traditional Knowledge to Participate in Global Climate Change Governance

General Secretary Xi Jinping emphasized the need for China to actively participate in global ecological and environmental governance to enhance its discourse power and influence (Xi, 2020a, 2020b). The Paris Agreement recognizes the value of diverse knowledge systems, including traditional knowledge, in climate change adaptation and establishes mechanisms for their participation (Sourgens, 2018). Within this framework, Chinese traditional knowledge under the ecological civilization knowledge system can contribute to global climate governance by providing strategies and pathways rooted in ecological wisdom, thereby strengthening China's voice in this arena. Although traditional knowledge is already embedded in international discourse through United Nations conventions, China's participation and contribution remain limited (Cheng et al., 2012). Thus, it is imperative to accelerate the construction of an international discourse system for Chinese traditional knowledge to expand its influence and articulate Chinese perspectives in global climate governance.

3.1 Aiming for Joint Global Ecological Civilization Construction

Xi Jinping's ecological civilization thought represents a significant theoretical innovation and ideological transformation in the history of human civilization, opening new horizons for sustainable development theory and practice. It presents a global initiative for ecological civilization construction, proposing that "China calls for all parties to join hands in tackling the challenge of climate change, working together to protect our common Earth, and upholding multilateral consensus. The United Nations Framework Convention on Climate Change (UNFCCC) and its Paris Agreement should serve as the fundamental legal basis for international cooperation in addressing climate change" (Dawson and Spannagle 2008). Therefore, building a discourse system for Chinese traditional knowledge in global climate change governance must actively serve this initiative. Guided by Xi Jinping's ecological civilization thought, it should adhere to the global win–win vision of ecological civilization and the concept of a shared human destiny. Constructing this discourse system involves thoroughly explaining the goals and significance of jointly pursuing global ecological civilization, clarifying the unique advantages and differences of China's ecological civilization concept compared to Western environmentalism, and elucidating the value and role of this pursuit in global climate governance. Furthermore, it should highlight its profound impact on promoting harmonious coexistence between humans and nature and constructing a global community of life on Earth, thereby providing strong support for the collective advancement of global ecological civilization (Fig. 1).

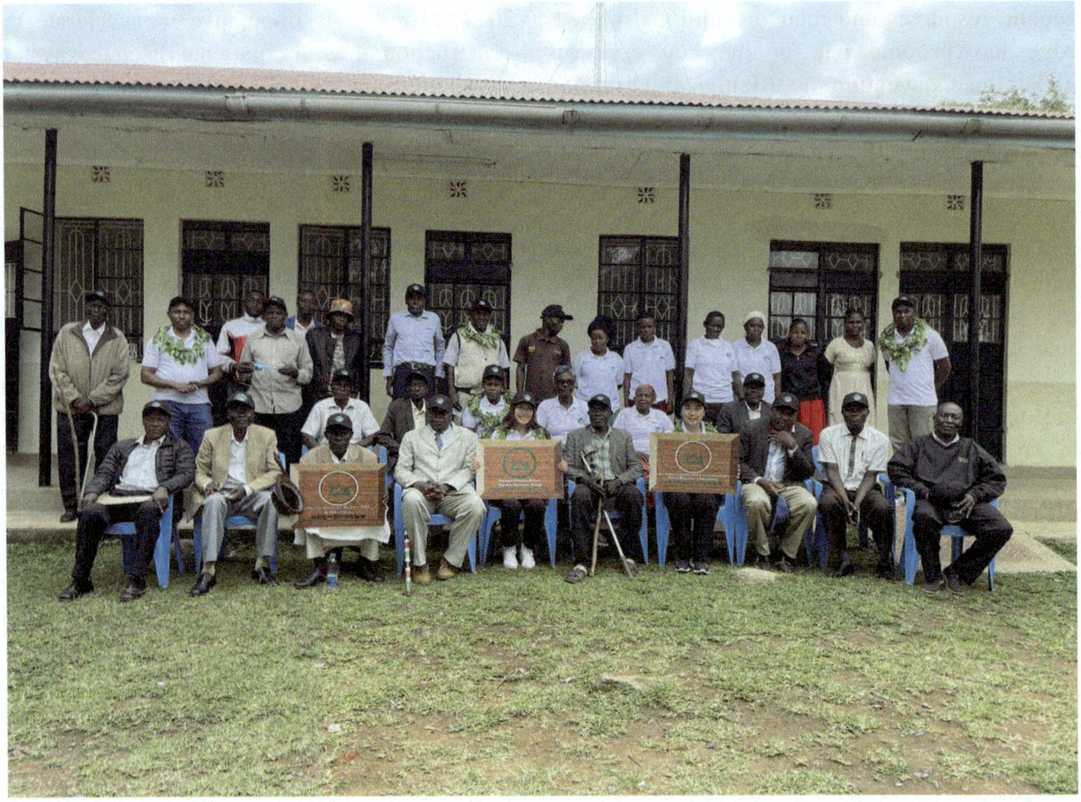

Fig. 1 Ecological civilization rural practice laboratory of the Ngany area in Kenya

3.2 Rooted in the Practical Applications of Chinese Traditional Knowledge in Climate Change Response

China's five-thousand-year-old civilization has cultivated profound ecological wisdom, with concepts such as "the unity of man and nature" (tian ren he yi) and "following the way of nature" (dao fa zi ran) forming the philosophical basis of traditional knowledge. As a practical science, this knowledge has vivid applications in both mitigation and adaptation to climate change. In terms of mitigation, forest-related traditional knowledge helps preserve ecosystem diversity, enhances carbon sequestration, and reduces greenhouse gas emissions, contributing to the achievement of China's "dual carbon" goals. In terms of adaptation, agricultural biodiversity knowledge enables the selection of suitable crops for changing conditions of soil, temperature, slope, orientation, and sunlight, while traditional techniques such as intercropping, relay cropping, and mixed cropping reduce climate risks in agriculture (Yin, 2024). Embedding these practices and the ecological wisdom behind them into China's discourse system not only demonstrates the dynamic role of traditional knowledge in addressing climate change but also provides compelling Chinese narratives for ecological civilization construction and global climate governance (Fig. 2).

4 Clarifying the Goals of Chinese Traditional Knowledge in Global Climate Change Governance

China's National Biodiversity Conservation Strategy and Action Plan proposes the urgent protection and transmission of relevant traditional knowledge, calling for surveys and cata-

Fig. 2 Ecological civilization rural practice laboratory in East Himalaya, Yunnan, China (photoed by Zhang Xiaohan)

loguing of such knowledge, and establishes the "Priority Action on the Protection and Inheritance of Traditional Knowledge" to enhance public cultural confidence in biodiversity-related traditional knowledge (Huang & Yin, 2024). The Ministry of Ecology and Environment issued the Technical Provisions on the Classification, Investigation, and Cataloguing of Biodiversity-Related Traditional Knowledge (Trial), which defines traditional knowledge as "knowledge, innovations, and practices created, inherited, and developed by various ethnic groups and local communities in the course of long-term traditional production and life practices, which are beneficial to biodiversity conservation and sustainable use" (Xue, 2021). On this basis, traditional knowledge is classified into five main categories: (1) knowledge related to the traditional breeding of agricultural genetic resources; (2) knowledge related to traditional medicine; (3) traditional technologies, production, and lifestyles related to the sustainable use of biological resources; (4) traditional cultures related to biodiversity; and (5) traditional knowledge related to products with traditional geographical indications. Among these, the third category—traditional cultures related to biodiversity—is further defined as "religious beliefs, traditional festivals, customary laws, and other cultural practices formed by various ethnic groups and local communities during long-term production and life practices that contribute to biodiversity conservation and sustainable use" (Yin, 2022). Specifically, this includes six subcategories: religious beliefs and ecological ethics, traditional festivals, customary laws, traditional literature and art, traditional food culture, and other traditional cultural practices.

China's definition of traditional knowledge largely aligns with the discourse of the Convention on Biological Diversity (CBD), reflecting its effort to integrate national policy with global governance frameworks. As host of CBD COP15, China proposed the vision of "Ecological Civilization: Building a Shared Future for All Life on Earth," and at UNFCCC COP28 it launched the Kunming–Montreal Global Biodiversity Framework Implementation Initiative. These initiatives show that China's concept of ecological civilization is increasingly embedded in UN environmental conventions, shaping issues such as traditional knowledge. Therefore, guided by Xi Jinping's thought on ecological civilization and culture, it is essential to clarify the goals of Chinese traditional knowledge in global climate change governance.

4.1 Co-building the Global Community of Life on Earth as the Core Goal

The core goal of Chinese traditional knowledge in participating in global climate change governance is to contribute to the co-building of the global community of life on Earth. Chinese traditional knowledge is characterized by three key traits: inclusiveness, unity, and innovation. From the perspective of inclusiveness, throughout the long history of adapting to and managing the ecological environment, as well as protecting and utilizing biodiversity, China's diverse ethnic groups have exchanged, absorbed, and integrated each other's ecological wisdom, forming the rich body of traditional knowledge. This inclusiveness enables harmony between human communities and nature, as well as fostering peaceful coexistence among various ethnic groups. From the perspective of unity, the different ethnic groups in China reside in similar or identical ecosystems, using comparable natural resources for production and daily life. This shared environment has led to the development of similar customs, beliefs, and worldviews, creating a unified cultural foundation for traditional knowledge. Finally, innovation is inherent in Chinese traditional knowledge, which is a dynamic process of accumulation and historical progress. Passed down through generations, this knowledge has continuously adapted to changing ecological and social environments, driven by innovation. This continuous adaptation provides traditional knowledge with natural ecological vitality and cultural energy. Under the core goal of co-building the global community of life on Earth, these three attributes of Chinese traditional knowledge will contribute the Chinese perspective, solutions, and contributions to the global governance of climate change.

4.2 A System Based on Chinese Traditional Knowledge

The inherent characteristics of Chinese traditional knowledge form the foundation of its goal system in participating in global climate change governance. Co-building the global community of life on Earth requires inclusivity, coexistence, and mutual learning among different civilizations and cultures. Over a long period of harmonious coexistence, the Chinese nation has developed a shared ecological view of "you within me, and I within you", which has evolved into an ecological symbiotic relationship of "we cannot exist without each other". This relationship has further developed into ecological economic links of "mutual benefit and support" and collaborative governance mechanisms of "mutual cooperation and joint management". Examples of these concepts include the shared ecological views of the Lancang River basin among the Han, Tibetan, Lisu, and Naxi ethnic groups, the shared views of the Cangshan and Jizu Mountains among the Han, Tibetan, and Bai peoples, the ecological symbiosis in rice cultivation among the Han, Dai, Blang, Hani, and Lahu peoples, the ecological economic links of tea-horse trade among the Han, Tibetan, Lisu, and Naxi peoples, and the collaborative water and forest resource management mechanisms among the Lahu, Hani, Blang, and Dai peoples. These examples reflect the vibrant practices of Chinese traditional knowledge in building a community of shared destiny (Yin, 2021). The shared ecological views, ecological symbiotic relationships, ecological economic connections, and collaborative governance mechanisms together define the characteristics of Chinese traditional knowledge. They offer Chinese wisdom in co-building the global community of life on Earth and form the goal system for Chinese traditional knowledge's participation in global climate change governance.

5 Pathways for the Participation of Chinese Traditional Knowledge in Global Climate Change Governance

5.1 International Mechanisms for the Participation of Chinese Traditional Knowledge in Climate Governance

General Secretary Xi Jinping emphasized: "We must deeply participate in global environmental governance, enhance China's voice and influence in the global governance system, actively guide the direction of international order transformation, and contribute solutions for global environmental protection and sustainable development" (Kitagawa, 2017). Accordingly, after constructing the discourse system and clarifying the goals, the next step is to gradually establish mechanisms for the participation of Chinese traditional knowledge in global climate change governance.

First, China can follow the Paris Agreement and subsequent Conference of the Parties (COP) decisions on traditional knowledge, promoting Chinese traditional knowledge in related discussions—for example, by establishing dedicated topics and sessions, and including more scholars and experts in traditional knowledge in Chinese delegations. Second, participation in the Lima–Paris Action Agenda can be strengthened by introducing a Chinese discourse system and goals for traditional knowledge, thereby expanding China's voice in these issues. Third, Chinese traditional knowledge should be shared in global policy discussions such as those within the Intergovernmental Panel on Climate Change (IPCC), highlighting stories, experiences, and successful practices. Fourth, drawing on the Paris Agreement's role as a platform for recording and sharing traditional knowledge and experiences, China can propose specific contributions across five dimensions: (1) phenological knowledge of climate change, (2) traditional natural resource use techniques for mitigation, (3) livelihood practices for adaptation, (4) ecosystem management mechanisms for reducing climate risks, and (5) cultural traditions and customs for understanding climate change. Finally, within the framework of the Paris Agreement, China can utilize existing traditional knowledge platforms to foster exchanges, share ecological wisdom and experiences, and collaborate with other countries to explore cases and roles of traditional knowledge in climate action. This would promote in-depth dialogue on traditional knowledge issues, enhance South–South cooperation on climate governance, and lead the formation of a global governance network on climate change with developing countries at its core.

5.2 Domestic Mechanisms for the Participation of Chinese Traditional Knowledge in Global Climate Change Governance

General Secretary Xi Jinping emphasized: "The 'second integration' represents a new intellectual liberation, allowing us to fully utilize the valuable resources of Chinese excellent traditional culture in a broader cultural space, and explore theoretical and institutional innovations for the future" (Li, 2021). To establish international mechanisms for the participation of Chinese traditional knowledge in global climate change governance, it is necessary to support these efforts with domestic theoretical, policy, legal, and institutional innovations. In this context, the National Biodiversity Conservation Strategy and Action Plan has established priority projects for the protection and inheritance of traditional knowledge, highlighting the value of traditional knowledge in adapting to climate change. It is recommended that the country incorporate traditional knowledge into its climate change policies, actions, and plans, and clarify the role and value of traditional knowledge in global climate change governance, in order to improve related climate policies and action plans. For instance, topics on traditional knowledge could be included in reports like the National Climate Change Assessment Report and China's Climate Change Blue Book, and traditional knowledge should be incorporated into national climate change action plans such as China's Policies and Actions for Addressing Climate Change and China's Regional Action on Climate Change during the 14th Five-Year Plan.

In terms of legal frameworks, formal mechanisms and channels should be established to integrate traditional knowledge into climate change laws and regulations, such as the Climate Change Response Law of the People's Republic of China. Additionally, the status of traditional knowledge in adaptation to climate change and green low-carbon development strategies should be clearly defined. Following the Paris Agreement's directive to "be based on and follow existing best science, as well as appropriate traditional knowledge, indigenous peoples' knowledge, and local knowledge systems," traditional knowledge should be integrated with scientific and technological advancements. This will support China's transition to green, low-carbon development and contribute to achieving the "dual carbon" targets (He et al., 2022).

6 The Vision of Chinese Traditional Knowledge Leading Global Climate Change Governance

In the global climate change discussions under the UNFCCC and the Paris Agreement, three main tendencies emerge regarding the integration of different knowledge systems. The first, led by Western developed nations, emphasizes existing science and technology, treating traditional and indigenous knowledge as supplementary. The second, represented by indigenous communities, views indigenous knowledge as not only an informational system but also as a cultural and value-based knowledge system that opposes "Western knowledge", highlighting the rights to land, resources, and autonomy. The third, represented by China and India, embraces an inclusive perspective, advocating for equal recognition of all cultures and knowledge systems. Both countries emphasize the ecological attributes of traditional knowledge and have developed national frameworks for its documentation and use. China, in particular, is developing a traditional knowledge database as part of its broader vision for ecological civilization, seeking to position itself as a leader in integrating traditional knowledge into global climate governance.

In the future, China can not only strengthen cooperation and exchanges with other countries in applying biodiversity-related traditional knowledge to climate change, thereby promoting the synergistic governance of biodiversity conservation and climate action, but also further integrate traditional knowledge into the ecological civilization knowledge system, achieving its elevation from traditional knowledge to ecological civilization knowledge. Within the framework of the Convention and the Paris Agreement, China may draw on the existing traditional knowledge platform mechanisms to lead the establishment of an "Ecological Civilization Knowledge and Climate Change Platform" and related mechanisms. This platform would facilitate the exchange of ecological civilization knowledge, as well as the sharing of experiences and wisdom (Hansen, et al, 2018). Based on its functions, a systematic classification and database of "ecological civilization knowledge related to climate change" can be developed. By jointly exploring with other countries the cases and roles of ecological civilization knowledge in addressing climate change, a global ecological civilization knowledge network for climate governance can be formed, fully leveraging its unique contribution to environmental protection and sustainable development.

At the domestic level, China should clearly establish the connection between the ecological civilization knowledge system and climate change in future climate policies and action plans, thereby enhancing relevant policies and action plans. For example, the National Climate Change Assessment Report should include a dedicated section on the ecological civilization knowledge system, and it should be incorporated into national climate change action plans. Simultaneously, formal mechanisms and channels should be established to involve the ecological civilization knowledge system in the policy-making process, integrating it into social, economic, and environmental policies aimed at transitioning to a green, low-carbon economy. In line with the

Paris Agreement, this approach should not only be guided by science and technology but also highlight the role and value of diverse knowledge systems, positioning ecological civilization knowledge as an essential component of climate change governance.

At the international level, China can follow the Paris Agreement and its subsequent COP decisions, proposing and exploring pathways for the participation of China's ecological civilization knowledge in global climate change governance. This would encourage the integration of Chinese ecological civilization knowledge into the global climate governance process, showcasing the value and role of China's ecological wisdom, culture, and practices in addressing climate change. China can also share its successful experiences in ecological civilization construction.

In conclusion, from traditional knowledge to ecological civilization knowledge, and from a participant to a leader, the ecological civilization knowledge system can form China's autonomous discourse in addressing climate change, demonstrating the value and contributions of China's ecological civilization construction and practices in global climate governance.

References

An, L. & Lin, Z. (2023). Creating an interdisciplinary discipline for ecological civilization construction and innovating China's independent ecological civilization knowledge system. *Urban and Environmental Research, 35*(2). (In Chinese).

Anja Byg, & Jan Salick (2009). Local Perspectives on a global phenomenon-Climate change in Eastern Tibetan villages. *Global Environmental Change, 19*, 156–166.

Cheng, G., Wang, C. & Xue, D. (2012). The attitude of international intergovernmental organizations towards traditional knowledge issues and China's countermeasures and suggestions. *Biodiversity Science, 15*(4). (In Chinese).

Couzin, J. (2007). Opening doors to native knowledge. *Science, 315*(5818), 1518–1519.

Crate, S. A., & Nuttall, M. eds. (2009). *Anthropology and climate change: From encounters to actions* (pp. 9–34). California: Left Coast Press, Inc.

Dawson, B. & Spannagle, M. (2008). *United Nations framework convention on climate change (Unfccc)[M]* (pp. 392–403). The complete guide to climate change. Routledge.

Fang, R. Y. (2006). Folk Characteristics and Comparative Study on 24 Solar Terms of Kazaks. *Journal of Xinjiang University (Philosophy and Social Sciences), 4*, 83–88. (In Chinese).

Fikret Berkes (2012). *Sacred Ecology (Third Edition)* (pp. 171–192). New York and London: Routledge.

Gao, Y. S., Sang, Y. Y, et al. (2004). The rising and declining of loulan and environmental changes and catastrophe. *Journal of Chengdu University (Natural Science), 23*(3), 50–52. (In Chinese).

Ge, Q. S., et al. (2010). *Climate change through every chinese dynasty*. Beijing: Science Press. (In Chinese).

Hansen, M. H., Li, H., & Svarverud, R. (2018). Ecological civilization: Interpreting the Chinese past, projecting the global future. *Global Environmental Change, 53*, 195–203.

Harris, M. (1968). *Rise of anthropological theory: A history of theories of culture* (pp. 41–42). New York: Harper and Row.

He, J., Li, Z., Zhang, X., et al. (2022). Towards carbon neutrality: A study on China's long-term low-carbon transition pathways and strategies. *Environmental Science and Ecotechnology, 9*, Article 100134.

Hongyi, L. (2021). The role and logic of nontraditional security in China's engagement in global governance mechanisms under Xi Jinping's Regime. *Journal of Chinese Political Science, 26*(3), 505–523.

Huang, C. (2018). On the theoretical origins of Xi Jinping's ecological civilization thought. *Urban and Environmental Studies, 43*(2). (In Chinese).

Huang, C. (2023). Five major relationships must be correctly handled in the construction of ecological civilization. *Frontline, 30*(11). (In Chinese).

Huang, Q. & Yin, L. (2024). Achievements and prospects of biodiversity conservation in Xishuangbanna Prefecture, Yunnan Province. *Western Forestry Science*, (4). (In Chinese).

Kitagawa, H. (2017). *Environmental policy under president Xi Jinping leadership: the changing environmental norms[M]* (pp. 1–15). Environmental policy and governance in China. Tokyo: Springer Japan.

Larry, C. P., & Gerald, H. H. (2005). Climate and the Collapse of Maya Civilization--A series of multi-year droughts helped to doom an ancient culture. *American Scientist, 93*, 322–329.

Li, H. Y. (2021). The role and logic of nontraditional security in China's engagement in global governance mechanisms under Xi Jinping's Regime. *Journal of Chinese political science, 26*(3), 505–523.

Liu, C. H., Cheng, G., Xue, D. Y., et al. (2013). An Overview on Traditional Knowledge with the Linkage of Climate Change. *Journal of Yunnan Agricultural University, 28*(4), 570–575. (In Chinese).

Luo, S. Z. (2006). Reflections on Qinghai's Meteorological Culture. *Journal of Qinghai Meteorology, 49*(F09), 42–44. (In Chinese).

Michael, W. B., Alan, L. K., Mark B., et al. (1997). Climate Variation and the Rise and Fall of an Andean Civilization. *Quaternary Research, 47*, 235–248.

Nan, W. Y. (2004). The Belief in the Eternal Sky. *CDMKFYJ, 1*, 56–57. (In Chinese).
Niu, Z. B., & Liu, H. Y., et al. (2011). *Introduction to Meteorological Philosophy* (pp. 24–26). Beijing: China Meteorological Press. (In Chniese).
Publicity Department of the Communist Party of China, Ministry of Ecology and Environment of the People's Republic of China. (2022). *An Outline for Studying Xi Jinping Thought on Ecological Civilization*. Beijing: Study Press; People's Publishing House. (In Chinese)
Richard, A. W. & William, E. R. (1983). Societal response to CO2 – induced climate change: Opportunities for research. *Climatic Change, 3*(4), 387–428.
Sarah, S., & Benjamin, S. O. eds. (2003). *Weather, Climate, Culture* (pp. 3–11). Oxford, New York, NY: Berg Publishers.
Sourgens, F. (2018). *Paris agreement: Regained or Lost?* Initial Thoughts. EJIL: Talk, 28.
Toni, H. & Poul, P. (1997). Meteorological knowledge and environmental ideas in traditional and modern societies: The case of tibet. *The Journal of the Royal Anthropological Institute, 3*(3), 577–597.
Vedwan, N. (2006). Culture, climate and the environment: Local knowledge and perception of climate change among apple growers in northwestern india. *Journal of Ecological Anthropology, 10*, 4–18.
Wang, W. R. (2012). A study on the origins of china's meteorological culture. *Journal of Shaanxi Meteorology, 5*, 50–54. (In Chinese).
Xi, J. (2017a). *The 19th CPC national congress report tutorial*. People's Publishing House. (in Chinese).
Xi, J. (2017b). *The governance of China* (Vol. 2). Foreign Languages Press. (In Chinese).
Xi, J. (2020a). Ideological and political courses are key courses for implementing the fundamental task of cultivating morality and educating people. *Qiushi*, (17). (In Chinese).
Xi, J. (2020b). *The governance of China* (Vol. 3). Foreign Languages Press. (In Chinese).
Xi, J. (2022). Outline for studying Xi Jinping's thought on ecological civilization. Learning Press, People's Publishing House. (In Chinese).
Xi, J. (2024). *Selected Important Speeches of Xi Jinping (2023 Combined Volume)*. Beijing: People's Publishing House. (In Chinese).
Xue, D. (2021). China enters a new era in implementing the convention on biological diversity. *Biodiversity Science, 29*(2). (In Chinese).
Yang, K. (2023). Construction of China's independent ecological civilization knowledge system. *Urban and Environmental Research*. (In Chinese).
Yang, W. H. (2009). Climate Resources and Belief-Traditional Knowledge of the Bai Ethnic People in Relation to the Change of Climate. *Academic Exploration, 2*, 89–93. (In Chinese).
Yin, S. T. (2010). *Collected papers on chinese culture and the environment*. Kunming, Yunnan people's Publishing House. (In Chinese).
Yin, L. (2021). National ecological cultural identity and the Chinese nation community. *Journal of Original Ecological National Culture, 32*(1). (In Chinese).
Yin, L. (2022). Ethnic ecological customary law and biodiversity protection: theory, value and approach. *Journal of Original Ecological Ethnic Culture, 22*(5). (In Chinese).
Yin, L. (2023). Traditional knowledge participation in global climate change governance under the framework of the paris agreement. *China Environment, 11*, 22–25. (In Chinese).
Yin, L. (2024). Theory and methods of climate anthropology. *Journal of Yunnan Normal University (Philosophy and Social Sciences Edition), 56*(2). (In Chinese).
Yin, L., & Xue, D. Y. (2013). Tibentans' belief in holy mountain and global climate change: The case of hongpo village of deqin county of yunnan province. *Journal of Yunnan Minzu University (Philosophy and Social Sciences Edition), 30*(3), 47–51. (In Chinese).
Zhang, Y. (2022). Constructing China's independent ecological civilization knowledge system. *People's Daily*. (In Chinese).
Zhang, Y. & Wang, Y. (2024a). Preliminary exploration of the construction of China's independent ecological civilization knowledge system-Interview with Professor Zhang Yunfei. *Journal of Poyang Lake Studies, 26*(1). (In Chinese).
Zhang, Y. F., & Wang, Y. X. (2024b). Preliminary exploration of China's independent knowledge system of ecological civilization: An interview with professor Zhang Yunfei. *Journal of Poyang Lake, 1*(48–80), 156–158. (In Chinese).
Zhuang, G. (2024). The internal logic and world significance of Xi Jinping's ecological civilization thought. *Contemporary World, 25*(2). (In Chinese).

Postscript

Climate is not only a natural phenomenon but also a social and cultural one. This work examines the relationship between traditional knowledge and climate change from theoretical, focal, developmental, field-based, and practical perspectives. Research in this area explores the cultural connections between humans and climate, the influence of climate change on human culture, and the corresponding adaptive strategies. Its aim is to provide accurate understandings and effective approaches for humanity to recognize, utilize, modify, adapt to, and sustain the climatic environment, as well as to respond to climate change and its related disasters. At the heart of this endeavor lies the cultural construction of climate. Interdisciplinary research, together with the development and innovation of theories and methods, represents a fundamental direction for future studies on traditional knowledge and climate change. Using Tibetan society in the Eastern Himalayan region of Yunnan Province, China, as a case study, this research moves beyond the earlier confines of climate studies restricted to the natural sciences. By adopting an interdisciplinary approach, it investigates the impacts of climate change on traditional knowledge and explores practical, localized strategies for adapting to climate change and its associated disasters through the lens of traditional knowledge.

This study encompasses three main aspects.

First, it examines traditional knowledge related to climate and climate change. Although climate models can project macro-level scenarios of climate change and estimate various trends and consequences for future human development, they cannot provide precise information about regional climate change. In recent years, there has been growing recognition of the value of traditional knowledge concerning climate and climate change. Rooted in observations of organisms and environments far more complex than scientific instruments can capture, this knowledge represents a valuable source of insight into regional climate change. The Tibetan people in Deqin County possess rich and diverse traditional knowledge of climate and climate change, which offers important perspectives for enhancing our understanding of climate change.

Second, this research examines the impact of climate change on traditional knowledge related to the conservation and sustainable use of biological resources. Beyond its influence on traditional climatic and phenological knowledge used to observe and predict weather, the most profound impact of climate change on traditional knowledge is found in the domain of knowledge concerning the conservation and sustainable use of biological resources. This book focuses on

Y. Lun, *Climate Change and Traditional Knowledge*, Sustainable Development Goals Series,
https://doi.org/10.1007/978-3-032-12015-1

analyzing how climate change affects traditional knowledge and, drawing on a five-category classification system of traditional knowledge, investigates the direct and indirect impacts of climate change and extreme climate events on the traditional knowledge of the Tibetans in Deqin County.

Third, this study considers traditional knowledge and its role in adaptation to climate change. Traditional knowledge holds significant potential and value for both mitigating and adapting to current and future climate change. Another central focus of this book is on strategies and practices of climate change adaptation grounded in traditional knowledge, as well as on how the integration of traditional knowledge into policy frameworks can enhance responses to the adverse impacts of climate change. Building on the classification system of traditional knowledge, this book examines adaptation strategies derived from traditional knowledge and their concrete implementation within the Tibetan community of Yunling Township, Deqin County.

The theoretical exploration of traditional knowledge and climate change constitutes one of the key innovations of this book, with several notable breakthroughs. First, it transcends the limitations of earlier climate change research that was confined purely to the natural sciences, where indigenous and ethnic minority traditional knowledge was often overlooked or undervalued. Second, it addresses the shortcomings of research within the social sciences, which frequently reduced traditional knowledge to little more than predictive understandings of climate and phenology or collections of agricultural proverbs. Building on a classification system of traditional knowledge and an analysis of the impacts of climate change and extreme climate phenomena on such knowledge, this study develops a theoretical framework for research on traditional knowledge and climate change. The establishment of this theory and its framework offer a solid basis and direction for future inquiry in the field.

The exploration of research methods for traditional knowledge and climate change constitutes another innovation of this book. Since traditional knowledge is primarily grounded in relative and localized experience and lacks standardized scientific benchmarks or measurement scales, achieving its quantitative analysis has long been an academic challenge. This limitation has, in turn, fostered skepticism within the scientific community regarding the credibility and value of traditional knowledge. To address this, the study investigates the transformation of traditional knowledge into indicator-based and data-driven forms, thereby establishing a framework for its quantitative analysis. This framework represents both an innovation and an initial attempt in the quantitative study of traditional knowledge, enhancing its credibility and value to some extent while also providing a referential model for its indicator-based and data-driven transformation.

This book also explores the classification of climate-related traditional knowledge: the direct and indirect observations and understandings of local climate and phenological phenomena accumulated by people in a specific region over generations, as well as the knowledge derived from these observations for predicting and responding to climate change and extreme climatic events. This system of knowledge is organized into five categories.

The purpose of advancing the theories and methods of research on traditional knowledge and climate change in this book is to encourage more scholars to engage with this field and to promote comprehensive interdisciplinary studies, thereby contributing to its further development.

The path of personal scholarship seems endless, continuing as long as life itself. In this sense, this postscript may also be read as a preface to my future academic journey.

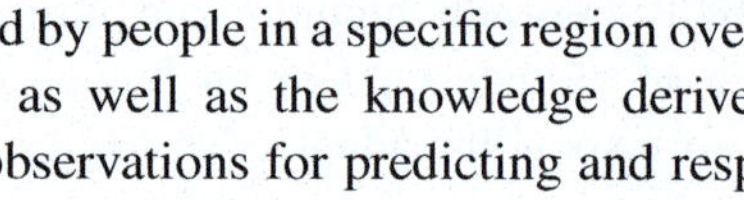

Lun Yin
Golden Horde, Kunming
September, 2025

Index

Y. Lun, *Climate Change and Traditional Knowledge*, Sustainable Development Goals Series,
https://doi.org/10.1007/978-3-032-12015-1

The manufacturer's authorised representative in the EU is Springer
Nature Customer Service Centre GmbH, Europaplatz 3, 69115 Heidelberg,
Germany. If you have any concerns regarding our products, please
contact ProductSafety@springernature.com

Printed and bound by CPI Group (UK) Ltd, Croydon, CR0 4YY
05/10/2026
02228021-0012